CISM COURSES AND LECTURES

The series presents lecture notes, monographs, edited works and proceedings in the field of Mechanics, Engineering, Computer Science and Applied Mathematics.
Purpose of the series in to make known in the international scientific and technical community results obtained in some of the activities organized by CISM, the International Centre for Mechanical Sciences.

INTERNATIONAL CENTRE FOR MECHANICAL SCIENCES

COURSES AND LECTURES - No. 339

EUROCODE '92
INTERNATIONAL SYMPOSIUM ON CODING THEORY AND APPLICATIONS

EDITED BY

P. CAMION
INRIA - ROCQUENCOURT

P. CHARPIN
INRIA - ROCQUENCOURT

S. HARARI
UNIVERSITY OF TOULON AND VAR - LA GARDE

SPRINGER-VERLAG WIEN GMBH

Le spese di stampa di questo volume sono in parte coperte da
contributi del Consiglio Nazionale delle Ricerche.

This volume contains 47 illustrations.

Originally published by Springer Verlag Wien-New York in 1993

In order to make this volume available as economically and as rapidly as possible the authors' typescripts have been reproduced in their original forms. This method unfortunately has its typographical limitations but it is hoped that they in no way distract the reader.

ISBN 978-3-211-82519-8 **ISBN 78-3-7091-2786-5 (eBook)**
DOI 10.1007/78-3-7091-2786-5

PREFACE

This book is made of the proceedings of EUROCODE 1992 which was held in Udine (Italy) at the Centre International des Sciences Mécaniques from October 23 through October 30,1992, where 53 conferences were selected for a gathering of about 90 participants from the academic and industrial worlds. It is a continuation of EUROCODE 1990 which was also held in Udine in 1990 and whose proceedings appeared as Lecture Notes in Computer Sciences, Volume 514 (P. CHARPIN, G. COHEN eds.).

The aim of EUROCODE 1992 was to attract high level research papers and to encourage interchange of ideas among the areas of coding theory and related fields which share the same tools for applications in the science of communications, computer science, software engineering and mathematics.

Each of the contributions, except for the five invited conferences, was refereed for an oral presentation on the basis of an extended summary. The submitted full papers were separately refereed for publication by at least two international referees. The topics of the conference were: Combinatorial Coding, Algebraic Coding, Convolutional Coding, Algorithms and applications, Security of transmissions, Authentication, Signature, Synchronisation and radio-localization, Source Coding, Modulation and Coding, Applications of Coding and Implementations.

There were five invited one hour conferences which were given by:

G. Battail (E.N.S.T. France): "We can think of good codes and even decode them"

J. Massey (University of Zurich): "From fields to rings to groups"

J. Pellikaan (University of Eindhoven): "On the efficient decoding of algebraic geometric codes"

V. Pless (University of Illinois at Chicago): "Duadic codes and generalizations"

C.P. Schnorr (University of Frankfurt): "On discrete log authentication and signature schemes".

The book contains three invited papers and thirty selected full papers. The papers are regrouped under several topics; they are presented in next paragraph.

The editors would like to thank the referees (see enclosed list) with a special mention to P.G. FARREL (University of Manchester) as well as C. GUIZIOU, secretary of "Projet CODES-INRIA".

CONTENTS

1 Algebraic Coding Theory

In her full-paper PLESS introduces the duadic codes and presents a survey on recent results related with these codes and their generalizations. CAMION, COURTEAU and MONTPETIT obtain the coset weight enumerator and some fundamental parameters for the Reed Muller, the quadratic residue code and the other three extremal doubly-even codes of length 32 given by Conway and Pless code. CARBONNE and THIONGLY improve the exponent in the Pellikaan algorithm for decoding algebraic geometric codes. COHEN, HONKALA and LITSYN show that finding M packings and M coverings amount to one problem. They give results on the existence of perfect q-ary linear (m_0,m_1)- coverings. LEVY-DIT-VEHEL examines the case of binary extended cyclic *affine-invariant codes of length 2^m whose zero's set can be defined by one cyclotomic coset; among those are some duals of extended BCH codes. MARTIN gives a method of construction of binary even length codes derived from shorter codes, which are better than odd length corresponding codes. MUNUERA and PELLIKAAN give a criterion for self duality of geometric codes. RODIER shows that a conjecture of Macwilliams and Sloane concerning the weights of BCH codes is false for some codes of minimum distance satisfying particular conditions. SEGUIN and WOUNGANG give a simple characterization of all the q^m-ary cyclic codes which possess a cyclic q-ary image. WOLFMANN estimates the weights of binary cyclic codes of primitive lengths by means of weights of irreducible cyclic codes.*

2 Mathematical Tools for Coding

CARLET presents results on the link between boolean functions and differential equations which define them. GILLOT improves a bound on the number of solutions of the trace equation and applies this result to improve the bound on the distance of certain cyclic codes. LANGEVIN studies the properties of generalized bent functions and answers the question of the distance property and degree of such functions. LEVENSHTEIN gives a bound on the size of antipodal codes of given minimum distance d and gives applications to the estimation of the minimum distance of self dual codes.

3 Cryptography

CHABAUD gives an asymptotic analysis of the problem of finding a codeword of given weight in a linear code. HARARI and LIARDET introduce formally markovian cryptosystems and study HARALIA which is an implementation of such a system. MASET and SGARRO study the rate regions for which there exist codes that are resistant to mixed attacks (deception an decryption) and give a theorem that shows that their construction is asymptotically the best possible. MORAIN gives a survey on recent results concerning pseudo-prime integers, including tables of recent pseudo-primes, and reports on a generalization of the Miller Rabin algorithm. VAUDENAY gives the description of a practical cryptosystem based on an authentication tree extending Merkle's previous results.

4 Decoding Block Codes

In his invited paper, PELLIKAAN gives a general survey on the decoding of algebraic-geometric codes. CHABANNE shows how Gröbner bases can replace generator polynomials for the study of abelian codes, and generalizes a permutation decoding algorithm to decode them. OLCAYTO and IRVINE study some multi-parameter burst error correcting codes whose parity check matrices are arrays of identity matrices. These codes are useful in situations where bursts of errors take a particular shape. SAKATA introduces 2D parallel decoding for 2D cyclic codes. SENDRIER introduces new tools to evaluate the performance of a decoding algorithm, in the case of complete error correction up to the minimum distance or incomplete error correction and error detection. SNYDERS exhibits codes for which a maximum likelihood decoding algorithm uses both code search and error search. TOLHUIZEN sharpens a bound on the number of miscorrected error patterns for Reed-Solomon codes.

5 Information Theory

FIORETTO, INGLESE and SGARRO study the zero-error capacity of multiple access channels. By using time sharing techniques they localize it and show that it is limited in surface.

6 Convolutional Coding, Synchronisation

DARNELL, GRAYSON and HONARY present a discussion on novel techniques for symbol and word synchronisation deriving information from the normal operational traffic, and the design of optimum preambles for word or block synchronisation. HONARY, MARKARIAN and DARNELL introduce a new algorithm which allows to construct the trellis diagram of array codes and a soft maximum likelihood decoding algorithm based upon this trellis diagram.

7 Coding Performance

In his invited paper BATTAIL proposes an alternative definition of a good code which relies on the closeness of its distance distribution with respect to that of random coding, instead of the largeness of its minimum distance. DUMER constructs suboptimal decoding algorithms for additive channels with linear output. This implies the generalization of the notion of weight. SWEENEY and SHIN study the problem of applying soft decision methods in decoding Reed-Solomon codes. To this end codes are defined as subfield subcodes of codes defined over a larger field and by a specific mapping. Trellis decoding can be applied and Viterbi decoding can be used to limit the complexity of trellis decoding and other search methods.

8 Applications of coding

CREUTZBURG, GEISELMANN and HEYL describe the VLSI implementation of a lossy image compression. The basic concept has been inspired by the fractal geometry.

Organization of the Conference

The conference had an Organizing Committee made of: P. Charpin (INRIA) Publications, S. Harari (GECT) Secretary, C. Goutelard (LETTI) Treasurer, G. Longo (CISM), L. Rizzi (DRET), A. Sgarro (Trieste), J. Stern (GRECC), J. Wolfmann (GECT) scholarships.

The Scientific Committee of the conference included P. Camion (CNRS-INRIA) as President and had as members G. Battail (Paris), V. Capellini (Florence), R. Capocelli (Rome), G. Chassé (Nantes), G. Cohen (Paris), J. Conan (Montréal), B. Courteau (Sherbrooke), J. L. Dornstetter (Paris), M. Elia (Turin), P. Farrell (Manchester), H. Fell (Boston), D. Haccoun (Montréal), G. Lachaud (Luminy), D. Lazard (Paris), S. Lebel (Paris), D. Lebrigand (Paris), A. de Luca (Rome), G. Norton (Bristol), P. Piret (Rennes), M. Tallini (Rome), A. Thiong Ly (Toulouse).

This colloquium was sponsored by DRET (Paris, France), INRIA (Rocquencourt, France) and by the CNRS of France (through the research groups Mathematics and Computer Science and MEDICIS).

List of referees

D. Augot, G. Battail, T. Berger, F. Blanchet, R. Calderbank, P. Camion, C. Carlet, G. Castagnoli, P. Charpin, G. Chassé, H. Chabanne, G. Cohen, J. Conan, B. Courteau, A. De Santis, J.L. Dornstetter, M. Elia, P.G. Farrell, M. Girault, D. Haccoun, S. Harari, C.R.P. Hartmann, T. Helleseth, A. Joux, S. Lebel, G. Lachaud, F. Laubie, D. Lazard, D. Le Brigand, F. Levy-dit-Vehel, G. Longo, P. Langevin, J.P. Martin, J. Massey, U. Maurer, A. Montpetit, F. Morain, G. Norton, P. Piret, J.J. Quisquater, F. Rodier, R. Sabin, N. Sendrier, J. Simonis, J. Stern, A. Thiong-ly, H.C.A. Van-Tilborg, J. Wolfmann, G. Zémor.

P. Camion
P. Charpin
S. Harari

CONTENTS

1

ALGEBRAIC CODING THEORY

DUADIC CODES AND GENERALIZATIONS

V. Pless
University of Illinois at Chicago, Chicago, IL, USA

Since their definition in 1984 [7] there has been much work on duadic codes with generalizations in various directions. We survey this for its own interest and as it might shed light on cyclic codes in general. Duadic codes are generalizations of quadratic residue codes. This tells us for which lengths n they exist and also gives us an easy way to construct their generating idempotents over fields of characteristic 2. As cyclic codes whose extensions are self-dual must be duadic, we also learn at which lengths such codes can exist. On the one hand duadic codes are generalized to triadic and polyadic codes, and on the other hand to codes generated by difference sets in abelian groups.

We can tell when the code generated by a cyclic projective plane is in a duadic code and give a coding theoretic proof that there cannot exist a cyclic projective plane of order $n \equiv 2 \pmod 4$ unless $n = 2$. The case of quaternary duadic codes is special as there are two inner products with the Hermitian one yielding more interesting self-dual codes. We get further non-existence criteria for cyclic projective planes. Codes generated by the incidence matrices of certain tournaments turn out to be binary or quaternary duadic codes.

As it is not difficult to compute generating idempotents or generating polynomials for duadic codes of lengths till 241, their minimum weights have been determined on computers. They are very good when the duadic codes have prime lengths. When the length is composite, but not a prime power, some codes are very good and some not. The worst case is when the length is a prime power greater than one. Similar computations were done for

m-adic residue codes and the same observation about lengths holds. We start with some facts we need about cyclic codes.

I. Cyclic Codes.

Our notation is as in [9]. As usual we consider cyclic codes of odd length n over $F = GF(q)$ where $\gcd(n,q) = 1$. We label coordinates with $0, 1, \ldots, n-1$. If F_n is the space of n-tuples over $GF(q)$, then C is a cyclic code if C is invariant under the coordinate permutation $i \longrightarrow i+1 \pmod{n}$. Let a vector $a = (a_0, \ldots, a_{n-1})$ in F_n correspond to a polynomial $a(x) = a_0 + a_1x + \ldots + a_{n-1}x^{n-1}$ in the ring $R_n = F[x]/(x^n - 1)$. We refer to either F_n or R_n. It is well-known that cyclic codes correspond to ideals of R_n.

Fact 1. Every cyclic code or ideal in R_n can be expressed uniquely as a sum of minimal ideals.

Associated with a cyclic code C are two distinguished polynomials, each of which generates C as an ideal of R_n. One of these is the **generator polynomial** $g(x)$ which is the unique monic divisor of $x^n - 1$ in C. The other is the **idempotent generator** $e(x)$ which is the multiplicative unit for C, that is $e(x)c(x) = c(x)$ for each $c(x)$ in C. Clearly then $e^2(x) = e(x)$. We write $C = \langle g(x) \rangle = \langle e(x) \rangle$. Let $h(x) = (1 + x + \ldots + x^{n-1})$. Then $h(x)$ is the generator polynomial for a minimal ideal M_0 of dimension 1 whose idempotent generator is $h'(x) = (1/n)h(x)$.

Fact 2. If $C_1 = \langle g_1(x) \rangle = \langle e_1(x) \rangle$ and $C_2 = \langle g_2(x) \rangle = \langle e_2(x) \rangle$, then $C_1 + C_2$ and $C_1 \cap C_2$ are cyclic codes. Further, $C_1 + C_2 = \langle \gcd(g_1(x), g_2(x)) \rangle = \langle e_1(x) + e_2(x) - e_1(x)e_2(x) \rangle$ and $C_1 \cap C_2 = \langle \mathrm{lcm}(g_1(x), g_2(x)) \rangle = \langle e_1(x)e_2(x) \rangle$.

Let a be an integer which is relatively prime to n. Then the mapping μ_a called a **multiplier** induces a coordinate permutation on F_n as follows.

$$\mu_a : i \longrightarrow ai \pmod{n} \qquad (i = 0, 1, \ldots, n-1)$$

If $w = (w_0, w_1, \ldots, w_{n-1})$ is in F_n, then $\mu_a(w)$ denotes the vector with w_i in position $ai \pmod{n}$. If C is a code, then $\mu_a(C)$ consists of all vectors $\mu_a(c(x))$ where $c(x)$ is in C. The following is one reason why multipliers are interesting for cyclic codes.

Fact 3. The multiplier μ_a sends a cyclic code $C = \langle e(x) \rangle$ onto another cyclic code $\mu_a(C) = \langle \mu_a(e(x)) \rangle$.

Proof. It can be seen computationally that $\mu_a(c(x)d(x)) = \mu_a(c(x))\mu_a(d(x))$ for any two polynomials $c(x)$ and $d(x)$ in R_n. Hence if $c(x)$ is in C, $\mu_a(c(x)) = \mu_a(e(x)c(x)) = \mu_a(e(x))\mu_a(c(x))$. □

From this fact we see that not only does a multiplier send a cyclic code onto another cyclic code but the entire image code is determined by the image of one vector, the idempotent generator. This is one advantage of the idempotent generator. Note that $\mu_a(g(x))$

need not be the generator polynomial of $\mu_a(C)$. Another advantage is that it is often possible, easily, to distinguish all inequivalent cyclic codes of a given length by looking at their idempotent generators.

Fact 4. Let n be such that $\gcd(n, \varphi(n)) = 1$. Then two cyclic codes of length n are equivalent iff they are equivalent by a multiplier.

Note that this includes any n which is a prime. For a proof of this see [2]. There is another condition in [2] ensuring that the above holds when n is a product of two distinct primes.

Consider cyclic codes over $GF(q)$. Then it can be shown that μ_q leaves any idempotent, hence any cyclic code, invariant. This is immediate for binary codes as then $\mu_2(e(x)) = e^2(x) = e(x)$. When the length of the code is a prime p, we let G be the multiplicative group of non-zero elements of $GF(p)$ and H the subgroup of G generated by μ_q. As G is cyclic, so is H. Let r be the index of H in G. Then in order to find codes equivalent to a given cyclic code, we need only consider $r - 1$ multipliers. For example, when $p = 7$ and $q = 2$, G has order 6 and H has order 3 so we need only look at one multiplier. We can often find, quickly, the number of inequivalent cyclic codes this way. Among multipliers, μ_{-1} is special.

Fact 5. If $C = \langle e(x) \rangle$, then $C^{\perp} = \langle 1 - \mu_{-1}(e(x)) \rangle$.

Proof. An easy calculation shows that $c(x)b(x) = 0$ iff $\mu_{-1}(b(x))$ is orthogonal to $c(x)$ and all its cyclic shifts. As $e(x)(1 - e(x)) = 0$, $1 - e(x)$ is the idempotent of the code consisting of the $b(x)$ so that $c(x)b(x) = 0$ for all $c(x)$ in C.

An important concept here is a generalization of the ideas of even and odd weight vectors for binary codes. A vector $a = (a_0, \ldots, a_{n-1})$ is called **even-like** if $\sum_{i=0}^{n-1} a_i \equiv 0$ (in F). Otherwise it is called **odd-like**. A code is called even-like if it only contains even-like vectors, otherwise it is called odd-like. Let **E** be the $n-1$ dimensional cyclic code of all even-like vectors. Then E is the dual of $\langle h'(x) \rangle$ so that $E = \langle 1 - h'(x) \rangle$.

Fact 6. A cyclic code $C = \langle g(x) \rangle$ is odd-like iff $h(x)$ is in C iff $(x-1)$ does not divide $g(x)$.

Proof. This is so as the sum of any odd-like vector and its cyclic shifts equals $\alpha h(x)$ for some $\alpha \neq 0$.

Fact 7. $a(x)$ is an even-like vector iff $a(x)h(x) = 0$.

$a(x)$ is an odd-like vector iff $a(x)h(x) = \alpha h(x)$, $\alpha \neq 0$.

Much information about cyclic codes can be determined from the easily computed cyclotomic cosets. We define the **q-cyclotomic coset** of i for n, where $0 \leq i \leq n$, to be the set

$$C_i^{(q)} = \{i, qi, q^2 i, q^3 i, \ldots\}$$

where the elements in the set are taken modulo n. The q-cyclotomic cosets partition the integers from 0 to $n-1$. We label each q-cyclotomic coset by its numerically smallest element, i. We omit the q when it is obvious.

Binary idempotents are very easy to compute from knowledge of the 2-cyclotomic cosets as a polynomial is an idempotent iff its exponents are unions of cyclotomic cosets. It is not much harder to determine quaternary idempotents. Let t be the order of 2 (mod n). If t is odd all quaternary idempotents are binary. This occurs for any length n whose prime factors are all $\equiv -1 \pmod 8$, and sometimes when prime factors are also $\equiv 1 \pmod 8$. When t is even a quaternary idempotent has the following form.

$$e(x) = \alpha + \sum_{\text{(some (or no))}\ j} \left(\sum_{i \in C_j} (\omega x^i + \bar{\omega} x^{2i}) \right) + \sum_{\text{(some (or no))}\ r} \left(\sum_{i \in C_r} (x^i + x^{2i}) \right) \quad \text{where } \alpha = 0, 1.$$

It is also possible to determine idempotents over $GF(2^s)$.

II. Duadic Codes.

Duadic codes were first defined in [7]. Then quaternary duadic codes were investigated in [10]. Many of the results there and much of the reasoning can be extended to general fields F. See [12, 15, 17] for their theory.

My favorite definition of duadic codes is in terms of generating idempotents. If $C_1 = \langle e_1 \rangle$ and $C_2 = \langle e_2 \rangle$ are even-like cyclic codes of length n, then they are **even-like duadic** codes if there is a multiplier μ_a with

1) $\mu_a(C_1) = C_2$ And $\mu_a(C_2) = C_1$ and

2) $e_1 + e_2 = 1 - h'$.

As e_1 and e_2 are even-like we see immediately from 2) that $e_1 e_2 = 0$ so that $C_1 \cap C_2 = \emptyset$ and $C_1 + C_2 = E$. Clearly they have the same dimension as they are equivalent. Hence $\dim C_1 = \dim C_2 = \frac{n-1}{2}$.

In terms of generator polynomials this means that if $C_1 = \langle g_1(x) \rangle$ and $C_2 = \langle g_2(x) \rangle$, then g_1 has roots α^i where i is in $\{0\} \cup S_1$, a union of cyclotomic cosets and g_2 has roots α^j for $j \in \{0\} \cup S_2$, another union of cyclotomic cosets with $S_1 \cap S_2 = \emptyset$, $S_1 \cup S_2 = \{1, \dots, n\}$ and $S_1 = \mu_a(S_2)$, $S_2 = \mu_a(S_1)$. Such an S_1 and S_2 is called a **splitting** given by μ_a.

If $C_1 = \langle e_1 \rangle$ and $C_2 = \langle e_2 \rangle$ are even-like duadic codes, then $D_1 = \langle h' + e_1 \rangle$ and $D_2 = \langle h' + e_2 \rangle$ are **odd-like duadic codes**. The following holds.

1) $D_1 = \langle 1 - e_2 \rangle$ and $D_2 = \langle 1 - e_1 \rangle$

2) $\mu_a(D_1) = D_2$ and $\mu_a(D_2) = D_1$

3) $C_1 \subset D_1$ and $C_2 \subset D_2$

4) $D_1 = C_1 + \langle h' \rangle$ and $D_2 = C_2 + \langle h' \rangle$

5) $D_1 + D_2 = F_n$ and $D_1 \cap D_2 = \langle h' \rangle$

6) $\dim D_1 = \dim D_2 = \frac{n-1}{2}$

These are all easy to demonstrate. For example as $e_1(h' + e_1) = e_1$, $C_1 \subset D_1$. Also $1 - e_2 = h' + e_1$ since $e_1 + e_2 = 1 - h'$.

From the above discussion we have the following theorem.

Theorem 1. Duadic codes exist iff a splitting exists.

Quadratic residue codes are duadic codes of prime length p with S_1 the set of non-zero quadratic residues in $GF(p)$, S_2 the set of non-residues with the multiplier given by any fixed non-residue.

Theorem 2. Duadic codes of length n over $GF(q)$ exist iff q is a square (mod n).

Proof. We give the proof for prime length p (as this is the most interesting case) and indicate why it is true for any n.

Assume that duadic codes of length p over $GF(q)$ exist. Let G and H be as before. Since we have a splitting, the number N of non-zero cyclotomic cosets must be even. Clearly $|G| = N|H|$. But if Q is the subgroup of G of quadratic residues, then $|G| = 2|Q|$ so that $N|H| = 2|Q|$. Hence $|H|$ divides $|Q|$. As a cyclic group contains precisely one subgroup of any order dividing its order, $H \subset Q$ so that q is a square (mod p).

Conversely suppose that q is a square (mod n). Then quadratic residue codes exist and these are duadic codes.

From the existence of a splitting for p, it is not difficult to construct a splitting for p^i. From the existence of splittings for two relatively prime numbers it is also straightforward to construct a splitting for their product. This indicates how to prove the theorem for any n.

Corollary. Binary duadic codes exist for all lengths $n = p_1^{a_1} p_2^{a_2} \dots p_r^{a_r}$ where each p_i is a prime $\equiv \pm 1 \pmod 8$.

Ternary duadic codes exist for all lengths $n = p_1^{a_1} p_2^{a_2} \dots p_r^{a_r}$ where each p_i is a prime $\equiv \pm 1 \pmod{12}$.

Quaternary duadic codes exist of any odd length n.

As already noted all quadratic residue codes are duadic codes. Quadratic residue codes exist only at prime lengths. Duadic codes exist at many composite lengths and at some prime lengths where quadratic residue codes exist there are additional duadic codes. Indeed, some of these might be better, as at length 113, where the odd-like binary quadratic residue code has minimum weight 15 but an odd-like duadic code has minimum weight 18 (yes, it can be even). Many Reed-Solomon and punctured Reed-Muller codes are duadic codes. We will see why below.

Theorem 3. An $(n, \frac{n-1}{2})$ cyclic code is self-orthogonal iff it is an even-like duadic code whose splitting is given by μ_{-1}.

Proof. Let $C_1 = \langle e \rangle$ be an even-like duadic code whose splitting is given by μ_{-1}. Then $C_2 = \langle \mu_{-1}(e) \rangle$ and $C_1 \subset D_1 = \langle 1 - \mu_{-1}(e) \rangle = C_1^\perp$. So C_1 is self-orthogonal.

Suppose $C_1 = \langle e \rangle$ is a self-orthogonal $(n, \frac{n-1}{2})$ code and let $C_2 = \langle \mu_{-1}(e) \rangle$. Now $C_1 \subset C_1^\perp = \langle 1 - \mu_{-1}(e) \rangle$. As C_1 is self-orthogonal, C_1 is an even-like code so that h' is in $C_1^\perp$. Since $\dim C_1^\perp = \frac{n+1}{2}$, $C_1^\perp = C_1 \perp \langle h' \rangle$ so that the generating idempotent of $C_1^\perp$ is $e + h' = 1 - \mu_{-1}(e)$. The equation $e + \mu_{-1}(e) = 1 - h'$ shows that C_1 and C_2 are duadic codes. Clearly the splitting is given by μ_{-1}.

Corollary. Cyclic $(n, \frac{n-1}{2})$ self-orthogonal codes over $GF(q)$ exist iff q is a square (mod n) and μ_{-1} gives a splitting iff extended cyclic $(n+1, \frac{n+1}{2})$ self-dual codes over $GF(q)$ exist.

Again it is easiest to tell when μ_{-1} gives a splitting for p a prime. As -1 has even order, it must act as an involution on cyclotomic cosets of odd size. Hence μ_{-1} gives a splitting for a prime p over $GF(q)$ iff the order of q (mod p) is odd. Further, if the order of q (mod p) is odd and -1 is not a square (mod p), then μ_{-1} gives every splitting. To see this let μ_a give a splitting. Then the order of a (mod p) must be even, say $2s$. As $a^s \equiv -1$ (mod p), s must be odd. Now μ_{a^s} gives the same splitting as μ_a for any odd s. So μ_a gives the same splitting as μ_{-1}.

The order of q is odd mod (n) iff it is odd for every prime factor of n. Now 2 has odd order mod p for all primes $p \equiv -1$ (mod 8) and for some primes $p \equiv 1$ (mod 8). But -1 is not a square when $p \equiv -1$ (mod 8). These observations applied to $q = 2$ prove the next theorem.

Theorem 4. A self-dual extended cyclic binary code of length $n+1$ exists iff $n = \Pi p_i^{a_i}$ where each prime p_i is either $\equiv -1$ (mod 8) or is $\equiv 1$ (mod 8) where $\text{ord}_2(p_i)$ is odd.

Suppose we have duadic codes C_1 and C_2 of prime length p given by a splitting μ_a. Then a has even order $2s$ (mod p). As above, if s is odd, μ_a gives the same splitting as μ_{-1}. If s is even, μ_{a^s} leaves the splitting invariant. As $a^s \equiv -1$ (mod p), μ_{-1} leaves C_1 and C_2 invariant. This situation is also interesting.

Theorem 5. If C_1 and C_2 are even-like duadic codes and D_1 and D_2 are the corresponding odd-like duadic codes, then $C_1 = D_2^\perp$ (and hence $C_2 = D_1^\perp$) iff $\mu_{-1}(C_1) = C_1$ (and so $\mu_{-1}(C_2) = C_2$).

Proof. Suppose that $C_1 = \langle e_1 \rangle = D_2^\perp = \langle h' + e_2 \rangle^\perp$. Then $h' + e_2 = 1 - \mu_{-1}(e_1)$ so that $\mu_{-1}(e_1) + e_2 = 1 - h'$ showing that $\mu_{-1}(e_1) = e_1$. As $e_1 + e_2 = 1 - h'$, $\mu_{-1}(e_1) = e_1$ iff $\mu_{-1}(e_2) = e_2$. The converse is clear.

We have a square-root bound for the minimum odd-like weight in a duadic code which is stronger for splittings given by μ_{-1}.

Theorem 6. Let D_1 and D_2 be odd-like duadic codes of length n and let d_0 be their minimum odd-like weight. Then

$$d_0^2 \geqq n.$$

If the splitting is given by μ_{-1} (or μ_{-2} for $q = 4$), then

$$d_0^2 - d_0 + 1 \geqq n.$$

If also $q = 2$, then either

$$d_0^2 - d_0 + 1 = n \text{ or } d_0^2 - d_0 + 1 \geqq n + 12.$$

If $d_0^2 - d_0 + 1 = n$, d_0 is the minimum weight and the minimum weight vectors are multiples of binary vectors. These binary vectors are the lines of a cyclic projective plane of order $d_0 - 1$.

Proof. The proof is the same as for quadratic residue codes. See [9, 10, 18].

The data in [10, 13] shows that the minimum weight of binary and quaternary duadic codes of prime lengths are very good; at composite lengths there seem to be both very good and not so good duadic codes. However, duadic codes of prime power lengths always have low weight vectors. From this data also we see that some binary odd-like duadic codes can have even minimum weight. This is not so for quadratic residue codes and the proof of this uses the fact that the extended quadratic residue codes have transitive groups.

In fact, we know in some situations whether the group of an extended duadic code can be transitive. If C is a code, let $\bar{C}$ denote its **extended code.**

Theorem 7. [3]. Let D be an odd-like binary duadic code of length n. If the group of $\bar{D}$ is transitive on the $n+1$ coordinate positions and contains no regular normal subgroup, then n equals a prime p and D is equivalent to a quadratic residue code of length p.

Ito used the classification of the 2-transitive groups without regular normal subgroups to show this.

The group of the (8,4) binary extended quadratic residue code has order 1,344. The group of the (24,12) extended quadratic residue code is the famous Mathieu group M_{24}. Any other binary extended quadratic residue code of length $p+1$ has $PSL_2(p)$ of order $\frac{(p_1)p(p-1)}{2}$ as its group.

We can prove another theorem easily without knowledge of regular normal subgroups.

Lemma. If $\bar{D}$ is an extended cyclic, binary self-dual code of length $n+1$, then all even weight vectors in D have weight $\equiv 0 \pmod 4$ and all odd weight vectors have weight $\equiv n \pmod 4$.

Proof. As duadic codes of length n exist, n must be $\pm 1 \pmod 8$. Now $D = C \perp \langle h \rangle$ where $C = \langle e \rangle$ is an even-like duadic code. If $n \equiv 1 \pmod 8$, $wt(e) \equiv (n-1)/2 \pmod 8$ which is divisible by 4. If $n \equiv -1 \pmod 8$, $wt(e) \equiv \frac{n+1}{2} \pmod 8$ which is also divisible by 4. As e and its shifts generate C and C is self-orthogonal, all weights in C are divisible by 4. Hence all odd weights in D are $\equiv n \pmod 4$.

Theorem 8. If $\bar{D}$ is an extended cyclic, self-dual code of length $n+1$ and $n \equiv 1 \pmod 8$, then the group of $\bar{D}$ is not transitive.

Proof. By the lemma we know that all odd weights in D are $\equiv 1 \pmod 4$. Suppose the group of $\bar{D}$ were transitive and puncture $\bar{D}$ on a position where an even weight vector has a non-zero component. Then the punctured code would be equivalent to D but it would contain an odd weight vector whose weight is $\equiv 3 \pmod 4$. This contradiction proves the theorem. □

There are several things about quaternary codes which are of additional interest. First is whether the duadic code has a binary idempotent. In this situation it was shown [10] that the minimum weight of the quaternary code is that of the binary code with the same idempotent. Further, all minimum weight vectors in the quaternary code are multiples of the binary minimum weight vectors. When this occurs, the quaternary code is not too good in terms of minimum weight. It is not difficult to show that a quaternary code C has a binary basis iff $\mu_2(C) = C$.

Another point of interest is that there are two inner products with respect to which orthogonality can be defined. Any cyclic (ordinary) self-orthogonal $(n, \frac{n-1}{2})$ code must be duadic with splitting given by μ_{-1} (as is true in general). However, the most interesting self-dual codes occur for the Hermitian inner product. As an even-like duadic code is self-orthogonal iff its splitting is given by μ_{-1}, an even-like duadic code is Hermitian self-orthogonal iff its splitting is given by μ_{-2} [10]. The following theorems are also analogous to the general case.

Theorem 9. [10] Every quaternary Hermitian self-dual, extended cyclic code is an extended odd-like duadic code whose splitting is given by μ_{-2}.

Theorem 10. [10]. Let D_1 and D_2 be odd-like duadic, quaternary codes and C_1 and C_2 be the associated even-like duadic codes. If the splitting can be given by either μ_2 or μ_{-1} then C_1 and D_2 (also C_2 and D_1) are Hermitian-dual codes.

There exist quaternary duadic codes of any odd length n. The situation for prime lengths p is given next. Analogous results for composite lengths can be found in [10].

Theorem 11. [10]. If p is an odd prime, D an odd-like and C its associated even-like quaternary code of length p, then the following hold.

a) If $p \equiv -1 \pmod 8$, $\mu_2(D) = D$ and D has a binary generating set. Any splitting can be given by either μ_{-1} or μ_{-2}.
   C is self-orthogonal (ordinary and Hermitian).

b) If $p \equiv 3 \pmod 8$, $\mu_{-2}(D) = D$ and any splitting can be given by either μ_{-1} or μ_2.
The binary subcode of D is $\langle h \rangle$.
C is self-orthogonal.

c) If $p \equiv -3 \pmod 8$, $\mu_{-1}(D) = D$ and any splitting can be given by either μ_{-2} or μ_2.
The binary subcode of D is $\langle h \rangle$.
C is Hermitian self-orthogonal.

d) If $p \equiv 1 \pmod 8$, we have the three situations above depending on the order of 2 (mod p). In all cases we have additional splittings to the ones given above.

Different things can be said about cyclic projective planes occurring as minimum weight vectors in binary duadic codes and in quaternary duadic codes as we see in the next section.

III. Duadic Codes and Designs.

We have already seen a relation between the minimum weight of duadic codes and lines of a cyclic projective plane. Consider the binary cyclic code D generated by lines of a cyclic projective plane of order n. Then D is a code of length $n^2 + n + 1$. When n is even, D must be $\langle h \rangle \perp C$ where C is self-orthogonal (since the extension of D is self-orthogonal). When $n \equiv 2 \pmod 4$ the method of invariant factors shows that $\dim(D) = \frac{n^2+2}{2} + 1$ so that $\dim C = \frac{n^2+n}{2}$ and $C = D^{\perp}$. Hence C and D are duadic codes. Since μ_2 leaves D fixed, it must send the lines, the only vectors of minimum weight, onto themselves. As μ_2 has a fixed point 0, it also has a fixed line which is then the generating idempotent e of D. By the equation which e and $\mu_{-1}(e)$ must satisfy, $wt(e) = \frac{n^2+n}{2} = n + 1$ so that $n = 2$. This demonstrates the following theorem.

Theorem 12. The only cyclic projective plane of order $n \equiv 2 \pmod 4$ is the plane of order 2.

If a cyclic projective plane of even order n exists, its quaternary code D has a self-orthogonal extension. Hence D must be contained in a duadic code whose splitting is given by μ_{-1} if such a code exists. Further non-existence criteria can be found since for many lengths when the splitting is given by μ_{-1}, it is also given by μ_2, and in the latter situation h is the only binary vector in the duadic code. As the lines are also binary vectors, this contradiction demonstrates the non-existence of a cyclic plane under conditions given in the following theorem.

Theorem 13. Let n be even and $n^2 + n + 1 = \Pi p_i^{a_i}$, for odd primes p_i. Then there is no cyclic projective plane of order n when either all $p_i \equiv 3 \pmod 8$ or all $p_i \equiv 5 \pmod 8$ and occasionally (see [10] for details) when also some of the $p_i \equiv 1 \pmod 8$.

Cyclic projective planes of certain orders must be contained in duadic codes over certain fields [10, 11, 16].

Theorem 14. Let P be the code over $GF(q)$ with $q = p^{2^t}$ generated by a cyclic projective plane of order p^s. Then there is a duadic code containing P when 2^t divides s but no higher power of 2 divides s. Further, the splitting is given by μ_{-1}.

If $q = 2$, a plane of order 2^s is contained in a binary duadic code iff s is odd.

If $q = 2^2$, a plane of order 2^s is contained in a quaternary duadic code iff either s is odd or $s \equiv 2 \pmod 4$.

For $n = 2$, there are two (equivalent) cyclic projective planes and they generate the two odd-like (7,4,3) binary duadic codes. For $n = 4$, there are two (equivalent) cyclic projective planes. They each generate a 10 dimensional cyclic code over $GF(4)$. Call them P_1 and P_2. Then P_1 is contained in two odd-like (21,11,5) quaternary duadic codes D_1 and D_1'. Similarly for P_2 and odd-like duadic codes D_2 and D_2'. Now D_1 and D_2 are paired odd-like duadic codes with splitting given by μ_{-1} [10]. Similarly for D_1' and D_2'. The group of these duadic codes was shown [8] to be the subgroup of index 2 in the group of the order 4 plane. Also μ_2, which sends the plane to itself, sends $D_1(D_1')$ to $D_2(D_2')$. The group of these codes is $PGL(3,4)$ of order $4^3(4^3-1)(4^2-1) = 60,480$.

For $n = 8$, there are 8 projective planes, each of which spans a cyclic code of dimension 28. There are two sets of 8 duadic codes containing all these planes [13]. The codes in each set of 8 are equivalent to each other, but the codes in the two sets are not equivalent. The codes in one set have as their group, the group of the order 8 plane, of order 49, 448, 448 [8]. The codes in the other set have as their group a group of order 657 [8].

In addition to vectors of minimum weight holding a projective plane, vectors of a fixed weight in several other duadic codes hold t-designs. The most famous of these are the 5-designs held by vectors of any given weight in either the binary extended quadratic residue (Golay) (24,12,8) code or the ternary extended quadratic residue (Golay) (12,6,6) code. Besides these we just mention the 5-designs held by words of weights 12,14,16 in the quaternary extended duadic (30,15,12) code or words of weights 8 and 10 in the quaternary extended duadic (18,9,8) code [10].

Duadic codes are also related to tournaments [4,8]. A **tournament** is a (0,1) matrix A satisfying

$$(1) \qquad A + A^t = J - I$$

where J is the all-one matrix.

A tournament is called **even** if every two rows of A are orthogonal over $GF(2)$. A **tournament is cyclic** if each row of A is a cyclic shift of the previous row. Let r be the weight of a row of a matrix A of order n. Then by (1), $n = 2r + 1$. It was shown in [4] that a cyclic even tournament can exist iff $r \equiv 0$ or $1 \pmod 4$. Note that μ_{-1} sends the rows of A onto rows of A^t. Hence when $r \equiv 0 \pmod 4$, (1) shows that the rows of A generate an even-like duadic code whose splitting is given by μ_{-1}. The first row of A (A^t) is the generating idempotent. Conversely, the idempotents of even-like duadic codes whose splitting is given by μ_{-1} and their shifts constitute rows of a cyclic even tournament of order $n \equiv 1 \pmod 8$. The next theorem follows from this.

Theorem 15. Cyclic even tournaments of order $n = \Pi\, p_i^{a_i}$, with even weight rows exist
iff $n \equiv 1 \pmod 8$ with $\mathrm{ord}_2(p_i)$ odd for each i
iff there are binary cyclic codes of length $n \equiv 1 \pmod 8$ with self-dual extensions
iff there are binary duadic codes of length $n \equiv 1 \pmod 8$ with splitting given by μ_{-1}.
See [4,8,13].

When r is odd and A is obtained from a normalized cyclic Hadamard matrix, then $A + wI$ and $A^t + \bar{w}I$ are quaternary duadic codes with splitting μ_{-1} [5,8]. These can only exist if the order n of A is $\equiv 3 \pmod 8$. As above it is possible to give conditions on the prime factors of n for the tournament to exist.

Much of the theory of duadic codes holds if we consider abelian group codes instead of cyclic codes. Then duadic codes are related to difference sets in an abelian group G [16]. A projective plane is one example of a difference set.

IV. Triadic and Polyadic Codes.

A different direction of generalization is to three or more codes. It was not immediately apparent how this should go. Let us start with triadic codes [14]. What is clear is that we need a multiplier μ_a which acts on three cyclic codes $C_1 = \langle e_1\rangle$, $C_2 = \langle e_2\rangle$, and $C_3 = \langle e_3\rangle$.

Distinct **even-like** cyclic codes C_1, C_2 and C_3 are **triadic codes** if there is a μ_a so that

(1) $\mu_a(C_1) = C_2$, $\mu_a(C_2) = C_3$ and $\mu_a(C_3) = C_1$, and

(2) $e_1 + e_2 + e_3 - 2e_1e_2e_3 = 1 - h'$.

It is easy to show from (2) that all pairwise intersections are equal, i.e. $C_i \cap C_j = C_1 \cap C_2 \cap C_3$ and $C_1 + C_2 + C_3 = E$. For general polyadic codes, this latter fact must be added as part of the definition [1]. Hence the even-like triadic codes C_1, C_2, and C_3 are equivalent by a multiplier, have equal pairwise intersections and sum to the $(n-1)$-dimensional space of all even-like vectors. They can be defined in this way. In fact, all polyadic codes are defined in three ways and all these definitions are equivalent. One is the idempotent definition, another gives conditions on ideals in R_n and the third is by generating polynomials.

Given three even-like triadic codes as above, we define three **odd-like triadic** codes to be $D_1 = \langle 1 - e_1\rangle$, $D_2 = \langle 1 - e_2\rangle$ and $D_3 = \langle 1 - e_3\rangle$. Each D_i is **complementary** to C_i, i.e. $D_i + C_i = R_n$ and $D_i \cap C_i = 0$. If we let $f_i = \langle 1 - e_i\rangle$, then the f_i satisfy the following conditions.

$$f_i + f_j - f_if_j \text{ are equal for all } i \neq j \text{ and}$$

$$f_1f_2f_3 = h'$$

This means that the pairwise sum of any two D_i is the same and the intersection of all three is $\langle h' \rangle$.

If we look at the codes $D_i' = \langle h' + e_i \rangle$, we see three more odd-like codes permuted by μ_a and $C_i' = \langle 1 - h' - e_i \rangle$ are three even-like codes also permuted by μ_a. The C_i and D_i defined above are called the even-like and odd-like triadic codes of class I. The C_i' and D_i' are the even-like and odd-like triadic codes of class II. There are two classes of polyadic codes also. These classes collapse in the duadic case.

The idempotents e_i' of the even-like triadic codes of class II satisfy the following.

$$e_i' + e_j' - e_i'e_j' \text{ is the same for all } i \neq j \text{ and}$$
$$e_1'e_2'e_3' = 0$$

Hence their pairwise sums are the same and their intersections are 0.

The idempotents f_i' of the odd-like codes of class II satisfy the following.

$$f_i'f_j' \text{ are the same for all } i \neq j \text{ and}$$
$$f_1' + f_2' + f_3' - 2f_1'f_2'f_3' = 1.$$

Hence their pairwise intersections are the same and their sum is the whole space.

Unlike the duadic case, the dimensions of triadic or polyadic codes are not completely determined. Let k be the dimension of an even-like code of class I and ℓ be the dimension of the intersection of two of them.

Then it can be shown that

$$n - 1 = 3k - 2\ell.$$

An analogous formula holds for $m \geqq 3$ codes in the polyadic case. The dimensions of all the other codes in class I and class II can be determined from k.

There is a bound on the minimum odd-like weight.

Theorem 16. If d_0 is the minimum odd-like weight in an odd-like triadic code of length n in class I, then

$$d_0^3 \geqq n.$$

Again the most interesting situation occurs for codes of prime length p. We again let $G = GF(p)^*$ and H the subgroup generated by q. In an analogous fashion as for duadic codes, it can be shown that triadic codes over $GF(q)$ of length p exist iff 3 divides the index of H in G.

Theorem 17. Triadic (or polyadic for m) codes of prime length p over $GF(q)$ exist iff q is a cubic (m-adic) residue mod p.

We can also generalize quadratic residue codes to cubic (and m-adic) residue codes of prime length p. In this situation ℓ, the dimension of the intersection of two even-like class I codes is 0. This implies that the sum of two odd-like class I codes is R_p, the union of two even-like class II codes is E, and the intersection of two odd-like class II codes is $\langle h' \rangle$. Again it is often possible to construct binary idempotents for m-adic residue codes. Further properties of m-adic residue codes and the minimum weights of many of the codes of modest length are in [6].

References.

1. R. A. Brualdi and V. Pless, "Polyadic Codes," *Discrete Applied Math.*, **25**(1989), 3-17.
2. W. C. Huffman, V. Job, and V. Pless, "Multipliers and Generalized Multipliers of Cyclic Objects and Cyclic Codes," *J. Comb. Theory A*, **63**(1993).
3. N. Ito, "A Characterization of Quadratic Residue Codes," *Mathematical Journal of Okayama University*, **28**(1986), 1-5.
4. N. Ito, "On Cyclic even tournaments," preprint.
5. N. Ito, "On Hadamard Tournaments," *Journal of Algebra*, **2**(1990), 432-443.
6. V. Job, "*M*-adic Residue Codes," *IEEE Trans. Inform. Theory*, **IT-38**(1992), 496-501.
7. J. S. Leon, J. M. Masley, V. Pless, "Duadic Codes," *IEEE Trans. Inform. Theory*, **IT-30**(1984), 709-714.
8. J. Nemoyer, Ph.D. Thesis, University of Illinois at Chicago, 1993.
9. V. Pless, "Introduction to the Theory of Error-Correcting Codes," second edition, John Wiley and Sons, New York, 1989.
10. V. Pless, "*Q*-Codes," *J.Comb. Theory A*, **43**(1986), 258-276.
11. V. Pless, "Cyclic Projective Planes and Binary, Extended Cyclic Self-Dual Codes," *J. Comb. Theory A*, **43**(1986), 331-333.
12. V. Pless, "Duadic Codes Revisited," *Congressus Numerantium*, **59**(1987), 225-233.
13. V. Pless, J. M. Masley, and J. S. Leon, "On Weights in Duadic Codes," *J. Comb. Theory A*, **44**(1987), 6-21.
14. V. Pless, J. J. Rushanan, "Triadic Codes," *Linear Algebra Appl.*, **98**(1988), 415-433.
15. J. J. Rushanan, "Generalized *Q*-Codes," Ph.D. Thesis, Caltech., 1986.
16. J. J. Rushanan, "Duadic Codes and Difference Sets," *J. Comb. Theory A*, **57**(1991), 254-261.
17. M. H. M. Smid, "Duadic Codes," *IEEE Trans. Inform. Theory*, **IT-33**(1987), 432-433.
18. H. C. A. van Tilborg, "On weights in codes," Department of Mathematics, Tech. Univ. of Eindhoven, #71, The Netherlands, 1971.

COSET WEIGHT ENUMERATORS OF THE EXTREMAL SELF-DUAL BINARY CODES OF LENGTH 32

P. Camion
INRIA, Rocquencourt, Le Chesnay, France

B. Courteau and A. Monpetit
University of Sherbrooke , Sherbrooke, Quebec, Canada

1 Introduction

Conway and Pless have enumerated in [6] the 85 non-equivalent self-dual doubly-even codes of length 32. From this enumeration it happened that five codes were extremal, having the largest possible minimum distance which is 8 in length 32. Two of these five codes were already known, the second order Reed-Muller code RM32 and the extended quadratic residue code QR32 of length 32. The other three denoted $2g_{16}$, $8f_4$ and $16f_2$ by Conway and Pless were new. In [8], Koch has given another more direct construction of these Conway–Pless codes denoted by him $F = 2g_{16}, U = 8f_4$ and $G = 16f_2$. Since the five extremal codes, though pairwise non-equivalent, have the same weight enumerator it is natural to try to calculate the coset weight enumerators and to ask for some parameters which may distinguish them.

In [3] we have introduced a new parameter, the regularity number $\bar{r}$ of a code, which is related to other fundamental parameters by the inequalities

$$e \leq \rho \leq t \leq \gamma \leq \bar{r}$$

where e is the error correcting capacity, ρ the covering radius, t the external distance (in the linear case t is the number of non-zero weight of the dual code) and γ is the number of distinct proper coset weight enumerators of the given code. In [4, 2, 9] we have developed theoretical tools based on the notion of partition design that permit us to calculate (in principle) the coset weight enumerators once a subgroup of the automorphism group of the code is given, more precisely when the orbit space of this group is computable. The situation appeared for example in [5] where the necessary notions and theorems are stated.

In this work we obtain the coset weight enumerators, the parameters γ and $\bar{r}$ and also the so-called regularity matrices $\overline{M}$ of the five extremal codes of length 32 obtained

*INRIA, Domaine de Voluceau — Rocquencourt, B. P. 105, 78153 Le Chesnay, France.
†Département de mathématiques et d'informatique, Université de Sherbrooke, Sherbrooke (Québec) J1K 2R1. This research has been supported by NSERC grant A5120.

by Conway and Pless. The coset weight enumerators for QR32 have already been calculated by Assmus and Pless in [1].

We have observed that the codes G and U have the same classical parameters e, ρ, t and γ, but different regularity numbers. So, the regularity number contains new information about a code not contained in the classical parameters.

2 Combinatorial matrix and partition designs

Let $\mathbb{F} = GF(2)$ and let C be a linear e-error correcting code of length n and dimension $n-k$ with $e \geq 1$. Let H be a parity check matrix for C and let $\Omega \subseteq \mathbb{F}^k$ be the set of columns of H. We shall denote C by $C(\Omega)$.

We recall in this particular case some notions and theorems of [4, 2, 3] necessary for our calculations.

Definition 1. *The combinatorial matrix A of the code C is the $(2^k \times \infty)$ matrix whose entry in position (h,j), $h \in \mathbb{F}^k$, $j \in \mathbb{N}$ is*

$$A(h,j) = \text{card}\left\{(a_1,\dots,a_j) \;\middle|\; h = \sum_{i=1}^{j} a_i, a_i \in \Omega\right\}.$$

The reduced distance distribution matrix [7] of C is the $2^k \times (n+1)$ matrix B whose entry in position (h,j), $h \in \mathbb{F}^k$, $j \in \{0,1,\dots,n\}$ is

$$B(h,j) = \text{card}\{y \in C | d(x,y) = j\}$$

where $x \in \mathbb{F}^n$ is such that $h = Hx^T$ is its syndrome.

Remark 1. *The distinct rows of B are the distinct coset weight enumerators of the code C.*

Remark 2. *If $x \in \mathbb{F}^n$ is such that $h = Hx^T$, then $A(h,j) = A(x,j) =$ number of paths of length j joining x to the code $C = C(\Omega)$ in the Hamming graph.*

Theorem 1. *$A = BS$ where S is an upper triangular $(n+1) \times \infty$ matrix with $S_{ii} = i!$ for $i \in \{0,1,\dots,n\}$. The matrix $S = (S_{ij})$ may be determined by the following recurrence over its columns: for $j \geq 1, i \geq 0$*

$$S_{ij} = iS_{i-1,j-1} + (n-i)S_{i+1,j-1}$$

with initial conditions

$$S_{0,0} = 1, S_{0,1} = 0, S_{1,1} = 1, \quad \text{and} \quad S_{i,j} = 0 \quad \text{for} \quad i > j \geq 0.$$

Definition 2. *A r-partition design on* $(\mathbb{F}^k, \Omega)$ *is a partition* $\pi = \{\Omega_0, \Omega_1, \ldots, \Omega_r\}$ *of* $\mathbb{F}^k$ *such that* $\Omega_0 = \{0\}$, $\Omega_1 = \Omega$ *and for* $u, v \in \{0, 1, \ldots, r\}$

$$m_{uv} = \mathrm{card}((h - \Omega) \cap \Omega_v)$$

is a constant for all $h \in \Omega_u$. *The code* $C = C(\Omega)$ *is said to admit the partition design* π. *The matrix* $M = (m_{uv})$ *is called the associate matrix of* π. *The regularity number is the least number* r *such that* C *admits a* r*-partition design and the regularity matrix which is unique up to a permutation of the indices [9] is the corresponding associate matrix* $\overline{M}$.

Geometrically, if $E_u = \{x \in \mathbb{F}^n \mid Hx^T \in \Omega_u\}$, then $\pi = \{E_u\}$ a partition of $\mathbb{F}^n$ such that $m_{uv} = \mathrm{card}\{y \in E_v \mid d(x,y) = 1\}$ is independent of the choice of x in E_u. This geometric property of the numbers $m_{u,v}$ means that the associate matrix M doesn't depend on the choice of the parity check matrix H of C. We shall also refer to this partition as a partition design of the surrounding space $\mathbb{F}^n$ admitted by the code $C = C(\Omega)$.

Proposition 1. *Let* G *be a subgroup of* $GL(k, 2)$. *If* G *acts transitively on* Ω, *then the set of orbits* $\{\Omega_0 = \{0\}, \Omega_1 = \Omega, \ldots, \Omega_r\}$ *under* G *is a partition design admitted by* $C = C(\Omega)$.

Theorem 2. *If* C *admits the* r*-partition design* $\pi = \{\Omega_0, \ldots, \Omega_r\}$ *then*

(1) $A_j(h) = A_j(g)$ *for any* $h, g \in \Omega_u$, $u \in \{0, \ldots, r\}$;

(2) *The numbers* $A_j(u) = A_j(h)$, $h \in \Omega_u$ *satisfy the recurrence relation*

$$A_j(u) = \sum_{v=0}^{r} m_{uv} A_{j-1}(v)$$

with $u \in \{0, \ldots, r\}$, $j \geq 1$. *By definition* $A_0(0) = 1$ *and* $A_0(u) = 0$ *for* $u \neq 0$;

(3) *The combinatorial partition* π_0 *defined by the equivalence relation*

$$h \equiv g \iff A_j(h) = A_j(g) \quad \text{for all } j \geq 0$$

is coarser than the partition design π. *If* $\gamma + 1$ *is the number of distinct rows in* A, *then* $\gamma \leq r$;

(4) *There is a* $\overline{r}$*-partition design* $\overline{\pi}$ *with least possible number* $\overline{r}+1$ *of classes which lies between the combinatorial partition (which is not in general a partition design) and the given partition design* π. *The corresponding associate matrix* $\overline{M}$ *is unique not depending on the choice of* π. $\overline{M}$ *is called the regularity matrix of the code* C.

To determine the regularity number $\overline{r}$ of a code C we have the following result. We shall use $\pi(g)$ to denote the class modulo π containing g and we shall write

$$m_{\pi(g),\pi(h)} = m_{uv}$$

when $\pi = \{\Omega_u \mid 0 \leq u \leq r\}$ and $\pi(g) = \Omega_u$, $\pi(h) = \Omega_v$.

Theorem 3. *Let π_0 be the combinatorial partition associated to C and let π be any partition design admitted by C with associate matrix $M = (m_{uv})$. Then the partition design $\overline{\pi}$ admitted by C and having the minimum number of classes and its associate matrix matrix $\overline{M}$ are obtained by the following algorithm:*

(1) *For $P \in \pi_0$ define on P the equivalence relation R_P as follows: for g, $h \in P$*

$$g\ R_P\ h \iff \sum_{\pi(\ell)\subseteq Q} m_{\pi(h),\pi(\ell)} = \sum_{\pi(\ell)\subseteq Q} m_{\pi(g),\pi(\ell)} \quad \text{for all } Q \in \pi_0.$$

Let π_1 be the refinement of π_0 so obtained. If π_1 is different from π_0, then set $\pi_0 := \pi_1$.

(2) If $\pi_1 = \pi_0$ (that means that π_0 is a partition design) then return $\overline{\pi} = \pi_1$ and $\overline{M} = (\overline{m}_{\pi_1(g),\pi_1(h)})_{\pi_1(g),\pi_1(h)\in\pi_1}$ with $\overline{m}_{\pi_1(g),\pi_1(h)} = \sum_{\pi(\ell)\subseteq\pi_1(h)} m_{\pi(g),\pi(\ell)}$, else set $\pi_0 := \pi_1$ and go to (1).

3 Results

To apply Proposition 1, we take a permutation group $G \subseteq S_n$ letting the code $C = \ker H$ invariant and we let it act on the syndrome space $\mathbb{F}^k$ as follows.

If $\sigma \in G$ and $h \in \mathbb{F}^k$, let x be any element in $\mathbb{F}^n$ such that $h = Hx^T$ is the syndrome of x. Then we define $\sigma(h) = H(\sigma x)^T$ where $\sigma x = \sigma(x_1, \ldots, x_n) = (x_{\sigma(1)}, \ldots, x_{\sigma(n)})$. This is well defined because if y is such that $Hy^T = Hx^T = h$ then $y - x \in C$, $\sigma(y) - \sigma(x) = \sigma(y - x) \in C$ and so $H(\sigma(y))^T - H(\sigma(x))^T = H(\sigma(y-x))^T = 0$.

We have written programs in the computer algebra system Maple that implement the above theorems. To obtain the orbit partition designs for the extended quadratic residue code and the second order Reed-Muller code we have taken the full automorphism groups $PSL_2(31)$ and $GA(5,2)$ respectively. For the code $G = 16f_2$, we have taken a subgroup of its automorphism group generated by 12 automorphisms fixing the set of glue components has given 316 orbits in the syndrome space. For the two last codes U and F we have taken subgroups of order 5376 and 9216 respectively given by Mrs. Vera Pless [10]. The results appear in Tables 1 to 8 where we have ordered the codes by increasing regularity number. We have given first the parameters d, ρ, t, γ, $\overline{r}$, then the $(\overline{r}+1)\times(\overline{r}+1)$ matrix $\overline{M}$ associated to the partition design $\overline{\pi}$ having the least possible number $\overline{r}+1$ of classes and finally the distinct coset weight enumerators.

These enumerators are symmetric because the all-one vector belongs to the codes considered. So we give only the first 17 components. As a consequence of Theorems 2 and 1, the matrix $\overline{M}$ contains all the information necessary to recover the matrix B.

The remarkable fact that all enumerators of the cosets of weight 1,2,3,5 and 6 are the same for the five codes has already been observed by Assmus and Pless [1]. We have observed that the codes G and U have the same parameters $d = 8$, $\rho = 6$, $t = 6$ and $\gamma = 11$ but different regularity numbers $\overline{r}(G) = 21$, $\overline{r}(U) = 19$. This shows that the regularity number contains new information about a code which is not contained in the classical parameters d, ρ, t or γ.

We thank Mrs. Vera Pless for her kind help in furnishing the basic data about the codes U and F.

References

[1] E.F. Assmus, V. Pless, *On the covering radius of extremal self-dual codes*, IEEE Trans. on Inform. Theory, **29** (3) (1983) 359–363.

[2] P. Camion, B. Courteau, P. Delsarte, *On r-partition designs in Hamming spaces*, Tech. Rep. 626, INRIA, 1987.

[3] P. Camion, B. Courteau and P. Delsarte, *On r-partition designs in Hamming spaces*, Applicable Algebra in Engineering, Commun. and Comput., **2** (1992) 147–162.

[4] P. Camion, B. Courteau, G. Fournier and V. S. Kanetkar, *Weight distributions of Translates of linear codes and generalized Pless identities*, J. Inform. Optim. Sci. **8** (1987) 1–23.

[5] P. Camion, B. Courteau and A. Montpetit, *Weight distribution of cosets of* 2-*error-correcting binary BCH codes of length* 15, 63 *and* 255, IEEE Trans. on Information Theory **38** (1992) 1353-1357.

[6] J.N. Conway, V. Pless, *On the enumeration of self-dual codes*, J. Combin. Theory, Ser. A **28** (1980) 26–53.

[7] P. Delsarte, *Four fundamental parameters of a code and their combinatorial significance*, Inform. and Control, 23 (1973) 407-438.

[8] H. Koch, *On self-dual, doubly even codes of length 32*, J. Combin. Theory, Ser. A **51** (1989) 63–76.

[9] A. Montpetit, *Codes dans les graphes réguliers*, Ph. D. thesis, Université de Sherbrooke, Canada, 1987.

[10] V. Pless, *personal communication*, jan. 1993.

Table 1: The second order Reed-Muller code RM(2,5)

$$d = 8, \rho = 6, t = 6, \gamma = 7, \overline{r} = 7$$

$$\overline{M} = \begin{bmatrix} 0 & 32 & 0 & 0 & 0 & 0 & 0 & 0 \\ 1 & 0 & 31 & 0 & 0 & 0 & 0 & 0 \\ 0 & 2 & 0 & 30 & 0 & 0 & 0 & 0 \\ 0 & 0 & 3 & 0 & 1 & 28 & 0 & 0 \\ 0 & 0 & 0 & 32 & 0 & 0 & 0 & 0 \\ 0 & 0 & 0 & 8 & 0 & 0 & 24 & 0 \\ 0 & 0 & 0 & 0 & 0 & 15 & 0 & 17 \\ 0 & 0 & 0 & 0 & 0 & 0 & 32 & 0 \end{bmatrix}$$

coset weight	coset weight enumerators 0	1	2	3	4	5	6	7	8	9	10	11	12	13	14	15	16	number of cosets
0	1	0	0	0	0	0	0	0	620	0	0	0	13888	0	0	0	36518	1
1	0	1	0	0	0	0	0	155	0	465	0	5208	0	8680	0	18259	0	32
2	0	0	1	0	0	0	35	0	240	0	2193	0	6720	0	14155	0	18848	496
3	0	0	0	1	0	7	0	84	0	892	0	3949	0	10507	0	17328	0	4960
4	0	0	0	0	2	0	24	0	324	0	1976	0	6878	0	14384	0	18360	17360
4	0	0	0	0	8	0	0	0	336	0	2048	0	6776	0	14336	0	18528	155
5	0	0	0	0	0	6	0	106	0	850	0	3934	0	10620	0	17252	0	27776
6	0	0	0	0	0	0	32	0	320	0	1952	0	6912	0	14400	0	18304	14756

Table 2: The Conway-Pless code F

$d = 8, \rho = 6, t = 6, \gamma = 9, \overline{r} = 12$

$$\overline{M} = \begin{bmatrix} 0 & 32 & 0 & 0 & 0 & 0 & 0 & 0 & 0 & 0 & 0 & 0 & 0 \\ 1 & 0 & 31 & 0 & 0 & 0 & 0 & 0 & 0 & 0 & 0 & 0 & 0 \\ 0 & 2 & 0 & 14 & 16 & 0 & 0 & 0 & 0 & 0 & 0 & 0 & 0 \\ 0 & 0 & 3 & 0 & 0 & 1 & 12 & 16 & 0 & 0 & 0 & 0 & 0 \\ 0 & 0 & 3 & 0 & 0 & 0 & 0 & 14 & 7 & 8 & 0 & 0 & 0 \\ 0 & 0 & 0 & 32 & 0 & 0 & 0 & 0 & 0 & 0 & 0 & 0 & 0 \\ 0 & 0 & 0 & 8 & 0 & 0 & 0 & 0 & 0 & 0 & 8 & 16 & 0 \\ 0 & 0 & 0 & 2 & 6 & 0 & 0 & 0 & 0 & 0 & 0 & 24 & 0 \\ 0 & 0 & 0 & 0 & 16 & 0 & 0 & 0 & 0 & 0 & 0 & 16 & 0 \\ 0 & 0 & 0 & 0 & 4 & 0 & 0 & 0 & 0 & 0 & 0 & 28 & 0 \\ 0 & 0 & 0 & 0 & 0 & 0 & 15 & 0 & 0 & 0 & 0 & 0 & 17 \\ 0 & 0 & 0 & 0 & 0 & 0 & 1 & 8 & 1 & 8 & 0 & 0 & 14 \\ 0 & 0 & 0 & 0 & 0 & 0 & 0 & 0 & 0 & 0 & 32 & 0 & 0 \end{bmatrix}$$

coset	coset weight enumerators																	number of
weight	0	1	2	3	4	5	6	7	8	9	10	11	12	13	14	15	16	cosets
0	1	0	0	0	0	0	0	0	620	0	0	0	13888	0	0	0	36518	1
1	0	1	0	0	0	0	0	155	0	465	0	5208	0	8680	0	18259	0	32
2	0	0	1	0	0	0	35	0	240	0	2193	0	6720	0	14155	0	18848	496
3	0	0	0	1	0	7	0	84	0	892	0	3949	0	10507	0	17328	0	4960
4	0	0	0	0	1	0	28	0	322	0	1964	0	6895	0	14392	0	18332	7680
4	0	0	0	0	2	0	24	0	324	0	1976	0	6878	0	14384	0	18360	10640
4	0	0	0	0	4	0	16	0	328	0	2000	0	6844	0	14368	0	18416	1680
4	0	0	0	0	8	0	0	0	336	0	2048	0	6776	0	14336	0	18528	35
5	0	0	0	0	0	6	0	106	0	850	0	3934	0	10620	0	17252	0	27776
6	0	0	0	0	0	0	32	0	320	0	1952	0	6912	0	14400	0	18304	12236

Table 3: The extended quadratic residue code QR(32)

$$d = 8, \rho = 6, t = 6, \gamma = 10, \overline{\tau} = 13$$

$$\overline{M} = \begin{bmatrix}
0 & 32 & 0 & 0 & 0 & 0 & 0 & 0 & 0 & 0 & 0 & 0 & 0 & 0 \\
1 & 0 & 31 & 0 & 0 & 0 & 0 & 0 & 0 & 0 & 0 & 0 & 0 & 0 \\
0 & 2 & 0 & 30 & 0 & 0 & 0 & 0 & 0 & 0 & 0 & 0 & 0 & 0 \\
0 & 0 & 3 & 0 & 9 & 6 & 3 & 10 & 1 & 0 & 0 & 0 & 0 & 0 \\
0 & 0 & 0 & 4 & 0 & 0 & 0 & 0 & 0 & 14 & 4 & 6 & 4 & 0 \\
0 & 0 & 0 & 12 & 0 & 0 & 0 & 0 & 0 & 6 & 0 & 6 & 8 & 0 \\
0 & 0 & 0 & 16 & 0 & 0 & 0 & 0 & 0 & 16 & 0 & 0 & 0 & 0 \\
0 & 0 & 0 & 8 & 0 & 0 & 0 & 0 & 0 & 12 & 4 & 4 & 4 & 0 \\
0 & 0 & 0 & 20 & 0 & 0 & 0 & 0 & 0 & 0 & 12 & 0 & 0 & 0 \\
0 & 0 & 0 & 0 & 11 & 1 & 1 & 6 & 0 & 0 & 0 & 0 & 0 & 13 \\
0 & 0 & 0 & 0 & 15 & 0 & 0 & 5 & 1 & 0 & 0 & 0 & 0 & 11 \\
0 & 0 & 0 & 0 & 9 & 3 & 0 & 6 & 0 & 0 & 0 & 0 & 0 & 14 \\
0 & 0 & 0 & 0 & 12 & 4 & 0 & 3 & 0 & 0 & 0 & 0 & 0 & 13 \\
0 & 0 & 0 & 0 & 0 & 0 & 0 & 0 & 0 & 18 & 6 & 6 & 2 & 0
\end{bmatrix}$$

Table 4: The extended quadratic residue code QR(32)

coset weight	coset weight enumerators																	number of cosets
	0	1	2	3	4	5	6	7	8	9	10	11	12	13	14	15	16	
0	1	0	0	0	0	0	0	0	620	0	0	0	13888	0	0	0	36518	1
1	0	1	0	0	0	0	0	155	0	465	0	5208	0	8680	0	18259	0	32
2	0	0	1	0	0	0	35	0	240	0	2193	0	6720	0	14155	0	18848	496
3	0	0	0	1	0	7	0	84	0	892	0	3949	0	10507	0	17328	0	4960
4	0	0	0	0	1	0	28	0	322	0	1964	0	6895	0	14392	0	18332	11160
4	0	0	0	0	2	0	24	0	324	0	1976	0	6878	0	14384	0	18360	6200
4	0	0	0	0	3	0	20	0	326	0	1988	0	6861	0	14376	0	18388	2480
4	0	0	0	0	4	0	16	0	328	0	2000	0	6844	0	14368	0	18416	930
4	0	0	0	0	5	0	12	0	330	0	2012	0	6827	0	14360	0	18444	248
5	0	0	0	0	0	6	0	106	0	850	0	3934	0	10620	0	17252	0	27776
6	0	0	0	0	0	0	32	0	320	0	1952	0	6912	0	14400	0	18304	11253

Table 5: The Conway-Pless code U

$d=8, t=6, \rho=6, \gamma=11, \overline{r}=19$

$$\overline{M}=\begin{bmatrix}
0&32&0&0&0&0&0&0&0&0&0&0&0&0&0&0&0&0&0&0\\
1&0&3&28&0&0&0&0&0&0&0&0&0&0&0&0&0&0&0&0\\
0&2&0&0&2&28&0&0&0&0&0&0&0&0&0&0&0&0&0&0\\
0&2&0&0&0&6&24&0&0&0&0&0&0&0&0&0&0&0&0&0\\
0&0&3&0&0&0&0&1&28&0&0&0&0&0&0&0&0&0&0&0\\
0&0&1&2&0&0&0&0&2&1&1&1&8&8&8&0&0&0&0&0\\
0&0&0&3&0&0&0&0&0&0&0&0&3&3&3&6&8&6&0&0\\
0&0&0&0&32&0&0&0&0&0&0&0&0&0&0&0&0&0&0&0\\
0&0&0&0&2&6&0&0&0&0&0&0&0&0&0&0&0&0&24&0\\
0&0&0&0&0&24&0&0&0&0&0&0&0&0&0&0&0&0&8&0\\
0&0&0&0&0&16&0&0&0&0&0&0&0&0&0&0&0&0&16&0\\
0&0&0&0&0&8&0&0&0&0&0&0&0&0&0&0&0&0&24&0\\
0&0&0&0&0&6&6&0&0&0&0&0&0&0&0&0&0&0&20&0\\
0&0&0&0&0&4&4&0&0&0&0&0&0&0&0&0&0&0&24&0\\
0&0&0&0&0&2&2&0&0&0&0&0&0&0&0&0&0&0&28&0\\
0&0&0&0&0&0&16&0&0&0&0&0&0&0&0&0&0&0&16&0\\
0&0&0&0&0&0&8&0&0&0&0&0&0&0&0&0&0&0&24&0\\
0&0&0&0&0&0&4&0&0&0&0&0&0&0&0&0&0&0&28&0\\
0&0&0&0&0&0&0&0&3&0&0&0&3&3&3&0&0&6&0&14\\
0&0&0&0&0&0&0&0&0&0&0&0&0&0&0&0&0&0&32&0
\end{bmatrix}$$

Table 6: The Conway-Pless code U

coset weight	coset weight enumerators 0	1	2	3	4	5	6	7	8	9	10	11	12	13	14	15	16	number of cosets
0	1	0	0	0	0	0	0	0	620	0	0	0	13888	0	0	0	36518	1
1	0	1	0	0	0	0	0	155	0	465	0	5208	0	8680	0	18259	0	32
2	0	0	1	0	0	0	35	0	240	0	2193	0	6720	0	14155	0	18848	496
3	0	0	0	1	0	7	0	84	0	892	0	3949	0	10507	0	17328	0	4960
4	0	0	0	0	1	0	28	0	322	0	1964	0	6895	0	14392	0	18332	10752
4	0	0	0	0	2	0	24	0	324	0	1976	0	6878	0	14384	0	18360	6888
4	0	0	0	0	3	0	20	0	326	0	1988	0	6861	0	14376	0	18388	1792
4	0	0	0	0	4	0	16	0	328	0	2000	0	6844	0	14368	0	18416	1428
4	0	0	0	0	6	0	8	0	332	0	2024	0	6810	0	14352	0	18472	56
4	0	0	0	0	8	0	0	0	336	0	2048	0	6776	0	14336	0	18528	1
5	0	0	0	0	0	6	0	106	0	850	0	3934	0	10620	0	17252	0	27776
6	0	0	0	0	0	0	32	0	320	0	1952	0	6912	0	14400	0	18304	11354

Table 7: The Conway-Pless code G

$d = 8, \rho = 6, t = 6, \gamma = 11, \overline{r} = 21$

$$\overline{M} = \begin{bmatrix}
0 & 32 & 0 & 0 & 0 & 0 & 0 & 0 & 0 & 0 & 0 & 0 & 0 & 0 & 0 & 0 & 0 & 0 & 0 & 0 & 0 & 0 \\
1 & 0 & 31 & 0 & 0 & 0 & 0 & 0 & 0 & 0 & 0 & 0 & 0 & 0 & 0 & 0 & 0 & 0 & 0 & 0 & 0 & 0 \\
0 & 2 & 0 & 8 & 12 & 8 & 2 & 0 & 0 & 0 & 0 & 0 & 0 & 0 & 0 & 0 & 0 & 0 & 0 & 0 & 0 & 0 \\
0 & 0 & 3 & 0 & 0 & 0 & 0 & 0 & 2 & 9 & 0 & 9 & 0 & 9 & 0 & 0 & 0 & 0 & 0 & 0 & 0 & 0 \\
0 & 0 & 3 & 0 & 0 & 0 & 0 & 0 & 0 & 8 & 1 & 11 & 1 & 8 & 0 & 0 & 0 & 0 & 0 & 0 & 0 & 0 \\
0 & 0 & 3 & 0 & 0 & 0 & 0 & 0 & 1 & 6 & 0 & 10 & 3 & 9 & 0 & 0 & 0 & 0 & 0 & 0 & 0 & 0 \\
0 & 0 & 3 & 0 & 0 & 0 & 0 & 0 & 0 & 0 & 0 & 14 & 7 & 8 & 0 & 0 & 0 & 0 & 0 & 0 & 0 & 0 \\
0 & 0 & 3 & 0 & 0 & 0 & 0 & 0 & 0 & 12 & 0 & 2 & 3 & 12 & 0 & 0 & 0 & 0 & 0 & 0 & 0 & 0 \\
0 & 0 & 0 & 8 & 0 & 12 & 0 & 0 & 0 & 0 & 0 & 0 & 0 & 0 & 12 & 0 & 0 & 0 & 0 & 0 & 0 & 0 \\
0 & 0 & 0 & 2 & 4 & 4 & 0 & 2 & 0 & 0 & 0 & 0 & 0 & 0 & 0 & 10 & 6 & 4 & 0 & 0 & 0 & 0 \\
0 & 0 & 0 & 0 & 24 & 0 & 0 & 0 & 0 & 0 & 0 & 0 & 0 & 0 & 0 & 0 & 0 & 0 & 8 & 0 & 0 & 0 \\
0 & 0 & 0 & 0 & 4 & 4 & 0 & 0 & 0 & 0 & 0 & 0 & 0 & 0 & 0 & 6 & 4 & 12 & 0 & 2 & 0 & 0 \\
0 & 0 & 0 & 0 & 2 & 8 & 4 & 2 & 0 & 0 & 0 & 0 & 0 & 0 & 0 & 0 & 0 & 16 & 0 & 0 & 0 & 0 \\
0 & 0 & 0 & 0 & 0 & 4 & 0 & 0 & 0 & 0 & 0 & 0 & 0 & 0 & 4 & 0 & 0 & 24 & 0 & 0 & 0 & 0 \\
0 & 0 & 0 & 0 & 0 & 0 & 0 & 0 & 1 & 0 & 0 & 5 & 0 & 15 & 0 & 0 & 0 & 0 & 0 & 0 & 0 & 11 \\
0 & 0 & 0 & 0 & 0 & 0 & 0 & 0 & 0 & 4 & 0 & 3 & 0 & 12 & 0 & 0 & 0 & 0 & 0 & 0 & 0 & 13 \\
0 & 0 & 0 & 0 & 0 & 0 & 0 & 0 & 0 & 3 & 0 & 6 & 0 & 9 & 0 & 0 & 0 & 0 & 0 & 0 & 0 & 14 \\
0 & 0 & 0 & 0 & 0 & 0 & 0 & 0 & 0 & 1 & 0 & 6 & 1 & 11 & 0 & 0 & 0 & 0 & 0 & 0 & 0 & 13 \\
0 & 0 & 0 & 0 & 0 & 0 & 0 & 0 & 0 & 0 & 1 & 0 & 0 & 24 & 0 & 0 & 0 & 0 & 0 & 0 & 0 & 7 \\
0 & 0 & 0 & 0 & 0 & 0 & 0 & 0 & 0 & 0 & 0 & 15 & 0 & 0 & 0 & 0 & 0 & 0 & 0 & 0 & 0 & 17 \\
0 & 0 & 0 & 0 & 0 & 0 & 0 & 0 & 0 & 0 & 0 & 9 & 1 & 8 & 0 & 0 & 0 & 0 & 0 & 0 & 0 & 14 \\
0 & 0 & 0 & 0 & 0 & 0 & 0 & 0 & 0 & 0 & 0 & 0 & 0 & 0 & 0 & 8 & 8 & 16 & 0 & 0 & 0 & 0
\end{bmatrix}$$

Table 8: The Conway-Pless code G

coset weight	coset weight enumerators 0	1	2	3	4	5	6	7	8	9	10	11	12	13	14	15	16	number of cosets
0	1	0	0	0	0	0	0	0	620	0	0	0	13888	0	0	0	36518	1
1	0	1	0	0	0	0	0	155	0	465	0	5208	0	8680	0	18259	0	32
2	0	0	1	0	0	0	35	0	240	0	2193	0	6720	0	14155	0	18848	496
3	0	0	0	1	0	7	0	84	0	892	0	3949	0	10507	0	17328	0	4960
4	0	0	0	0	1	0	28	0	322	0	1964	0	6895	0	14392	0	18332	11040
4	0	0	0	0	2	0	24	0	324	0	1976	0	6878	0	14384	0	18360	6060
4	0	0	0	0	3	0	20	0	326	0	1988	0	6861	0	14376	0	18388	2880
4	0	0	0	0	4	0	16	0	328	0	2000	0	6844	0	14368	0	18416	750
4	0	0	0	0	5	0	12	0	330	0	2012	0	6827	0	14360	0	18444	160
4	0	0	0	0	6	0	8	0	332	0	2024	0	6810	0	14352	0	18472	60
5	0	0	0	0	0	6	0	106	0	850	0	3934	0	10620	0	17252	0	27776
6	0	0	0	0	0	0	32	0	320	0	1952	0	6912	0	14400	0	18304	11321

ZETA FUNCTIONS OF SOME CURVES AND MINIMAL EXPONENT FOR PELLIKAAN'S DECODING ALGORITHM OF ALGEBRAIC-GEOMETRIC CODES

Ph. Carbonne and A. Thiong Ly
University of Toulouse le Mirail, Toulouse, France

Abstract.
It is shown that the Pellikaan's decoding algorithm of some families of Goppa codes $C_\Omega(D,G)$ needs at most $(g+1)$ effective divisors if the degree of G is odd and at most $\lfloor g/2 \rfloor+1$ effective divisors if the degree is even, where g is the genus of the curve used.

0 INTRODUCTION.

Let X be a plane ,projective,nonsingular ,absolutely irreducible curve over the finite field GF(q) q a power of a prime.We suppose that the genus of X satisfies $g \geqslant 3$.We denote by J(X) the jacobian and by D_k the set of the positive divisors of X with degree k.The cardinalities of J(X) and D_k are denoted by h and by a_k.For an integer $s \geqslant 2$ the mapping Ψ_k^s from D_k^s to $J(X)^{s-1}$ is defined by:

$$\Psi_k^s(F_1,F_2,\ldots,F_s) = (\overline{F_2-F_1},\ldots,\overline{F_s-F_{s-1}})$$

where $\overline{K}$ is the class of a zero degree divisor K in the jacobian.
In this paper we determine for some families of curves the smallest integers s_1 and s_2 such that: $a_{g-1}^{s_1} < h^{s_1-1}$ and $a_{g-2}^{s_2} < h^{s_2-1}$.
In [3] R.Pellikaan describes an algorithm for decoding a residue Goppa code $C_\Omega(D,G)$ defined by the divisors D and G.See also[4].When the degree of G is odd (resp even) one must find an integer s such that the mapping Ψ_{g-1}^s (resp Ψ_{g-2}^s) is nonsurjective.The condition $s \geqslant s_1$ (resp $s \geqslant s_2$) is a sufficient condition to satisfy the nonsurjectivity.
The zeta function of X over GF(q) is denoted by Z(X,T,GF(q)).

Recall that: $$Z(X,T,GF(q)) = \frac{P(T)}{(1-T)(1-qT)} = \sum_{k=0}^{\infty} a_k T^k$$

where : $$P(T) = A_0 + A_1T + \ldots + A_{2g}T^{2g} = \prod_{j=1}^{g} (1-\alpha_j T)(1-\overline{\alpha_j} T)$$

It is known([2]) that: - $A_0=1$ and $A_{2g-j}=q^{g-j}A_j$ for $j=0,\dots,g$
- $|\alpha_j|=\sqrt{q}$ for $j=0,\dots,g$
- $h=P(1)$
- $(\sqrt{q}-1)^{2g}\leqslant h\leqslant(\sqrt{q}+1)^{2g}$

It is proved in [3] that:

$$a_{g-1}=\sum_{j=0}^{g-1}\frac{q^{g-j}-1}{q-1}A_j = \sum_{0\leqslant j+m\leqslant g-1} q^m A_j$$

$$a_{g-2}=\sum_{j=0}^{g-2}\frac{q^{g-j-1}-1}{q-1}A_j = \sum_{0\leqslant j+m\leqslant g-2} q^m A_j .$$

In the following we consider the curves X over GF(q) such that:

-$P(T)=(1+\sqrt{q}T)^{2g}$ (maximal curves reaching the Weil bound)

-$P(T)=(1-\sqrt{q}T)^{2g}$ ("minimal" curves reaching the lower Weil bound)

-$P(T)= 1+q^gT^{2g}$ (supersingular curves)

-$P(T)=(1+aT+qT^2)^g$ with $a\geqslant0$

For these curves and under some condition about g and q, the values s_1 and s_2 are given.

We obtain bounds for these values when studying a polynomial P(T) the coefficients of which are all positive.

1. CURVES SUCH THAT THE COEFFICIENTS OF P(T) ARE POSITIVE.

Lemma 1.

If the coefficients of the polynomial P(T) are positive, then:

$$a_{g-t}<\frac{h-q^{t-1}}{q^{t-1}(q-1)} \qquad 1\leqslant t<g.$$

Proof.

$$a_{g-t}=\sum_{i+j=g-t}\frac{q^{i+1}-1}{q-1}A_j$$

$$=\sum_{j=0}^{g-t}\frac{q^{g-j-t+1}-1}{q-1}A_j$$

$$=\frac{1}{q^{t-1}(q-1)}\sum_{j=0}^{g-t}A_{2g-j}-\frac{1}{q-1}\sum_{j=0}^{g-t}A_j$$

Since all the coefficients A_j are positive and since $h=P(1)$, $A_0=1$ we conclude that:

$$a_{g-t}\leqslant\frac{h}{q^{t-1}(q-1)}-\frac{1}{q-1}$$ □

From the Lemma 1 we deduce that:

$$a_{g-1}<\frac{h}{q-1} \qquad (1)$$

$$a_{g-2}<\frac{n}{q(q-1)} \qquad (2)$$

Proposition 1

If $q \geqslant g^2$ and $g \geqslant 3$, and if the coefficients of the polynomial $P(T)$ are positive, then: $s_1 \leqslant g+1$ and $s_2 \leqslant \lfloor g/2 \rfloor +1$ where $\lfloor x \rfloor$ denotes the integer part of x.

Proof.

We first prove the two inequalities:

$$q-1 \geqslant (\sqrt{q}+1)^{2g/(g+1)} \qquad (3)$$

$$q(q-1)>(\sqrt{q}+1)^{4g/(g+2)} \qquad (4)$$

The first one is equivalent to: $(\sqrt{q}-1)^{g+1} \geqslant (\sqrt{q}+1)^{g-1}$.

If $g=3$ this is equivalent to $(\sqrt{q}-1)^2 \geqslant \sqrt{q}+1$, which is true for $q \geqslant 9=g^2$.

If $g \geqslant 4$ and $q \geqslant g^2$, we have: $\sqrt{q}-1 \geqslant 4-1=3$ and $\dfrac{\sqrt{q}+1}{\sqrt{q}-1} \leqslant \dfrac{g+1}{g-1}$.

Since: $\dfrac{g-1}{2}\operatorname{Log}\left(\dfrac{g+1}{g-1}\right)=\dfrac{g-1}{2}\operatorname{Log}\left(1+\dfrac{2}{g-1}\right) \leqslant \dfrac{g-1}{2}.\dfrac{2}{g-1}=1$

We have: $\left(\dfrac{\sqrt{q}+1}{\sqrt{q}-1}\right)^{(g-1)/2} \leqslant \left(\dfrac{g+1}{g-1}\right)^{(g-1)/2} \leqslant e<3 \leqslant \sqrt{q}-1$.

Therefore: $\sqrt{q}+1)^{g-1} \leqslant (\sqrt{q}-1)^2(\sqrt{q}-1)^{g-1}=(\sqrt{q}-1)^{g+1}$.

The secund inequality may be deduced from the first one.

Indeed: $q(q-1)>(q-1)^2 \geqslant (\sqrt{q}+1)^{4g/(g+1)}> (\sqrt{q}+1)^{4g/(g+2)}$.

By (1) we have:

$$h^{g/(g+1)}-a_{g-1}>h^{g/(g+1)}-\frac{h}{q-1}=\frac{h^{g/(g+1)}}{q-1}(q-1-h^{1/(g+1)}).$$

Since $h \leqslant (1+\sqrt{q})^{2g}$ we have:

$$h^{g/(g+1)}-a_{g-1}>\frac{h^{g/(g+1)}}{q-1}(q-1-(1+\sqrt{q})^{2g/(g+1)})$$

So by (3): $h^g>a_{g-1}^{g+1}$, hence $s_1 \leqslant g+1$.

By using (2) and (4) we prove in a similar manner that $s_2 \leqslant \lfloor g/2 \rfloor+1$.□

3)The inequality $h^{g/2}>a_{g-2}^{g/2+1}$ is proved in a similar manner.■.

2.CURVES SUCH THAT $P(T)=(1+aT+qT^2)^g$, $a \geqslant 0$.

Proposition 2.

If $q \geqslant g^2$ and $g \geqslant 3$, and if $P(T)=(1+aT+qT^2)^g$ with $a \geqslant 0$, then:

$$s_1=g+1 \quad \textit{and} \quad s_2=\lfloor g/2 \rfloor+1.$$

Proof.

1)We have: $$P(T)=\sum_{i+j+k=g}\frac{g!}{i!j!k!}a^i q^j T^{i+2j}$$

Hence: $$A_t=\sum_{\substack{i+j+k=g\\ i+2j=t}}\frac{g!}{i!j!k!}a^i q^j.$$

All the coefficients are positive.So,by Proposition 1 we have:

$s_1\leqslant g+1$ and $s_2\leqslant\lfloor g/2\rfloor+1$.

2)We have : $$a_{g-1}=\sum_{0\leqslant m+t\leqslant g-1}q^m A_t=\sum_{\substack{i+j+k=g\\ 0\leqslant i+2j+m\leqslant g-1}}\frac{g!}{i!j!k!}a^i q^{j+m}.$$

On the other hand,we have:

$$h^{(g-1)/g}=(1+a+q)^{g-1}=\sum_{i+j+n=g-1}\frac{(g-1)!}{i!j!n!}a^i q^n.$$

In $h^{(g-1)/g}$ the term $a^i q^n$ has the coefficient:

$$H=\sum_{i+j+n=g-1}\frac{(g-1)!}{i!j!n!}.$$

In a_{g-1} the term for which $j+m=n$ has the coefficient:

$$K=\sum_{\substack{i+j+k=g\\ 0\leqslant i+j+n\leqslant g-1}}\frac{g!}{i!j!k!}$$

We consider in K the terms such that k=n+1.We have:

$i+j+n=i+j+k-1=g-1$ so, $n+1\leqslant g$.

The sum K' of these terms satisfies:

$$K'=\sum_{i+j+n=g-1}\frac{(g-1)!}{i!j!n!}.\frac{g}{n+1}\geqslant\sum_{i+j+n=g-1}\frac{(g-1)!}{i!j!n!}=H$$

So we have: $H\leqslant K'\leqslant K$ and we can write $h^{g-1}\leqslant a^g_{g-1}$.

Therefore $s_1>g$.By Prop 1 we conclude that $s_1=g+1$.

3)We have: $$a_{g-2}=\sum_{0\leqslant m+t\leqslant g-2}q^m A_t=\sum_{\substack{i+j+k=g\\ 0\leqslant i+2j+m\leqslant g-2}}\frac{g!}{i!j!k!}a^i q^{j+m}.$$

On the other hand we have:

$$h^{(g-2)/g}=(1+a+q)^{g-2}=\sum_{i+j+n=g-2}\frac{(g-2)!}{i!j!n!}a^i q^n$$

Let L be the coefficient of $a^i q^n$ in $h^{(g-1)/g}$.

In a_{g-2} the term for which $j+m=n$ has the coefficient:

$$M=\sum_{\substack{i+j+k=g\\ 0\leqslant i+j+n\leqslant g-2}}\frac{g!}{i!j!k!}$$

We consider in M the terms such that k=n+2.We have:

$i+j+n=i+j+k-2=g-2$ and so $n+2\leqslant g$.

The sum M' of these terms satisfies:

$$M' = \sum_{i+j+n=g-2} \frac{(g-2)!}{i!j!n!} \cdot \frac{g(g-1)}{(n+1)(n+2)} \geq \sum_{i+j+n=g-2} \frac{(g-2)!}{i!j!n!} = L$$

Hence we have: $L \leq M' \leq M$. This implies that $h^{g/2-1} \leq a^{g/2}_{g-2}$.

So $s_2 > g/2$. With 1) we can conclude that $s_2 = \lfloor g/2 \rfloor + 1$. □.

By applying the Proposition 2 with the particular value $a=2\sqrt{q}$,we obtain the following corollary:

Corollary 1.

Let X be a maximal curve over GF(q);so $P(T)=(1+\sqrt{q}T)^{2g}$.

If $g \geq 3$ *and if* $q \geq g^2$, *then* $s_1=g+1$ *and* $s_2 = \lfloor g/2 \rfloor + 1$.

3. CURVES SUCH THAT $P(T)=(1+qT^2)^g$.

We can apply Proposition 2 with $a=0$. But as shown below the condition $q \geq g^2$ may be improved and replaced by $q \geq g+1$.

Proposition 3.

Suppose $g \geq 8$ *and* $q \geq g+1$. *Let X be a curve over GF(q) with genus g and such that* $P(T)=(1+qT^2)^g$. *Then:* $s_1=g+1$ *and* $s_2=\lfloor g/2 \rfloor+1$.

Proof

We first prove the two inequalities:

$$(q-1) > (q+1)^{g/(g+1)} \tag{5}$$

$$q(q-1) > (q+1)^{g/\lfloor g/2 \rfloor+1} \tag{6}$$

Let $\varphi(q) = g\mathrm{Log}(q+1) - (g+1)\mathrm{Log}(q-1)$.

Then $\varphi(g+1) = g\mathrm{Log}(g+2) - (g+1)\mathrm{Log}(g)$

$$= g\mathrm{Log}(1+2/g) - \mathrm{Log}(g) < 2 - \mathrm{Log}(g).$$

Hence $\varphi(g+1) < 0$ if $g > e^2$, that is if $g \geq 8$.

On the other hand one verifies that : $\varphi'(q) < 0$.

So if $q \geq g+1$ we have: $\varphi(q) \leq \varphi(g+1) < 0$. This proves (5).

Now since $q(q-1) > (q-1)^2$ we have by using (5):

$$q(q-1) > (q+1)^{2g/(g+1)}.$$

Clearly we have: $\dfrac{2g}{g+1} \geq \dfrac{g}{\lfloor g/2 \rfloor+1}$.

Therefore: $q(q-1) > (q+1)^{g/\lfloor g/2 \rfloor+1}$

This proves (6).

The relations (5) and (6) imply respectively that:

$$a^{g+1}_{g-1} < h^g \text{ and } a^{\lfloor g/2 \rfloor+1} < h^{\lfloor g/2 \rfloor}.$$

Indeed, since $a_{g-1} < \dfrac{h}{q-1}$,the first inequality is obtained if we have $\left(\dfrac{h}{q-1}\right)^{g+1} \leq h^g$, which is equivalent to $q-1 \geq (1+q)^{g/(g+1)}$. Similarly, since

$a_{g-2} < \frac{h}{q(q-1)}$, we obtained the second inequality if we have

$$\left(\frac{h}{q(q-1)}\right)^{\lfloor g/2\rfloor+1} \leqslant h^{\lfloor g/2\rfloor}$$

but this condition is equivalent to the inequality (6).

Hence: $s_1 \leqslant g+1$ and $s_2 \leqslant \lfloor g/2\rfloor+1$.

On the other hand the inequalities $s_1 > g$ and $s_2 > g/2$ was proved (see the proof of the Proposition 2)without any condition about g and q. Hence if $g \geqslant 8$ and $q \geqslant g+1$ we have: $s_1 = g+1$ ang $s_2 = \lfloor g/2\rfloor+1$.□.

4.CURVES SUCH THAT $P(T)=1+q^g T^{2g}$.

1) We have: $h^{g-1} = (1+q^g)^{g-1} = \sum_{j=0}^{g-1} C^j_{g-1} q^{gj}$

$$a^g_{g-1} = (q^{g-1}+\ldots+1)^g = \sum_{j_1+\ldots+j_g=g} \frac{g!}{j_1!\ldots j_g!} q^{(g-1)j_1+\ldots+2j_{g-2}+j_{g-1}}.$$

The highest degree term in these expressions is $q^{g(g-1)}$.

For $j=0,\ldots,g-2$ the term in a^g_{g-1} obtained by taking $j_1=j, j_2=g-j$, $j_3=\ldots=j_g=0$ has for coefficient $C^j_g \geqslant C^j_{g-1}$ and has for exponent:

$$\mu=(g-1)j+(g-2)(g-j)=gj+(g-2)(g-1-j)+(g-2-j) \geqslant gj.$$

SO we have: $h^{g-1} \leqslant a^g_{g-1}$.This proves without any condition on q that: $s_1 > g$.

2) We have: $h^{g-2} = (1+q^g)^{g-2} = \sum_{j=0}^{g-2} C^j_{g-2} q^{gj}$

and:

$$a^g_{g-2} = (q^{g-2}+\ldots+1)^g = \sum_{j_1+\ldots+j_{g-1}=g} \frac{g!}{j_1!\ldots j_{g-1}!} q^{(g-2)j_1+\ldots+j_{g-2}}$$

In these both expressions the highest degree term is $q^{g(g-2)}$.

For $j=0,\ldots,g-3$ the term in a^g_{g-2} obtained by taking $j_1=j, j_2=g-j$, $j_3=\ldots=j_{g-2}=0$ has for coefficient $C^j_g \geqslant C^j_{g-2}$ and has for exponent:

$$\mu'=(g-2)j+(g-3)(g-j)=gj+(g-3)(g-j-2)+2(g-j-3) \geqslant gj.$$

So we have : $h^{g-2} \leqslant a^g_{g-2}$ or $h^{g/2-1} \leqslant a^{g/2}_{g-2}$.

This proves without any condition on q that: $s_2 > g/2$.

3) By Proposition 1 we can conclude that for $q \geqslant g^2$, $s_1 = g+1$ and $s_2 = \lfloor g/2\rfloor+1$.But we will show that the condition $q \geqslant g^2$ may be improved by $q \geqslant g+1$.

Proposition 4.

Suppose that $g \geqslant 3$ and $q \geqslant g+1$.Let X be a curve over GF(q) with genus *g and such that* $P(T)=1+q^g T^{2g}$. *Then:* $s_1=g+1$ *and* $s_2=\lfloor g/2\rfloor+1$.

Proof.

Similar to the proof of Proposition 3, using the following inequalities:

$$q-1>q^{g/(g+1)} \quad \text{and} \quad q(q-1)>q^{g/\lfloor g/2\rfloor+1} \quad \square.$$

5. CURVES SUCH THAT $P(T)=(1-\sqrt{q}T)^{2g}$.

We have: $h=(\sqrt{q}-1)^{2g}$

$$a_{g-1}=\sum_{j=0}^{g-1}(-1)^j C_{2j}^j q^{j/2}(1+\ldots+q^{g-j-1})$$

and

$$a_{g-2}=\sum_{j=0}^{g-2}(-1)^j C_{2g}^j q^{j/2}(1+\ldots+q^{g-j-2}).$$

Proposition 5.

Suppose that $g\geqslant 3$ *and* $q\geqslant 2^{4g}$. *Let X be a curve over GF(q) with genus g and such that* $P(T)=(1-\sqrt{q}T)^{2g}$. *Then* $s_1=g$ *and* $s_2=\lfloor (g-1)/2\rfloor+1$.

Proof.

For $j\geqslant 1$, let μ,λ and λ' such that:

$$0\leqslant\mu\leqslant 1 \quad , \quad 0\leqslant\lambda\leqslant g-1-j \quad , \quad 0\leqslant\lambda'\leqslant g-j.$$

Define $K_{\lambda,\lambda',j}$ and $L_{\lambda,\lambda',j}$ by:

$$K_{\lambda,\lambda',j}=(C_{2g}^1+\ldots+C_{2g}^{2\lambda+1}-C_{2g-2j}^{2\lambda+1})\sqrt{q}-(C_{2g}^0+\ldots+C_{2g}^{2\lambda'})+C_{2g-2j}^{2\lambda'}$$

and $L_{\lambda,\lambda',j}=(C_{2g}^0+\ldots+C_{2g}^{2\lambda'}-C_{2g-2j}^{2\lambda'}\mu)\sqrt{q}-(C_{2g}^1+\ldots+C_{2g}^{2\lambda+1})+C_{2g-2j}^{2\lambda+1}\mu$

Then we have: $K_{\lambda,\lambda',j}>0$ and $L_{\lambda,\lambda',j}>0$.

Indeed:

$$K_{\lambda,\lambda',j}\geqslant\sqrt{q}-(1+1)^{2g}+C_{2g}^{2g}=\sqrt{q}-2^{2g}+1\geqslant 1>0. \qquad (7)$$

$$L_{\lambda,\lambda',j}\geqslant\sqrt{q}-(1+1)^{2g}+C_{2g}^{0}=\sqrt{q}-2^{2g}+1\geqslant 1>0. \qquad (8)$$

Now we prove that $s_1=g$.

1) For $s\geqslant g$ we have:

$$\omega=h^{(s-1)/s}-a_{g-1}\geqslant h^{(g-1)/g}-a_{g-1}=(\sqrt{q}-1)^{2(g-1)}-a_{g-1}=\varphi$$

where

$$\varphi=\sum_{j=0}^{2(g-1)}C_{2g-2}^j(-1)^j q^{j/2}-\sum_{j=0}^{g-1}C_{2g}^j(-1)^j q^{j/2}(1+\ldots+q^{g-j-1})=\sum_{j=0}^{g-1}\varphi_j q^j.$$

We have: $\varphi_{g-1}=0$

$\varphi_j=K_{j,j,1}$ for $0\leqslant j\leqslant\lfloor g/2\rfloor+1$

$\varphi_j=K_{g-j-2,g-j-1,1}$ for $\lfloor g/2\rfloor\leqslant j\leqslant g-2$.

By (7) we have $\varphi>0$. Hence: $h^{s-1}>a_{g-1}^s$.

2) We have: $2g.\frac{g-2}{g-1}=2g-2-\frac{2}{g-1}$.

So:

$$\omega'=\ h^{(g-2)/(g-1)}-a_{g-1}=(\sqrt{q}-1)^{2g-2-2/(g-1)}-a_{g-1}=\mu(\sqrt{q}-1)^{2g-2}-a_{g-1}$$

where

$$\mu=(\sqrt{q}-1)^{-2/(g-1)}>0 \text{ and } \frac{1}{\mu}=(\sqrt{q}-1)^{2/(g-1)}\geqslant(2^{2g}-1)^{2/(g-1)}\geqslant 16.$$

Therefore : $0<\mu<1$.

On the other hand we can write $\omega'=\omega'_0+\sum_{j=1}^{g-1} q^{j-1/2}\omega'_j$

We have:

$$\omega'_0=\mu-1<0$$

$$\omega'_j=-L_{j-1,j,1} \qquad \text{for } 1\leqslant j\leqslant\lfloor(g-1)/2\rfloor$$

$$\omega'_j=-L_{g-j-1,g-j-1,1} \qquad \text{for } \lfloor(g-1)/2\rfloor+1\leqslant j\leqslant g-1.$$

By (8) we have $\omega'<0$. Hence : $h^{g-2}<a_{g-1}^{g-1}$.

We can conclude that $s_1=g$.

Now we prove that $s_2=\lfloor(g-1)/2\rfloor+1$.

3) Let $g'=\lfloor(g-1)/2\rfloor$.

For $g=2g'+2$ we have:

$$2g\frac{g'}{g'+1}=2g\frac{2g'}{g}=4g'=2(g-2)$$

For $g=2g'+1$ we have:

$$2g\frac{g'}{g'+1}=2g\frac{2g'}{2g'+2}=2g\frac{g-1}{g+1}=2g(1-\frac{2}{g+1})\geqslant 2(g-2).$$

In both cases we obtained for $s\geqslant g'+1$:

Let $\theta=h^{(s-1)/s}-a_{g-2}$

We have:

$$\theta\geqslant h^{g'/(g'+1)}-a_{g-2}=(\sqrt{q}-1)^{2gg'/(g'+1)}-a_{g-2}\geqslant(\sqrt{q}-1)^{2(g-2)}-a_{g-2}=\psi$$

where

$$\psi=\sum_{j=0}^{2(g-2)} C^j_{2g-4}(-1)^j q^{j/2}-\sum_{j=0}^{g-2} C^j_{2g}(-1)^j q^{j/2}(1+\ldots+q^{g-j-2})\sum_{j=0}^{g-2}\psi_j q^j.$$

We can write:

$$\psi_j=K_{j,j,2} \qquad \text{for } 0\leqslant j\leqslant g'-1$$

$$\psi_j=K_{g-j-3,g-j-2,2} \qquad \text{for } g'\leqslant j\leqslant g-3$$

$$\psi_{g-2}=0.$$

By (7) we have $\theta\geqslant\psi>0$. Therefore : $h^{s-1}>a^s_{g-2}$.

4) If $g=2g'+2$, we have: $2g\frac{g'-1}{g'}=2(g-\frac{g}{g'})=2(g-2-\frac{2}{g'})$.

If $g=2g'+1$, we have: $2g\frac{g'-1}{g'}=2(g-\frac{g}{g'})=2(g-2-\frac{1}{g'})$.

In both cases we can write:

$$\theta'=h^{(g'-1)/g'}-a_{g-1}=(\sqrt{q}-1)^{2g(g'-1)/g'}-a_{g-2}\leqslant(\sqrt{q}-1)^{2(g-2-1/g')}-a_{g-2}=\psi'$$

We have $\psi'=(\sqrt{q}-1)^{2(g-2)}\mu-a_{g-2}$

where $\mu=(\sqrt{q}-1)^{-2/g'}>0$ and $\frac{1}{\mu}=(\sqrt{q}-1)^{2/g'}\geqslant(2^{2g}-1)^{2/g'}>1$.

So: $0<\mu<1$.

We can write: $\psi'=\psi'_0+\sum_{j=0}^{g-2} q^{j-1/2}\psi'_j$

with: $\psi'_0=\mu-1<0.$

$\psi'_j=-L_{j-1,j,2}$ for $1\leqslant j\leqslant \lfloor g/2\rfloor-1$

$\psi'_j=-L_{g-2-j,g-2-j,2}$ for $\lfloor g/2\rfloor\leqslant j\leqslant g-2$.

By (8) we have: $\theta'\leqslant\psi'<0$.

So: $h^{g'-1}<a^{g'}_{g-2}$.

We can conclude that: $s_2=g'+1=\lfloor(g-1)/2\rfloor+1$.∎.

We gather the previous results in the following theorem:

Theorem.

Let X be a nonsingular ,absolutely irreducible curve over GF(q) with genus g and let P(T) be the numerator of the zeta function of X over GF(q).

1)If $q\geqslant g^2$ and $g\geqslant 3$,and if $P(T)=(1+aT+qT^2)^g$ with $a\geqslant 0$, then:

$s_1=g+1$ and $s_2=\lfloor g/2\rfloor+1$.

2)If $q\geqslant g^2$ and $g\geqslant 3$,and if X is a maximal curve over GF(q), then:

$s_1=g+1$ and $s_2=\lfloor g/2\rfloor+1$.

3)If $q\geqslant g+1$ and $g\geqslant 8$,and if $P(T)=(1+qT^2)^g$ then:

$s_1=g+1$ and $s_2=\lfloor g/2\rfloor+1$.

4)If $q\geqslant g+1$ and $g\geqslant 3$,and if $P(T)=1+q^gT^{2g}$ then:

$s_1=g+1$ and $s_2=\lfloor g/2\rfloor+1$.

5)If $q\geqslant 2^{4g}$ and $g\geqslant 3$,anf if $P(T)=(1-\sqrt{q}T)^{2g}$ then:

$s_1=g$ and $s_2=\lfloor(g-1)/2\rfloor+1$.

6.EXAMPLES.

6.1 Klein quartic over finite fields with caracteristic 3.

The equation of the curve is $x^3y+y^3z+z^3x=0$.

The genus is g=3.Let $q=3^r$.One can verify ([1]) that there are six expressions for the polynomial P(T):

$P(T)=(1+qT^2)^3$ if $r\equiv 3 \bmod 6$ (1)

$P(T)=(1+\sqrt{q}T)^6$ if $r\equiv 6 \bmod 12$ (2)

$P(T)=(1-\sqrt{q}T)^6$ if $r\equiv 0 \bmod 12$ (3)

$P(T)=(1+q^3T^6)$ if $r\equiv\pm 1 \bmod 6$ (4)

$P(T)=(1+q^{3/2}T^3)^2$ if $r\equiv\pm 2 \bmod 12$ (5)

$P(T)=(1-q^{3/2})^2$ if $r\equiv\pm 4 \bmod 12$ (6)

The exact values of s_1 and s_2 may be given for cases (1) to (4).

By the proposition1 we have for the case (5): $s_1\leqslant 4$ and $s_2\leqslant 2$.

Case (6) won't be studied in this paper.

6.2 Snyder quintic over finite fields with caracteristic 2.

The equation of the curve is: $x^4y+y^4z+z^4x=0$.

The genus is 6.Let $q=2^r$.One verifies ([1])that there are eight expressions for the polynomial P(T):

$P(T)=(1+qT^2)^6$ if $r\equiv6 \mod 12$ (1)

$P(T)=(1+\sqrt{q}T)^{12}$ if $r\equiv12 \mod 24$ (2)

$P(T)=(1-\sqrt{q}T)^{12}$ if $r\equiv0 \mod 24$ (3)

$P(T)=1+q^6T^{12}$ if $r\equiv\pm1 \mod 6$ (4)

$P(T)=(1+q^3T^6)^2$ if $r\equiv\pm2 \mod 12$ (5)

$P(T)=(1+q^2T^4)^3$ if $r\equiv\pm3 \mod 12$ (6)

$P(T)=(1+q^{3/2}T^3)^4$ if $r\equiv\pm4 \mod 24$ (7)

$P(T)=(1-q^{3/2}T^3)^4$ if $r\equiv\pm8 \mod 24$ (8)

For cases (1) to (4) the exact values of s_1 and $s_2$may be given. For cases (5),(6),(7) (polynomials with coefficients $\geqslant0$) we have $s_1\leqslant7$ and $s_2\leqslant3$.The last case won't be studied in this paper.

6.3 The Fermat curve of degree 5 over GF(16).

The equation of the curve is: $x^5+y^5+z^5=0$.

The genus is g=6.Over GF(16) we obtain the polynomial

$P(T)=(1+\sqrt{q}T)^{2g}=(1+4T)^{12}$.

We have: $h=5^{12}$ and $a_5=15896673$.

But here we cannot apply our results since the conditions $q\geqslant g^2$ or $g\geqslant8$ and $q\geqslant g+1$ are not satisfied.By direct calculation we obtain $s_1=8=g+2$ instead of $s_1=g+1=7$.This example shows that the exact values of s_1 and s_2 depend crucially of some relation between q and g.

7.CONCLUSION.

It appears that for many curves and in particular for maximal curves,the decoding of a code $C_\Omega(D,G)$ over such curves may be easier for an even degree divisor G than for an odd degree divisor: one has to search at most $\lfloor g/2\rfloor+1$ effective divisors.Besides the conjecture given in [3] ("is it always true that $\Psi^{g+1}_{g-1}$is not surjective?") is shown to be true for the curves studied but under some conditions about g and q.

References.

[1]**P.Carbonne**.Calcul de quelques fonctions Zeta. Preprint.

[2]**C.J.Moreno**.Algebraic curves over finite fields.
Cambridge Tracts in Mathematics 97,Cambridge University Press 1991.

[3]**R.Pellikaan**.on a decoding algorithm for codes on maximal curves.
IEEE Trans Info Theory Vol 35,6 (1989),1228-1232.

[4]**S.G.Vladuts**. On the decoding of Algebraic Geometric Codes over
F_q for q≥16. IEEE Trans Info Theory Vol 36,6 (Nov 1990),1461-1463.

ON WEIGHTED COVERINGS AND PACKINGS WITH DIAMETER ONE

G.D. Cohen
Ecole Nationale Supérieure des Télécommunications, Paris, France

I.S. Honkala
University of Turku, Turku, Finland

S.N. Litsyn
Tel Aviv University, Ramat Aviv, Israel

ABSTRACT

We discuss a connection between weighted coverings and packings, study perfect weighted coverings (PWC) with diameter one and determine all the pairs (m_0,m_1) for which there exists a perfect q-ary linear (m_0,m_1)-covering.

RESUME

Nous discutons des liens entre recouvrements et pavages pondérés, étudions les recouvrements pondérés parfaits de diamètre 1 et dêterminons toutes les paires (m_0,m_1) pour lesquelles des (m_0,m_1) recouvrements q-aires parfaits linéaires existent.

1. INTRODUCTION AND BASIC DEFINITIONS

Covering codes have been extensively studied in recent years. Suppose q is a prime power and $\mathbb{F}_q$ is the finite field with q elements, and that C is a code over $\mathbb{F}_q$, i.e., $\emptyset \neq C \subseteq \mathbb{F}_q^n$. We say that C has covering radius R if every element in $\mathbb{F}_q^n$ is within Hamming distance R from at least one codeword of C and R is the smallest integer with this property. If a code $C \subseteq \mathbb{F}_q^n$ has covering radius R, then the Hamming spheres of radius R cover the whole space $\mathbb{F}_q^n$. However, it is sometimes natural to think that a codeword c covers a point x the more strongly the smaller their Hamming distance d(c,x), which leads us to the definition of *weighted coverings* [1], [2].

We denote

$$A_i(x) = |\{c \in C \mid d(c,x) = i\}|,$$

so that $A(x) = (A_0(x), A_1(x), \dots, A_n(x))$ is the weight distribution of C with respect to x. Given an (n+1)-tuple $M = (m_0, m_1, \dots, m_n)$ of rational numbers in the interval [0,1], the so-called *weights,* we define the M-density of the code $C \subseteq \mathbb{F}_q^n$ at a point $x \in \mathbb{F}_q^n$ as

$$\vartheta(x) := \sum_{i=0}^{n} m_i A_i(x) = \langle M, A(x) \rangle.$$

We say that the code C is an *M-covering,* if

$$\vartheta(x) \geq 1 \quad \text{for every } x \in \mathbb{F}_q^n,$$

and C is an *M-packing,* if

$$\vartheta(x) \leq 1 \quad \text{for every } x \in \mathbb{F}_q^n.$$

In particular, the code C is a *perfect M-covering (M-packing)* if

$$\vartheta(x) = 1 \quad \text{for every } x \in \mathbb{F}_q^n.$$

The largest i for which $m_i \neq 0$ is called the *diameter* δ of the M-covering. In general, we would like to know the smallest (largest) possible cardinality of an M-covering (M-packing) in $\mathbb{F}_q^n$.

In the definition we do not assume that $m_0 \geq m_1 \geq \dots \geq m_n$, but instead consider the general case. However, to avoid trivial cases we usually assume that $m_i = 0$ for $i \geq n/2$, i.e., $\delta < n/2$.

As an application of weighted packings we consider list decoding where the output of the decoder is a list with a given maximal size and correct decoding means that the transmitted codeword appears in the list. The size of the list could be an increasing function of the Hamming distance between the received word and the code. This kind of a criterion could be useful for example in spelling checkers.

As an application of weighted coverings we consider the football pool problem. A player wishes to forecast the outcome of n football matches (each with three possible outcomes) and is allowed to make several guesses (each guess being a word in $\mathbb{F}_3^n$). A guess with exactly k wrong outcomes wins the player a $(k+1)^{st}$ prize. If the player thinks that he knows the outcome of some $i \geq 1$ matches he may wish to fix an integer R and construct a set of forecasts in such a way that no matter what the outcomes in the remaining n-i matches are he will win at least an $(R+1)^{st}$ prize (provided that he is right about the i matches). In other words, he wishes to construct a code C $\subseteq \mathbb{F}_q^{n-i}$ with covering radius at most R. If he instead uses a μ-fold covering code (i.e. an M-covering with $\delta = R$ and $M = (1/\mu, 1/\mu, \ldots, 1/\mu)$) he will be guaranteed to win the $(R+1)^{st}$ prize at least μ times. However, from the practical point of view what the player really wants to do is to guarantee that he wins a certain amount of money, and in that sense it is natural to attach different weights to different prizes and construct a weighted covering instead. Since the Rth prize is usually several times the $(R+1)^{st}$ prize, say at least μ times, one natural example is to construct an M-covering with $\delta = R$ and $M = (1,1,\ldots,1,1/\mu)$, which will guarantee the player the $(R+1)^{st}$ prize at least μ times or at least one better prize, which alone will be worth at least the same (in [5] such M-coverings are called *multiple coverings of the farthest-off points,* or MCF's for short). For numerical tables for binary multiple coverings and MCF's, see [4] and [5].

The best currently known method of obtaining lower bounds on the cardinality of binary codes with a given length and covering radius, see [9], [10], [6], is based on showing that such a code is an M-covering for a suitable M and then using the sphere-covering lower bound for M-coverings.

We discuss a connection between weighted coverings and packings, study perfect weighted coverings (PWC) with diameter one and determine all the pairs (m_0,m_1) for which there exists a perfect q-ary linear (m_0,m_1)-covering. We finally give a simple non-existence result for perfect nonlinear binary $(1,1/t)$-coverings.

2. A CONNECTION BETWEEN WEIGHTED PACKINGS AND COVERINGS

It turns out that in this general set-up, the dual problems of finding all the best M-packings and M-coverings are really one and the same problem.

Suppose that $C \subset \mathbb{F}_q^n$ is an M-covering, where $M = (m_0, m_1, \ldots, m_\delta)$, and denote $C' := \mathbb{F}_q^n \setminus C \neq \emptyset$. If we denote

$$v = \sum_{i=0}^{\delta} \binom{n}{i} (q-1)^i m_i,$$

then clearly $v > 1$ and

$$\langle A(x),M\rangle + \langle A'(x),M\rangle = v$$

holds for all $x \in \mathbb{F}_q^n$, where $A'(x)$ is the weight distribution of C' with respect to x. Therefore the condition

$$\langle A(x),M\rangle \ge 1 \quad \text{for all } x \in \mathbb{F}_q^n$$

is equivalent to the condition

$$\langle A'(x),M'\rangle \le 1,$$

where $M' = (m_0/(v-1),\ m_1/(v-1),\ \ldots,\ m_\delta/(v-1))$. Similary C is an M-packing if and only if C' is an M'-covering.

A similar observation is already true for multiple coverings and packings. If $M = (m_0,m_1,\ldots,m_\delta)$ and $m_i = 1/\mu$ for all $i = 0, 1, \ldots, \delta$, and we denote by $K_q(n,\delta,\mu)$ (resp. $A_q(n,\delta,\mu)$) the smallest (largest) possible number of codewords in any M-covering (M-packing) in $\mathbb{F}_q^n$, then

$$K_q(n,\delta,\mu) = q^n - A_q(n,\delta,V_q(n,\delta)-\mu),$$

where

$$V_q(n,\delta) = \sum_{i=0}^{\delta} \binom{n}{i}(q-1)^i.$$

3. ON PERFECT Q-ARY LINEAR WEIGHTED COVERINGS WITH DIAMETER ONE

In [1] and [2] many results about perfect binary weighted coverings are given, see also [8]. The following two theorems will be needed to construct perfect q-ary linear weighted coverings, cf. [7] and [3].

Theorem 1. Assume that the codes $C(\alpha) \subseteq \mathbb{F}_q^n$, $\alpha \in \mathbb{F}_Q$, are (perfect) (m_0,m_1)-coverings and that they form a partition of $\mathbb{F}_q^n$. Assume further that $D \subseteq \mathbb{F}_Q^N$ is a (perfect) (M_0,M_1)-covering. Then the code

$$\bigcup_{(a_1,\ldots,a_N) \in D} C(a_1) \oplus \ldots \oplus C(a_N)$$

is a (perfect) $(M_0 - NM_1(1-m_0),\ m_1M_1)$-covering in $\mathbb{F}_q^{Nn}$.

Proof. Suppose $\mathbf{z} = (z_1, z_2, \ldots, z_N) \in \mathbb{F}_q^{Nn}$ where each z_i has length n. Because the union of the codes $C(\alpha)$ is the whole space $\mathbb{F}_q^n$, there exists a word $\mathbf{b} = (b_1,b_2,\ldots,b_N) \in \mathbb{F}_Q^N$ such that $\mathbf{z} \in C(b_1) \oplus \ldots \oplus C(b_N)$.

Assume first that $d(\mathbf{b},D) = 1$. Then there are at least $1/M_1$ words $\mathbf{a} = (a_1,\ldots,a_N) \in D$ such that $d(\mathbf{b},\mathbf{a}) = 1$ and for each such word $\mathbf{a}$ there are at least $1/m_1$ words in $C(a_1) \oplus \ldots \oplus C(a_N)$ that have distance 1 to $\mathbf{z}$.

Assume then that $\mathbf{b} \in D$. Then $\mathbf{z} \in C(b_1) \oplus \ldots \oplus C(b_N)$ and in each $C(b_i)$ there are at least $(1-m_0)/m_1$ words that have distance 1 to $\mathbf{z}_i$ and therefore together there are at least $N(1-m_0)/m_1$ words in $C(b_1) \oplus \ldots \oplus C(b_N)$ that have distance 1 to $\mathbf{z}$. Furthermore, there are at least $(1-M_0)/M_1$ words $\mathbf{a} \in D$ such that $d(\mathbf{b},\mathbf{a}) = 1$ and for each such word $\mathbf{a}$ there are again at least $1/m_1$ words in $C(a_1) \oplus \ldots \oplus C(a_N)$ that have distance 1 to $\mathbf{z}$.

In both cases it is easy to check that the $(M_0 - NM_1(1-m_0), m_1M_1)$-density at $\mathbf{z}$ is at least 1.

If the codes $C(\alpha)$ and D are perfect then in the previous discussion the estimated numbers of words are exact and in both cases the $(M_0 - NM_1(1-m_0), m_1M_1)$-density at $\mathbf{z}$ equals 1.■

Theorem 2. If C is a perfect q-ary (m_0,m_1)-covering then the code $C \oplus \mathbb{F}_q$ is a perfect $(m_0 - (q-1)m_1, m_1)$-covering provided that $m_0 \geq (q-1)m_1$.

Proof. Let $(x,\alpha) \in \mathbb{F}_q^n \times \mathbb{F}_q$ be arbitrary and denote $D = C \oplus \mathbb{F}_q$.

If $x \in C$, then $(x,\alpha) \in D$ and the words of D that have distance 1 to (x,α) are the q-1 words $(x,\beta) \in D$, $\beta \neq \alpha$, and the $(1-m_0)/m_1$ words (y,α) where $y \in C$ and $d(y,x) = 1$.

If $d(x,C) = 1$, then also $d((x,\alpha),D) = 1$ and the words of D that have distance 1 to (x,α) are precisely the $1/m_1$ words (y,α) where $y \in C$ and $d(y,x) = 1$.■

We now consider the existence of perfect q-ary linear coverings with diameter one. If $C \subseteq \mathbb{F}_q^n$ is a perfect linear (m_0,m_1)-covering with K codewords, then according to the sphere covering equality we have

$$(m_0 + (q-1)nm_1)K = q^n.$$

The numbers m_0, m_1 are rational numbers in the interval [0,1], w.l.o.g. $m_0 = a/t$ and $m_1 = b/t$ for some integers $t > 0$, $0 \leq a \leq t$, $0 \leq b \leq t$. Assume now that $C \neq \mathbb{F}_q^n$. Then there exists a point $x \notin C$ and $\vartheta(x) = 1$ implies that some multiple of b/t equals one, and hence b divides t. On the other hand, if $x \in C$ then $\vartheta(x) = 1$ implies that some multiple of b/t equals 1 - a/t and hence b divides a. We can therefore w.l.o.g. assume that b = 1.

Theorem 3. A perfect q-ary linear (m_0,m_1)-covering $C \neq \mathbb{F}_q^n$ of length n exists if and only if

$$m_0 = a/t, \quad m_1 = 1/t \quad \text{for some integers } t > 0,\ 0 \leq a \leq t$$

and $a \equiv t \pmod{q-1}$, and for some integer $s > 0$

$$n = \frac{tq^s - a}{q-1}.$$

Proof. We have already seen that m_0 and m_1 are as described in the theorem. By the sphere covering equality we have

$$(a/t + (q-1)n/t)K = q^n$$

and because C is linear, its cardinality K is a power of q, say q^{n-s}, $0 < s \le n$, and then $n = (tq^s - a)/(q-1)$. In particular $a \equiv t \pmod{q-1}$.

To construct such codes assume first that a = t. Then we can in Theorem 1 choose

$$C(\alpha) = \{(x_1, x_2, \ldots, x_t) \in \mathbb{F}_q^t \mid x_1 + x_2 + \ldots + x_t = \alpha\}$$

for $\alpha \in \mathbb{F}_q$ and

D = the q-ary linear Hamming code of length $(q^s-1)/(q-1)$.

Then each $C(\alpha)$ is a perfect (1, 1/t)-covering and D is a perfect (1,1)-covering. Then the construction of Theorem 1 yields a perfect (1, 1/t)-covering of length $t(q^s-1)/(q-1)$. By the construction and the definitions of $C(\alpha)$ and D it is clear that this code is linear. If a is smaller than t then $a = t - (q-1)j$ for some integer $j \ge 1$ and we obtain the required perfect (a/t, 1/t)-covering of length $j + t(q^s-1)/(q-1)$ by applying Theorem 2 to this code j times.■

4. A NON-EXISTENCE RESULT FOR PERFECT BINARY (1,1/t)-COVERINGS

In the binary non-linear case we have the following non-existence result.

Theorem 4. If a perfect (1,1/t)-covering $C \subset \mathbb{F}_2^n$ exists and $n \ne t$ then $n \ge 2t+1$.

Proof. W.l.o.g. the all-zero word does not belong to C and there are exactly t codewords of C of weight 1. Suppose x is any other word of weight 1. Then x is covered by t codewords of C of weight 2 and none of them have any 1's in common with the codewords of weight 1. Therefore $n \ge t + (t+1) = 2t+1$.■

Example. If i is odd and $t = (2^i+1)/3$, then $n = 2^i - t$ and $K = t2^{n-i}$ satisfy the equality $K(1 + n/t) = 2^n$. However, because $n = 2t-1$, no perfect binary (1,1/t)-covering of length n exists by the previous theorem.

ACKNOWLEDGMENT

The second author would like to thank Heikki Hämäläinen and Markku Kaikkonen for useful discussions.

REFERENCES:

1. Cohen, G.D., S.N. Litsyn and H.F. Mattson, Jr.: Binary perfect weighted coverings, in: Sequences '91, Positano, 17-22 June 1991, *Springer-Verlag Lecture Notes in Computer Science,* to appear.
2. Cohen, G.D., S.N. Litsyn and H.F. Mattson, Jr.: On perfect weighted coverings with small radius, in: First French-Soviet Workshop in Algebraic Coding, Paris, 22-24 July 1991, *Springer-Verlag Lecture Notes in Computer Science* 573, Berlin - Heidelberg, 1992, 32-41.
3. Etzion, T. and G. Greenberg: Constructions for perfect mixed codes and other covering codes, IEEE Trans. Inform. Theory, to appear.
4. Hämäläinen, H.O., I.S. Honkala, M.K. Kaikkonen and S.N. Litsyn: Bounds for binary multiple coverings, Designs, Codes and Cryptography, to appear.
5. Hämäläinen, H.O., I.S. Honkala and S.N. Litsyn, Bounds for binary codes that are multiple coverings of the farthest-off points, in preparation.
6. Li, D. and W. Chen: New lower bounds for binary covering codes, IEEE Trans. Inform. Theory, submitted.
7. Östergård, P.R.J.: Further results on (k,t)-subnormal covering codes, IEEE Trans. Inform. Theory 38 (1992) 206-210.
8. van Wee, G.J.M., G.D. Cohen and S.N. Litsyn: A note on perfect multiple coverings of Hamming spaces, IEEE Trans. Inform. Theory 37 (1991) 678-682.
9. Zhang, Z., Linear inequalities for covering codes: part I - pair covering inequalities, IEEE Trans. Inform. Theory 37 (1991) 573-582.
10. Zhang, Z. and C. Lo, Linear inequalities for covering codes: part II - triple covering inequalities," IEEE Trans. Inform. Theory 38 (1992) 1648-1662.

ON DUALS OF BINARY PRIMITIVE BCH CODES

F. Levy-dit-Vehel
INRIA, Rocquencourt, Le Chesnay, France

Abstract

We treat binary extended cyclic codes of length 2^m over $\mathbf{F}_2$. We introduce a class, denoted by $\{C^{(t)}\}_{2\leq t\leq m-1}$, of such codes, whose defining set is characterized by only one cyclotomic coset. We prove that they are duals of extended BCH codes. We study the divisibility of the $C^{(t)}$'s, and show that it determines the divisibility of all duals of extended BCH codes. Next we obtain a lower bound on their minimum distance, that yields results for several affine-invariant codes. In particular, it gives a bound for all duals of extended BCH codes, which is interesting especially when the Carlitz-Uchiyama bound is negative.

1 Preliminaries

A binary cyclic code C of length $n = 2^m - 1$ is an ideal in the ring $\mathbf{F}_2[Z]/(Z^n - 1)$. Let α be a primitive root in $\mathbf{F}_{2^m}$. Then, if $g(Z)$ is the generator polynomial of C, α^i is a zero of the code C if, and only if, $g(\alpha^i) = 0$. The set $T = \{i \in [0, n],\ \alpha^i \text{ is a zero of } C\}$, is the defining set of C. If $0 \notin T$, we can extend C and obtain a linear code C_e of length 2^m, so-called the extension of C, by adding a parity check symbol to each word of C:

$$c \in C,\ c = (c_0, \ldots, c_{n-1}) \Longleftrightarrow c' \in C_e,\ c' = (c_\infty, c_0, \ldots, c_{n-1}) \text{ where } c_\infty = \sum_{i=0}^{n-1} c_i.$$

The code C_e can be considered as a code of the group algebra $\mathcal{A} = \mathbf{F}_2[\{\mathbf{F}_{2^m}, +\}]$, which is the set of formal polynomials $x = \sum_{g\in\mathbf{F}_{2^m}} x_g X^g$, $x_g \in \mathbf{F}_2$.

And this because a word $c' = (c_\infty, c_0, \ldots, c_{n-1})$ in C_e can be represented by the polynomial

$$c'(X) = c_\infty X^0 + \sum_{i=0}^{n-1} c_i X^{\alpha^i}.$$

We say that $T_e = T \cup \{0\}$ is the *defining set* of C_e, and $[0,n] \setminus T_e$ is the *parity set* of C_e.
Recall that the BCH code of length n and designed distance d over $\mathbb{F}_2$ is the cyclic code with defining set : $T(d) = \cup_{1 \le s < d}\, cl(s)$, where $cl(s)$ is the cyclotomic coset of 2 modulo n containing s. We denote by $B(d)$, $EB(d)$, and $DEB(d)$ respectively the BCH code of designed distance d, its extension, and the dual of this extension.
We let $w(x)$ denote the Hamming weight of a vector x of length n or $n+1$, that is, the number of its non-zero components.

Definition 1 *Let $S = [0,n]$. The 2-ary expansion of an element $s \in S$ is : $s = \sum_{i=0}^{m-1} s_i 2^i$, $s_i \in \{0,1\}$, and the 2-weight of s is $\omega_2(s) = \sum_{i=0}^{m-1} s_i$.*

Note : We identify an element s in $[0,n]$ with its binary representation, that is, if $s = \sum_{i=0}^{m-1} s_i 2^i$, we shall write $s = (s_{m-1}, \ldots, s_0)$.

Definition 2 *We denote by $\ll$, the partial order on S defined as follows:*

$$s, t \in S,\ s \ll t \Longleftrightarrow s_i \le t_i,\ i \in [0, m-1]$$

When $s \ll t$, s is said to be a descendant of t. We say that s and t are not related when $s \not\ll t$ and $t \not\ll s$. An antichain of $(S, \ll)$ is a subset of S of non-related elements.
Let Δ be the map :

$$\Delta : I \subset [0,n] \mapsto \Delta(I) = \cup_{t \in I}\{s \in S,\ s \ll t\} = \cup_{t \in I} \Delta(\{t\})$$

A code of the algebra $\mathcal{A}$ (ie a $\mathbb{F}_2$-subspace of $\mathcal{A}$) is called affine-invariant if its automorphism group contains the group GA(m) of affine permutations of $\mathbb{F}_{2^m}$:

$$p_{u,v} : \sum_{g \in \mathbb{F}_{2^m}} x_g X^g \to \sum_{g \in \mathbb{F}_{2^m}} x_g X^{ug+v},\ u, v \in \mathbb{F}_{2^m},\ u \neq 0$$

Theorem 1 *([Ka,Li,Pe 2],[Ch.1]) Let C_e be an extended cyclic code of $\mathcal{A}$, with defining set T_e. Then, C_e is affine-invariant if, and only if :*
$\Delta(T_e) = T_e$, or C_e is an ideal of $\mathcal{A}$.

An element s is maximal (resp. minimal) in I if its strict ascendants (resp. descendants) are not in I.

Definition 3 *([Ch.1],[Ch.2]) Let C_e be an extended cyclic code with defining set T_e. The set of minimal elements of $[0,n] \setminus T_e$ is called the border of C_e.*

Note : Let C_e be an affine-invariant code with border F. Then the set
$\{n - f,\ f \in F\}$ is the set of maximal elements of the defining-set of the dual of C_e [Ch.1].

Let C be a binary cyclic code and C_e its extension. Denote by d_{min}, d_{odd}, and d_{even}, respectively the minimum weight, the minimum odd weight, and the minimum even weight of C. If C_e is affine-invariant, then $d_{min} = d_{odd} = d_{even} - 1$.

A binary linear code C is said to be k-divisible, $k > 1$, if the weights of all its words are divisible by k, and if there exists a word of C whose weight is not divisible by $l > k$. If $2 \not| k$, then the code is equivalent to a degenerate code (in which every symbol is repeated k times) [Wa]. We thus consider the case $k = 2^a$, $a \in \mathbb{N}^*$.
The divisibility of Reed-Muller codes is known [Mc El]. For every extended cyclic code can be located among the RM-codes, it has at least the divisibility of the smallest RM-code it is included in. We shall call this divisibility the minimum divisibility of the code.
We recall the theorem of Mac Eliece, adapted here to the context of binary extended cyclic codes :

Theorem 2 : *Let C_e be an extended cyclic code of length 2^m over $\mathbb{F}_2$, with defining set T. Then, C_e is 2^λ-divisible, where $\lambda = \omega - 1$, and*

$$\omega = min\{r, \prod_{i=1}^{r} \alpha^{u_i} = 1, \ u_i \in [0, n[\setminus T\}$$

α being a primitive root in $\mathbb{F}_{2^m}$.

The reader will find ampler proofs and more numerical results in the INRIA report [Le].

2 A class of duals of primitive extended BCH codes

We are studying a class of primitive extended cyclic codes over $\mathbb{F}_2$, defined as follows :

Definition 4 *Let $n = 2^m - 1$ and $t \in [2, m-1]$.*
The $C^{(t)}$ code is the affine-invariant code of length 2^m, whose border is $F_t = cl(2^t - 1)$.

Proposition 1 *The defining set of $C^{(t)}$ is*

$$D_t = \{u \in [0, n], \ u \text{ has no run of } t \text{ consecutive ones}\}$$

Note : An element $x \in [0, n]$ has a run of t consecutive ones if, and only if, a cyclic shift of x has a run of t consecutive ones. For instance, $(110\ldots0\underbrace{1\ldots1}_{t-2}) \in cl(2^t - 1)$ is such an element.

Proof . The defining-set I of an affine-invariant code with border F is given by

$$I = \bigcap_{f \in F} N(f),$$

where $N(f)$ is the set of non-majorants of f in S.

So the defining set of $C^{(t)}$ is :

$$D_t = \cap_{f \in cl(2^t-1)} N(f) = \cap_{f \in cl(2^t-1)} \{u \in [0,n] \setminus \{f\},\ u \ll f \text{ or } f \text{ and } u \text{ are not related}\}$$

Let $s(t) = 2^t - 1$. $s(t) = (0\dots0\underbrace{1\dots1}_{t})$ and $\omega_2(s(t)) = t$.
Let $u \in \cap_{f \in cl(s(t))} N(f)$. Then, if $u \ll f$ with $u \neq f$ for all f in F_t, we have $w(u) < t$.
If, for all f in F_t, u and f are not related, then u has not t consecutive ones. Indeed, if $u = (\epsilon\dots\epsilon\underbrace{1\dots1}_{t}\underbrace{\epsilon\dots\epsilon}_{z})$, where $\epsilon \in \{0,1\}$, then $2^z s(t)$ is a descendant of u.
Moreover, $w(u) < t \Rightarrow u$ has not t consecutive ones. So $D_t \subset \{u \in [0,n],\ u$ has not t consecutive ones$\}$.
Conversely, if $u \in [0,n]$ has not t consecutive ones, then either $\omega_2(u) < t$, or u is unrelated with every element of F_t, so u belongs to D_t. Thus,

$$D_t = \{u \in [0,n],\ u \text{ has not } t \text{ consecutive ones}\}$$

□

We denote by $\lceil x \rceil$, (resp. $\lfloor x \rfloor$), the least (largest) integer greater (smaller) than or equal to x.

Theorem 3 *The code $C^{(t)}$ is the dual of an extended BCH code. More precisely, $C^{(t)} = DEB(d_t)$, where d_t is the smallest element of $[0,n]$ having no run of t consecutive zeroes.*
If $t < \lceil \frac{m}{2} \rceil$, then $d_t = \sum_{i=1}^{a} 2^{m-it} + (1 - \delta_{r,0})$, where $m = at + r$, $r < t$, and $\delta_{r,0}$ is the Kronecker symbol.
If $t \geq \lceil \frac{m}{2} \rceil$, $d_t = 2^{m-t} + 1$

Proof . First, if $t \geq \lceil \frac{m}{2} \rceil$, then $m - t \leq \lfloor \frac{m}{2} \rfloor \leq t$, and so

$$2^{m-t} + 1 = (\underbrace{0\dots0}_{t-1}1\ \underbrace{0\dots0}_{m-t-1<t}\ 1)$$

is the smallest element of $[0,n]$ having no run of t consecutive zeroes.
If $t < \lceil \frac{m}{2} \rceil$, let $m = at + r$, $r < t$; (Note that in this case, $a \geq 2$). Then, the element

$$(\underbrace{\overbrace{0\dots0}^{t-1}1\dots\overbrace{0\dots0}^{t-1}1}_{a\ blocks}\underbrace{0\dots01}_{r<t})$$

is the smallest element of $[0,n]$ having no run of t consecutive zeroes.
Besides, we clearly have : $2^{m-t} - 1 < d_t$.
$d_t - (2^{m-t} - 1) = \sum_{i=1}^{a} 2^{m-it} + (1 - \delta_{r,0}) - (2^{m-t} - 1) \geq \sum_{i=2}^{a} 2^{m-it} + 1 > 0$.
To prove the theorem, it suffices to show that the defining set, say $D_t^{\perp}$, of the dual of $C^{(t)}$, is the defining set of the extended BCH code of designed distance d_t, that is :

$$D_t^{\perp} = \cup_{0 \leq s < d_t} cl(s)$$

We have : $D_t^\perp = \{n - s,\ s \notin D_t\} = \{n - s, s \text{ has at least } t \text{ consecutive ones}\}$, or :
$D_t^\perp = \{u \in [0, n],\ u \text{ has at least } t \text{ consecutive zeroes}\}$.
Recall that $\{n - f,\ f \in F_t\} = cl(2^{m-t} - 1)$, is the set of maximal elements of $D_t^\perp$. It means that the maximal elements of $D_t^\perp$ are those having exactly t consecutive zeroes. Thus we immediately have :

$$\bigcup_{0 \le s \le 2^{m-t}-1} cl(s) \subset D_t^\perp.$$

Moreover, $D_t^\perp$ cannot contain a class with smallest element strictly larger than $2^{m-t} - 1$, because it would also be a maximal element of $D_t^\perp$. Thus, $D_t^\perp$ is the defining-set of an extended BCH code. Its designed distance is the smallest element which is both the smallest element of its cyclotomic coset, and strictly greater than $2^{m-t} - 1$, that is, the smallest element not having t consecutive zeroes. Thus, $D_t^\perp = \cup_{0 \le s < d_t} cl(s)$. □

Proposition 2 *For $t \ge \lceil \frac{m}{2} \rceil$, $dim\, C^{(t)} = m\, 2^{m-t-1} + 1$.*
For $t < \lceil \frac{m}{2} \rceil$, we deduce from [Ma] that :

$$dim\, C^{(t)} = m \sum_{k=1}^{\lfloor \frac{m}{t+1} \rfloor} \frac{(-1)^{k-1}}{k} \binom{m - kt - 1}{k - 1} 2^{m-k(t+1)} + 1$$

The proof of this proposition can be found in [Le].

Example : Let $m = 8$, $t = 3$; $C^{(3)}$ is the affine-invariant code with border

$$F_3 = cl(7) = \{7, 14, 28, 56, 112, 224, 193, 131\}$$

It is the dual of the extended BCH code with designed distance
$d_3 = \sum_{i=1}^{2} 2^{8-3i} + (1 - \delta_{2,0}) = 37$.
The corresponding defining-set D_3 is the union of the following cyclotomic cosets :
$\{0\}$, $cl(1)$, $cl(3)$, $cl(5)$, $cl(9)$, $cl(13)$, $cl(17)$, $cl(19)$, $cl(21)$, $cl(25)$, $cl(27)$, $cl(37)$, $cl(43)$, $cl(45)$, $cl(51)$, $cl(53)$, $cl(85)$, $cl(91)$.
That corresponds to all elements of $[0, n]$ having not 3 consecutive ones in their binary representation. The dimension of the code is 125.

3 A bound on the minimum distance of $C^{(t)}$:

We denote by $RM(r, m)$, (or simply $RM(r)$ when there cannot be any ambiguity on m), the Reed-Muller code of length 2^m and order r, that is, the code with defining-set $\{s \in [0, n],\ \omega_2(s) < m - r\}$, where $\omega_2(s)$ is the 2-weight of the element s (see def.1). Since $C^{(t)} \subset RM(m - t)$, we immediately have $d_{min} C^{(t)} \ge 2^t$. But we shall see that, for almost all t's in the range $[2, m - 1]$, we can refine this bound.

lemma 1 *Let C be an affine-invariant code of length $n + 1$, with defining set I, and suppose*

$$[0, d[\subset I, \text{ and } d \notin I$$

Then

1. *Any interval included in I is of length $\leq d$.*

2. *If there exists s and z such that $(z,n)=1$, and*

$$[0,d[,\ [z,z+d[,\ldots,\ [(s-1)z,(s-1)z+d[$$

are included in I, then $d_{min}C \geq d+s$.

Proof . *1.* It suffices to show that, for all k in I, there exists an integer k' such that : $k'>k$, $k'\notin I$, and $k'-k\leq d$. Then, any interval with k as first element and incuded in I will be contained in $[k,k'[$ and thus will be of length smaller than or equal to d. Let k belong to I. If $0\leq k<d$, then any interval included in I, with k as first element is of length less than d, because $d\notin I$. We can thus suppose $k>d$.
Let $k=\sum_{i=0}^{m-1}k_i2^i$ and $d=\sum_{i=0}^{m-1}d_i2^i$. As $k>d$, there exists a j such that $k_j>d_j$ and, for all l, $j<l<m$, $k_l\geq d_l$. Let j_0 be the minimum of those j's. Then, $j_0\neq 0$ otherwise k would be an ascendant of d, and, as C is affine-invariant, k would not belong to I. Let $k'=\sum_{i=0}^{j_0-1}d_i2^i+\sum_{i=j_0}^{m-1}k_i2^i$.
By construction, $d\ll k'$, so that $k'\notin I$. Moreover, we have $k'>k$. But

$$k'-k\leq\sum_{i=0}^{j_0-1}d_i2^i\leq d$$

2. Let C^* be the cyclic code whose extension is C. Then, the defining-set of C^* is $I^*=I\backslash\{0\}$. Thus the following $(s-1)$ intervals belong to I^* :

$$[z,z+d[,\ldots,\ [(s-1)z,(s-1)z+d[$$

Recall that the Hartmann-Tzeng bound [Sl.] states that, if the elements $b+i_1c_1+i_2c_2$ belong to the defining-set of a cyclic code of length n, where c_1 and c_2 are relatively prime to n, $0\leq i_1\leq\delta-2$ and $0\leq i_2\leq r$, then the minimum distance of the code is at least $\delta+r$. We can apply this bound to I^*, with $c_1=1$, $c_2=z$, $b=z$, $\delta=d+1$, $r=s-2$. We obtain : $d_{min}(C^*)\geq\delta+r=d+1+s-2=d+s-1$. But, as C is affine-invariant, $d_{min}(C)=d_{min}(C^*)+1$, so that $d_{min}(C)\geq d+s$. □

Applying this lemma to the $C^{(t)}$'s, we obtain :

Theorem 4 *For $2\leq t<m-2$, the minimum weight δ_t of $C^{(t)}$ satisfies: $\delta_t\geq 2^{t+1}-2^{t-1}$.*

Proof . Let $2\leq t<m-2$. Then, the intervals $[0,2^t-1[$ and $[2^{t+1},2^{t+1}+2^t-1[$ belong to the defining-set D_t of $C^{(t)}$. With $z=2^{t+1}$ prime to n, we can apply lemma 1 and we find : $d_{min}C^{(t)}\geq 2^t-1+2=2^t+1$. For $C^{(t)}\subset RM(m-t)$, the smallest possible weight for $C^{(t)}$ is $2^{t+1}-2^{t-1}$, according to a theorem of Kasami & al. ([Ka,To],th.2). □

Remark : For $t=m-2$ or $t=m-1$, $C^{(t)}$ is the dual of the extended two or one error correcting BCH code, so that the minimum weight of $C^{(t)}$ is known: It is $2^{m-1}-2^{\lfloor\frac{m}{2}\rfloor}$ and 2^{m-1} respectively.

4 Divisibility properties

Theorem 5 *The code $C^{(t)}$ is $2^{\lceil \frac{m}{m-t} \rceil -1}$-divisible.*

Remark : The theorem means that $C^{(t)}$ has the minimum divisibility.

Proof . The code $C^{(t)}$ is at least $2^{\lceil \frac{m}{m-t} \rceil -1}$-divisible, as a subcode of RM$(m-t)$. It remains to prove that $C^{(t)}$ has exactly this divisibility. According to the theorem of Mac Eliece [Mc El], it suffices to show that there exists $\lceil \frac{m}{m-t} \rceil$ elements of $[0,n[\backslash D_t$ whose sum is n. We distinguish two cases :
Case 1 : $t \le \lfloor \frac{m}{2} \rfloor$:
Then $C^{(t)}$ is exactly 2-divisible because $s = 2^{\lfloor \frac{m}{2} \rfloor} - 1$ and $n - s$ both belong to $[0,n[\backslash D_t$.
Case 2 : $t \ge \lceil \frac{m}{2} \rceil$:
Let C be the code whose defining set can be deduced from the one of $C^{(t)}$ by changing the primitive root α into $\beta = \alpha^{-1}$. Those two codes are equivalent and so have the same divisibility. The defining set of C is $\{n-s,\ s \in D_t\}$. And the parity set, say $\mathcal{U}$, of C is $[0,n] \setminus \{n-s,\ s \in D_t\} = \{n-s,\ s \notin D_t\}$. It is worth to notice here that the parity set of C is $D_t^{\perp}$. But $D_t^{\perp}$ admits $cl(2^{m-t}-1)$ as maximal elements, that is :

$$D_t^{\perp} = \cup_{i=0}^{m-1} \Delta(2^i(2^{m-t}-1))$$

So $cl(2^{m-t}-1) \subset \mathcal{U}$.
Case 2.1 : $m-t \,|\, m$.
$m = (m-t)\,a$. Then, the a elements $2^{i(m-t)}(2^{m-t}-1)$, $0 \le i \le a-1$, belong to $\mathcal{U}$ and :

$$\sum_{i=0}^{a-1} 2^{i(m-t)}(2^{m-t}-1) = (2^{m-t}-1)\frac{1-2^m}{1-2^{m-t}} = n$$

Case 2.2 : $m-t \not| \, m$.
$m = (m-t)\,a + r$, $r < m-t$ and $\lceil \frac{m}{m-t} \rceil = a+1$. We have :

$$\sum_{i=0}^{a-1} 2^{i(m-t)}(2^{m-t}-1) = (\underbrace{0\ldots0}_{r}\underbrace{\overbrace{1\ldots1}^{m-t}\ldots\overbrace{1\ldots1}^{m-t}}_{a \text{ blocks}})$$

Let $b = (\underbrace{1\ldots1}_{r}0\ldots0) = 2^m - 2^{m-r}$. b belongs to $\mathcal{U}$ because $r < m-t$ so that $b \ll 2^t(2^{m-t}-1)$. The $a+1$ elements b, $2^{i(m-t)}(2^{m-t}-1)$, $0 \le i \le a-1$, belong to $\mathcal{U}$ and :

$$b + \sum_{i=0}^{a-1} 2^{i(m-t)}(2^{m-t}-1) = b + (2^{m-t}-1)\frac{1-2^{(m-t)a}}{1-2^{m-t}} = b - 1 + 2^{(m-t)a} = b - 1 + 2^{m-r} = n$$

There exists $\lceil \frac{m}{m-t} \rceil$ elements of the parity set of C whose sum is n. So C is $2^{\lceil \frac{m}{m-t} \rceil -1}$-divisible, and therefore $C^{(t)}$ also. □

Corollary 1 Divisibility of a code situated between two consecutive $C^{(t)}$'s
Let $t > 2$, and C be an affine-invariant code with defining set D, such that :

$$C^{(t)} \subseteq C \subset C^{(t-1)} \tag{1}$$

Then, C is $2^{\lceil \frac{m}{m-t} \rceil - 1}$-divisible, ie. the divisibility of C is the one of $C^{(t)}$.

1. *If $2 < t < m-2$, and the elements $u_0 = 2^t + 2^{t-1} - 1$ and $v_0 = 2^{t+1} + 2^{t-1} - 1$ belong to D,*

$$d_{min}(C) \geq 2^{t+1} - 2^{t-1}$$

2. *If $2 < t < m-2$, and either u_0 or v_0 does not belong to D,*

$$d_{min}(C) \geq 2^t$$

3. *If $t = m-2$,*

$$d_{min}(C) \geq 2^{m-1} - 2^{\lfloor \frac{m+2}{2} \rfloor}$$

4. *If $t = m-1$, $d_{min}(C) = 2^{m-1}$.*

Proof . By (1), we have : $D_{t-1} \subset D \subseteq D_t$. Thus

$$D = (D_{t-1}) \cup I$$

where $I \neq \emptyset$, and $I \subset \{s \in [0, n[, s \text{ has } t-1 \text{ consecutive ones but not } t\}$.
We claim that $cl(2^{t-1} - 1) \subset I$:
Indeed, let $s \in I$. Then s is of the following form : $(\epsilon \ldots \epsilon 0 \underbrace{1 \ldots 1}_{t-1} \underbrace{0 \epsilon \ldots \epsilon}_{z})$ where $\epsilon \in \{0,1\}$.
Then, $2^z(2^{t-1} - 1) \ll s$ and as D is let fixed by the Δ map (ie. C is affine-invariant), $2^z(2^{t-1} - 1)$ belongs to D. C being a code over F_2, $cl(2^{t-1} - 1) \subset D$.
Let now $s \in [0, n[$, with $\omega_2(s) = t - 1$.
-If $s \in cl(2^{t-1} - 1)$, then $s \in I$.
-If s and $cl(2^{t-1} - 1)$ are not related (ie. $\forall u \in cl(2^{t-1} - 1), s \not\ll u$ and $u \not\ll s$), s has not $t-1$ consecutive ones, so s is in $D_{t-1} \setminus \{0\}$.
In both cases, s belongs to D, ie D contains all the elements of 2-weight $t-1$; Thus C is necessarily a subcode of RM$(m-t)$. The exact divisibility of C follows immediately.

1. Suppose $2 < t < m-2$, and u_0, v_0 belong to D. We shall show that the intervals $[0, 2^t - 1[$ and $[2^{t+1}, 2^{t+1} + 2^t - 1[$ are included in D, and then apply lemma 1. First, as the elements of 2-weight less than or equal to $t-1$ all belong to D, we immediately have : $[0, 2^t - 2] \subset D$. Now what are the elements of $[2^{t+1}, 2^{t+1} + 2^t - 1[$ that have $t-1$ consecutive ones in their binary representation? They are exactly v_0 and $2u_0$. (The other members of this interval have no $t-1$ consecutive ones, so that they belong to $D_{t-1} \subset D$.) As v_0 and u_0 belong to D, $cl(u_0)$ and $cl(v_0)$ also, and thus $[2^{t+1}, 2^{t+1} + 2^t - 1[\subset D$. By the same reasoning as in the proof of theorem 4, we conclude that the minimum weight of C is at least $2^{t+1} - 2^{t-1}$.
2. Suppose $2 < t < m-2$ and either one or both u_0 and v_0 does not belong to D. Then

$[2^{t+1}, 2^{t+1}+2^t-1[$ is not a subset of D, and it is easy to see that there are no intervals of length 2^t-1 included in D except $[0, 2^t-1[$. Thus applying lemma 1 does not give a better estimate than the one we have by inclusion of C in RM$(m-t)$.
3. If $t = m-2$, then (1) implies that the defining-set, say $D^{\perp}$, of $C^{\perp}$ is such that

$$\{0\} \cup cl(1) \cup cl(3) \subseteq D^{\perp} \subset \{0\} \cup cl(1) \cup cl(3) \cup cl(5) \cup cl(7)$$

But, as C is affine-invariant, we can only have two possibilities for $D^{\perp}$, namely :
$D^{\perp} = \{0\} \cup cl(1) \cup cl(3)$, in which case $C = DEB(5)$, or,
$D^{\perp} = \{0\} \cup cl(1) \cup cl(3) \cup cl(5)$, and then $C = DEB(7)$.
In both cases, the minimum weight of C is at least the minimum weight of the dual of the extended 3-error correcting BCH code, that is $2^{m-1} - 2^{\lfloor \frac{m+2}{2} \rfloor}$.
4. If $t = m-1$, then C is the Reed-Muller code of order one, so that $d_{min}(C) = 2^{m-1}$. □

The next corollary requires a theorem of Carlitz-Uchiyama, that we recall now :

Theorem 6 *[Ca, Uc] If C is the dual of the BCH code of length $2^m - 1$ and of designed distance $d = 2r+1$, with $2r-1 < 2^{\lceil \frac{m}{2} \rceil} + 1$, then the weight of every nonzero codeword of C satisfies :*

$$2^{m-1} - (r-1)2^{m/2} \leq w \leq 2^{m-1} + (r-1)2^{m/2}$$

In terms of duals of extended BCH codes, the above corollary becomes

Corollary 2 Divisibility of the dual of an extended BCH code
Let $d \geq 3$, and l_d such that :

$$2^{l_d} + 1 \leq d < 2^{l_d+1} + 1 \tag{2}$$

Then, $DEB(d)$ is $2^{\lceil \frac{m}{l_d} \rceil - 1}$-divisible.
Let $t_d = m - l_d$.

1. *If $\lfloor \frac{m}{2} \rfloor \leq l_d < m-2$, and $d < 2^{l_d+1} - 3$,*

$$d_{min}(DEB(d)) \geq 2^{t_d+1} - 2^{t_d-1} \tag{3}$$

2. *If $\lfloor \frac{m}{2} \rfloor \leq l_d < m-2$, and $d \geq 2^{l_d+1} - 3$,*

$$d_{min}(DEB(d)) \geq 2^{t_d}$$

3. *If $l_d = m-2$ and $m \geq 4$, let d_0 denote the largest designed distance strictly less than $2^{m-1} - 1$, that is, the largest integer d such that $d < 2^{m-1} - 1$ and d is the smallest element of its cyclotomic coset. then :*
If m is odd and $d \leq d_0$, $d_{min}(DEB(d)) \geq 6$.
If m is even and $d < d_0$, $d_{min}(DEB(d)) \geq 6$.

4. *If $l_d < \lfloor \frac{m}{2} \rfloor$, a lower bound on $d_{min}(DEB(d))$ is given by the Carlitz-Uchiyama bound (th.6).*

Proof . By (2), and according to the characterization of the $C^{(t)}$ codes in terms of duals of BCH codes, we deduce :

$$\forall d \geq 3,\ C^{(t_d)} \subseteq DEB(d) \subset C^{(t_d-1)} \tag{4}$$

where $t_d = m - l_d$ and, if $l_d \leq \lfloor \frac{m}{2} \rfloor$, then $C^{(t_d)} = DEB(2^{l_d}+1)$, and if $l_d > \lfloor \frac{m}{2} \rfloor$, then $C^{(t_d)} = DEB(\sum_{i=1}^{a} 2^{m-it_d} + 1 - \delta_{r,0})$, with $m = at_d + r$, $r < t_d$.
The divisibility of $DEB(d)$ is then an immediate consequence of corollary 1.
Note that we distinguish the cases $l_d < m-2$ and $l_d = m-2$ because, if $l_d < m-2$, then $2^{l_d+1}-3$ is the smallest element of its cyclotomic coset, so that $2^{l_d+1}-3$ is the designed distance of a BCH code. Indeed, $2^{l_d+1}-3 = (\underbrace{0..0}_{\geq 2}\underbrace{1\ldots1}_{l_d-1}01)$. and any cyclic shift of $2^{l_d+1}-3$ has strictly less than two consecutive zeroes in the leftmost part of its binary representation. But if $l_d = m-2$, $2^{l_d+1}-3$ is not the smallest element of its cyclotomic coset, so that the condition $d < 2^{l_d+1}-3$ does not ensure that $2^{l_d+1}-3$ does not belong to $\cup_{s<d} cl(s)$.
1. Let $\lfloor \frac{m}{2} \rfloor \leq l_d < m-2$, and suppose $d < 2^{l_d+1}-3$.
In the same spirit as in the proof of part one of corollary 1, we want to show that $u_{0,d}$ and $v_{0,d}$ belong to the defining-set $U(d)$ of $DEB(d)$, where $u_{0,d} = 2^{t_d} + 2^{t_d-1} - 1$ and $v_{0,d} = 2^{t_d+1} + 2^{t_d-1} - 1$.
First, if $x \in [0,n]$, we shall denote by $(x)_0$, the smallest element of the cyclotomic coset of 2 modulo n containing x. We have :

$$n - v_{0,d} = 2^m - 2^{t_d+1} - 2^{t_d-1} = 2^{t_d-1}(2^{m-t_d+1} - 5)$$

so that

$$(n - v_{0,d})_0 = 2^{m-t_d+1} - 5 = 2^{l_d+1} - 5$$

But $d < 2^{l_d+1}-3$, that is $d \leq 2^{l_d+1}-5$, and so

$$(n - v_{0,d})_0 \notin \cup_{s<d} cl(s)$$

or

$$(n - v_{0,d})_0 \in [0,n] \setminus \cup_{s<d} cl(s)$$

thus

$$v_{0,d} \in \{n - z, z \in [0,n] \setminus \cup_{s<d} cl(s)\} = U(d)$$

The same reasoning holds for $u_{0,d}$:

$$n - u_{0,d} = 2^m - 2^{t_d} - 2^{t_d-1} = 2^{t_d-1}(2^{m-t_d+1} - 3)$$

$$(n - u_{0,d})_0 = 2^{m-t_d+1} - 3$$

$d < 2^{m-t_d+1} - 3$ so that $u_{0,d} \in U(d)$.

2. Let $\lfloor \frac{m}{2} \rfloor \leq l_d < m-2$. If $d = 2^{l_d+1}-3$, then $2^{l_d+1}-5 = (n - v_{0,t})_0$ belongs to $\cup_{s<d} cl(s)$, and so $v_{0,d}$ does not belong to $U(d)$.
If $d = 2^{l_d+1}-1$, then both $(n - v_{0,t})_0$ and $(n - u_{0,d})_0$ belong to $\cup_{s<d} cl(s)$, so that neither $v_{0,d}$ nor $u_{0,d}$ belong to $U(d)$.

3. If $l_d = m-2$ and m is odd, it is easy to see that $[0,2]$ and $[2^{\frac{m+1}{2}}, 2^{\frac{m+1}{2}}+2]$ are included in $U(d)$, the defining-set of $DEB(d)$, whenever $d \le d_0$. Applying lemma 1 with $d=3$, $z = 2^{\frac{m+1}{2}}$ and $s=2$ gives the result.
If m is even and $d < d_0$, $[0,2]$ and $[2^{\frac{m}{2}}, 2^{\frac{m}{2}}+2]$ are included in $U(d)$ (see [Le]).

4. If $l_d < \lfloor \frac{m}{2} \rfloor$, then $l_d + 1 \le \lfloor \frac{m}{2} \rfloor$, so $d < 2^{\lfloor \frac{m}{2} \rfloor} + 1$. We can then apply the Carlitz-Uchiyama bound, which is applicable here because if C is a cyclic code with defining set T, whose extension C_e is affine-invariant, then C_e has the same minimum weight as the cyclic code with defining set $T \cup \{0\}$. □

Note : Corollary 2 states that the dual of a BCH code has the divisibility of the largest $C^{(t)}$ code it contains (here $t = t_d$).

Remark : If m is even and $d = 2^{\frac{m}{2}} + 1$, then the bound given by (3) is better than the Carlitz-Uchiyama bound. Indeed, the C.U. bound in this case gives $d_{min}DEB(d) \ge 2^{\frac{m}{2}}$, and (3) gives $d_{min}DEB(d) \ge 2^{\frac{m}{2}+1} - 2^{\frac{m}{2}-1}$.

The following table gives the divisibility of all duals of extended BCH codes of designed distance d, of length 256. It also gives the lower bound on their minimum distance given by corollary 2. The star following some d's means that the corresponding codes are $C^{(t)}$ codes.

d	divisibility	lower bound
3*	128	128
5*	8	112
7	8	96
9*	4	80
11	4	64
13	4	48
15	4	32
17*, 19, 21, 23, 25, 27	2	24
29, 31	2	16
37*, 39, 43, 45, 47, 51, 53, 55, 59	2	12
61, 63	2	8
85*, 87, 91, 95, 111	2	6
119, 127	2	4

5 The $C^{(t)}$ codes as ideals of the modular algebra $\mathcal{A}$

Recall that $\mathcal{A}$ is the modular group algebra $\mathsf{F}_2[\{\mathsf{F}_{2^m}, +\}]$. By theorem 1, an affine invariant code is an extended cyclic code which is an ideal of $\mathcal{A}$.
If P denotes the ideal of $\mathcal{A}$ consisting of elements $x = \sum_{g \in \mathsf{F}_{2^m}} x_g X^g$ such that $\sum_{g \in \mathsf{F}_{2^m}} x_g = 0$, the product ideal P^j is the Reed Muller code of order $m-j$. The sequence $(P^j)_j$, is a decreasing sequence of ideals of $\mathcal{A}$. The $C^{(t)}$ codes are particular ideals of $\mathcal{A}$, as we shall now see. (The reader can refer to [Ch.1]).

Definition 5 *Let U be an ideal of $\mathcal{A}$.*

1. *The depth of x is the integer j such that : $x \in P^j$ and $x \notin P^{j+1}$. In the same way, the depth of U is the integer j such that : $U \subset P^j$ and $U \not\subset P^{j+1}$.*
2. *$x = \{x_1, \ldots, x_k\}$, $x_i \in \mathcal{A}$, is a generating system of U if $U = x_1\mathcal{A} + \ldots + x_k\mathcal{A}$. x is a minimal generating system of U, if the cardinality of every generating system of U is greater than or equal to k. k is then called the size of U.*
3. *U is a constant depth type ideal if every minimal generating system of U consists of elements of the same depth.*

The following result is due to P. Charpin [Ch.1] :

Theorem 7 *Let U be an affine-invariant code. Let j be the depth of U, and F, its border. Then :*

$$U \text{ is a constant depth type ideal } \Leftrightarrow (f \in F \Rightarrow \omega(f) = j)$$

From this theorem, we can easily derive the following corollary :

Corollary 3 *$\forall t \in [2, m-1]$, $C^{(t)}$ is a constant depth type ideal, and its size is m.*

So $C^{(t)}$ can be expressed as a sum of principal ideals of $\mathcal{A}$, in the following way :

$$C^{(t)} = x_1\mathcal{A} + \ldots + x_m\mathcal{A}, \quad x_i \in C^{(t)} \setminus P^{t+1}.$$

For example, x_i can be chosen to be the extension of the word $Z^{i-1}g_t(Z)$, where $g_t(Z)$ is the generator polynomial of the punctured $C^{(t)}$ code.

6 Numerical results

The following table sums up the parameters of the $C^{(t)}$ codes for several lengths. In the first column, we have the bound given by th.4. In the second one, we have the Carlitz-Uchiyama bound, which is negative for the first values of t. The divisibility of $C^{(t)}$ is that of theorem 5. Its depth is defined in def. 5. For the dimension of $C^{(t)}$, see prop. 2. "best low.b." is the best lower bound on the minimum distance of $C^{(t)}$. For $t \leq \lfloor \frac{m}{2} \rfloor$, it is the one of theorem 4; for $t \geq \lceil \frac{m}{2} \rceil$, when m is odd, and $t > \frac{m}{2}$ when m is even, using the Carlitz-Uchiyama bound and the divisibility yields a better result. (For m even and $t = \frac{m}{2}$, see the remark following the proof of corollary 2). The asterisque means that low.b. is the true minimum distance.

Parameters of the $C^{(t)}$ codes for the length 2^9.

t	b.of th.4	CU b.	divisibility	depth	dim.	best low.b.
2	6		2	2	436	6
3	12		2	3	271	12
4	24		2	4	145	24
5	48	97.61	4	5	73	100
6	96	188.12	4	6	37	192
7		233.37	16	7	19	240*
8		256	256	8	10	256*

Parameters of the $C^{(t)}$ codes for the length 2^{10}.

t	b.of th.4	CU b.	divisibility	depth	dim.	best low.b.
2	6		2	2	901	6
3	12		2	3	581	12
4	24		2	4	316	24
5	48	32	2	5	161	48
6	96	288	4	6	81	288
7	192	416	8	7	41	416
8		480	16	8	21	480*
9		512	512	9	11	512*

Aknowledgement

The author thanks P. Charpin without whom this paper wouldn't have been written.

References

[Ca,Uc] Carlitz L., Uchiyama S.: Bounds for exponential sums, Duke Math. Journal 24, pp.37-41, 1957.

[Ch.1] Charpin P.: Codes cycliques étendus affine-invariants et antichaines d'un ensemble partiellement ordonné. Discrete Mathematics 80, North-Holland, 1990, p.229-247.

[Ch.2] Charpin P.: Some applications of a classification of affine- invariant codes. Lect. Notes in Comp. Sci.356, Proceedings of AAECC 5, Springer-Verlag 1987.

[Ka,Li,Pe] Kasami T., Lin S., Peterson W.W.: Some results on cyclic codes which are invariant under the affine group and their applications. Info. and Control,vol 11, p. 475-496, 1967.

[Ka,To] Kasami T., Tokura N.:On the weight structure of Reed Muller codes. IEEE Trans. on Inf. theory, vol IT 16,no 6, nov. 1970, p.752-759.

[Le] Levy-dit-Vehel F. : On duals of primitive binary BCH codes. INRIA Report, to appear.

[Li,Mo] Litsyn S., Moreno C.J., Moreno O. : Divisibility properties and new bounds for cyclic codes and exponential sums in one and several variables, preprint.

[Ma] Mann H.B.: On the number of information symbols in Bose-Chauduri codes. Information and control 5, 1962, p.153-162.

[Mc El] Mc Eliece R.J.: Weight congruences for p-ary cyclic codes. Discrete Math. 3 1972, p.177-192.

[Sl] Sloane N.J.A., Mc Williams J.: The theory of error correcting codes. North Holland.

[Wa] Ward H.N.: Divisible codes. Archiv der Mathematik, vol. 36, 1981, fasc. 6, p.485-494.

Parameters of the $C^\perp$ codes for the length $2^m - 1$

[illegible]	[illegible]	[illegible]	[illegible]	[illegible]	[illegible]	m
	901	2	2		1	2
12	581	2			12	4
24	316	14	2		24	
48	161	5	4	32	48	5
256	81	6	4	248	96	6
160	41	7	8	16	96	7
460+	21	8	10	480		8
512+	11	9	512	32		9

Acknowledgements

The author thanks P. Charpin without whom this paper wouldn't have been written.

References

[Ca.De] Carlitz L., Uchiyama S., Bounds for exponential sums. Duke Math. Journal 24, pp.37-41, 1957

[Ch.1] Charpin P., Codes cycliques étendus affines-invariants et antichaînes d'un ensemble partiellement ordonné. Discrete Mathematics 80, North Holland, 1990, p.229-247

[Ch.2] Charpin P., Some applications of a classification of affine-invariant codes. Lect. Notes in Comp. Sc. 356, Proceedings of AAECC 5, Springer-Verlag, 1987

[Ka.Li.Pe] Kasami T., Lin S., Peterson W.W., Some results on cyclic codes which are invariant under the affine group and their applications. Inform. and Control vol 11, p.475-496, 1967

[Ka.To] Kasami T., Tokura N., On the weight structure of Reed-Muller codes. IEEE Trans. on Inf. Theory, vol IT-16, no 6, nov. 1970, p.752-759.

[La] Augot-Vallet D., On duals of primitive binary BCH codes. INRIA Report, to appear

[Mo] Moreno O., Moreno C.J., Divisibility properties and new bounds for cyclic codes and exponential sums in one and several variables. preprint

[Ma] Mann H.B., On the number of information symbols in Bose-Chaudhuri codes. Information and control 5, 1962, p.153-162

[McE] McEliece R.J., Weight congruences for p-ary cyclic codes. Discrete Math. 3, 1972, p.177-192

[SM] Sloane N.J.A., Mac Williams F., The theory of error correcting codes. North Holland

[Wa] Ward H.N., Divisible codes. Archiv der Mathematik vol. 36, 1981, fasc. 5, p.485-494

CONSTRUCTION OF THE BEST BINARY CYCLIC CODES OF EVEN LENGTH

J.-P. Martin
G.E.C.T., University of Toulon and Var, La Garde, France

Abstract – **This paper gives a method of construction for binary cyclic codes of even length out of shorter ones of odd length. This construction enables us to establish the set of the minimal distances reached by even length codes, and for each of these distances, to determine the associated generator polynomial of lowest degree. Thanks to this result we construct the complete table of the best possible binary cyclic codes on each minimal distance up to the length 64, finishing off in that way the table of the binary cyclic codes of odd length compiled by CHEN and lying in [5]. It can be noticed that VAN LINT found similar results but for only few specific values, see [3].**

1. Introduction

A cyclic code C of length N on the finite field with q elements $GF(q)$, consists of all multiples modulo $x^N - 1$ of a generator polynomial. C may have several different generator polynomials but among them there exists a unique nonzero monic one of lowest degree which is called *"the" generator polynomial.* This generator $g(x)$ is necessarily a divisor of $x^N - 1$ [1, p.190] ; so out of the splitting of $x^N - 1$ into irreducible factors in $GF(q)$, every possible generator polynomial and so every cyclic code of length N can be constructed.

$$x^N - 1 = \prod_{i=1}^{\ell} m_i(x) \quad \Rightarrow \quad g(x) = \prod_{\substack{i \in I \\ I \subseteq \{1..\ell\}}} m_i(x)$$

We can remark that if the characteristic p of $GF(q)$ is prime to N then the irreducible factors $m_i(x)$ are all distinct and so are the roots of $g(x)$ in the appropriate splitting field $GF(q^s)$. A code whose generator polynomial is such a $g(x)$ will be called a *simple-root cyclic code*.

Conversely, when p divides N we obtain a *repeated-root cyclic code*. As a matter of fact the following equality can be easily checked

$$\left.\begin{array}{l} N = p^\delta n \\ \gcd(n,p) = 1 \end{array}\right| \Rightarrow x^N - 1 = (x^n - 1)^{p^\delta}$$

and so we can deduce the general form of the generator polynomial

$$x^n - 1 = \prod_{i=1}^{\ell} m_i(x) \quad \Rightarrow \quad x^N - 1 = \prod_{i=1}^{\ell} m_i^{p^\delta}(x) \quad \Rightarrow \quad g(x) = \prod_{i=1}^{\ell} m_i^{e_i}(x) \qquad 0 \le e_i \le p^\delta$$

The multiplicities e_i of the roots of $g(x)$ make the study of such code by the classic theory ineffectual. For example the calculation of the BCH bound requires distinct roots and so does the decoding. It appeared that an efficient tool for such codes was the Hasse-derivative [4]. Massey, Castagnoli, Schoeller and Von Seeman succeeded thanks to this tool in expressing the minimal distance of a repeated-root cyclic code in terms of minimal distances of a corresponding set of simple-root cyclic codes [2]. Up to the present nothing has been told in a general point of view about the reduction of this set. The innovation in this paper is to do that work in the binary case : we determine the exact minimal distance of any binary repeated-root cyclic code out from its generator polynomial. Furthermore we give a universal method to construct for any even length and any possible minimal distance, the "best" repeated-root cyclic code i.e. the code whose dimension is the largest. We emphasize here that the term of "best" is slightly ironic because very local and concerns no other interesting property but the capacity of the code.

2. Castagnoli, Massey, Schoeller and Von Seemann's formula

We note :

- $\mathbf{F}_q = GF(q)$ the finite field of characteristic p.
- $A_q(N) = \mathbf{F}_q[x]/x^N - 1$ the algebra of polynomials modulo $(x^N - 1)$ over $\mathbf{F}_q$, i.e. the residue classes of $\mathbf{F}_q[x]$ modulo $x^N - 1$.
- $C_q(N)$ the set of the cyclic codes of length N on $\mathbf{F}_q$.
- $N = p^\delta n$ with $\delta \in \mathbf{N}$ and $\gcd(p,n) = 1$.
- $x^N - 1 = (x^n - 1)^{p^\delta} = \prod_{i=1}^{\ell} m_i^{p^\delta}(x)$ the splitting of $x^N - 1$ into irreducible factors in $\mathbf{F}_q$.
- $C(N,k,d) =< g(x) >$ the code of $C_q(N)$ generated by $g(x)$, whose dimension and minimal distance are respectively k and d. As $g(x)$ is a divisor of $x^N - 1$ it will be noted $\prod_{i=1}^{\ell} m_i^{e_i}(x)$ where $e_i \in \{0,1,\dots p^\delta\}$.
- $w_H(g(x))$ the Hamming weight of $g(x)$, i.e. the number of nonzero components of $g(x)$.

In their article of March 1991 [2] the authors showed that to a repeated-root cyclic code C in $C_q(p^\delta n)$ there corresponds a set of simple-root cyclic codes in $C_q(n)$ noted $\bar{C}_{\bar{t}}$ where $\bar{t}$ lies in T a subset of $\{0, 1, \dots p^\delta\}$. More exactly for all natural t, $\bar{C}_t$ is defined as hereunder

$$C(p^\delta n, k, d) = < g(x) > \qquad \text{with } g(x) = \prod_{i=1}^{\ell} m_i^{e_i}(x)$$

Implies

$$\bar{C}_t(n, \bar{k}, \bar{d}) = < \bar{g}_t(x) > \qquad \text{with} \quad \left| \begin{array}{l} \bar{g}_t(x) = \prod_{i \in I_t} m_i(x) \\ I_t = \{i \in \{1, \dots \ell\} \mid e_i > t\} \end{array} \right.$$

Thanks to this definition it was shown [2, Theorem 1] the following correspondence :

THEOREM 2.1. *Let T denote the set of values*

$$(p-1)p^{\delta-1} + (p-1)p^{\delta-2} + \cdots + (p-1)p^{\delta-(j-1)} + rp^{\delta-j}$$

with $r \in \{1, \cdots p-1\}$ and $j \in \{1 \cdots \delta\}$ adding to T the null value 0. Then there exists $\bar{t} \in T$ such that

$$d_{\min}(C) = P_{\bar{t}} \cdot d_{\min}(\bar{C}_{\bar{t}}) \qquad \textit{where } P_{\bar{t}} = p^{j-1}(r+1).$$

Proof – The proof of this theorem is the main purpose of the article [2]. for all natural t, P_t is originally introduced as the Hamming weight of $(x-1)^t$. Its value stems from the radix-p expansion of t which is $\sum_i p^i t_i$ where $0 \leq t_i \leq p-1$. We have then $P_t = \prod_i (t_i + 1)$. For the elements $\bar{t}$ of T we obtain the more concise form $P_{\bar{t}} = p^{j-1}(r+1)$. □

Problem – If we know the generator polynomial of a code C in $C_q(p^\delta n)$ and the minimal distances of the codes in $C_q(n)$, we can settle the possible field where $d_{\min}(C)$ lies :

$$d_{\min}(C) \in \mathcal{D} = \{P_{\bar{t}} \cdot d_{\min}(\bar{C}_{\bar{t}}) \mid \bar{t} \in T\}.$$

From the definition of T we calculate easily the cardinality of $\mathcal{D}$: $(p-1)\delta + 1$, and so we ascertain the incapability of determining the exact value of $d_{\min}(C)$.

3. Improvement of the formula in the binary case

In the binary case we are going to show that the exact minimal distance can be exactly calculated. From now on, as +1=-1 in $\mathbf{F}_2$, $A(n)$ will denote $A_2(n) = \mathbf{F}_2[x]/x^n + 1$.

THEOREM 3.1. *$C(2^\delta n, k, d) = < (x^n+1)^t v(x) >$ where x^n+1 does not divide $v(x)$. Then*

$$d = \min_{\bar{t}\geq t}\{P_{\bar{t}}\cdot d_{\min}(\bar{C}_{\bar{t}})\}$$

Proof – Let $\prod_{i=1}^{\ell} m_i(x)$ be the splitting in irreducible factors of x^n+1 and let $g(x)$ be the generator polynomial of $C(2^\delta n, k, d)$. Since $g(x)$ is a divisor of $x^{2^\delta n}+1$ it can be written $\prod_{i=1}^{\ell} m_i^{e_i}(x)$ with $0 \leq e_i \leq 2^\delta$. If we set $t = \inf\{e_i \mid 1 \leq i \leq \ell\}$ we can write :

$$g(x) = (x^n+1)^t v(x) \tag{1}$$

where x^n+1 does not divide $v(x)$. Show that for all $\bar{t} \geq t$ there exists a codeword in C whose weight is $P_{\bar{t}}\cdot d_{\min}(\bar{C}_{\bar{t}})$; in other words, show that every weight $P_{\bar{t}}\cdot d_{\min}(\bar{C}_{\bar{t}})$ with $\bar{t} \geq t$ is actually reached by codewords of C :
Let $\bar{g}_{\bar{t}}(x)$ be the generator polynomial of $\bar{C}_{\bar{t}}(n, \bar{k}, \bar{d})$. We have

$$\bar{g}_{\bar{t}}(x) = \prod_{i\in I} m_i(x) \quad \text{where } I = \{i \in \{1,\dots\ell\} \mid e_i > \bar{t}\}. \tag{2}$$

If $m(x)$ denotes a minimum weight nonzero code polynomial of $\bar{C}_{\bar{t}}$, there exists $k(x)$ unique with $\deg k(x) < n - \deg \bar{g}_{\bar{t}}(x)$ satisfying

$$m(x) = k(x)\bar{g}_{\bar{t}}(x) \tag{3}$$

Show that the code polynomial

$$\theta(x) = (x^n+1)^{\bar{t}} m^{2^\delta-\bar{t}}(x) \tag{4}$$

is divisible by $g(x) = (x^n+1)^t v(x)$ and so belongs to the code C :
Let $\bar{v}_{\bar{t}}(x)$ denote $\bar{g}_{\bar{t}}^{e_i-\bar{t}}(x)$. Since $\bar{t} \geq t$, $(x^n+1)^t$ divides $(x^n+1)^{\bar{t}}$ whereas $e_i - t \geq e_i - \bar{t}$ implies that $v(x)$ divides $\bar{v}_{\bar{t}}(x)$. Therefore we obtain $g(x)$ divides $(x^n+1)^{\bar{t}}\bar{v}_{\bar{t}}(x)$. In the same way $\bar{v}_{\bar{t}}(x)$ divides $\bar{g}_{\bar{t}}^{2^\delta-\bar{t}}(x)$ and so divides $m^{2^\delta-\bar{t}}(x)$ too (3). We finally obtain $g(x)$ divides $\bar{g}_{\bar{t}}^{2^\delta-\bar{t}}(x)$ which divides $\theta(x)$.
We are now going to study the weight of $\theta(x)$. By definition (Theorem 2.1.) we have for $p = 2$:

$$T = \{2^{\delta-1} + 2^{\delta-2} + \dots 2^{\delta-j} \mid 1 \leq j \leq \delta\} \cup \{0\} \tag{5}$$

Using the following equality

$$2^{\delta-1} + 2^{\delta-2} + \dots 2^{\delta-j} = 2^{\delta-j}(2^{j-1} + \dots 1) = 2^{\delta-j}(2^j-1) = 2^\delta - 2^{\delta-j} \tag{6}$$

T can be rewritten

$$T = \{2^\delta - 2^{\delta-j} \mid 1 \leq j \leq \delta\} \cup \{0\} = \{2^\delta - 2^{\delta-j} \mid 0 \leq j \leq \delta\} \tag{7}$$

For all $\bar{t}$ in T we have

$$(x^n+1)^{\bar{t}} = (x^n+1)^{2^{\delta-j}(2^j-1)} = (x^{2^{\delta-j}n}+1)^{2^j-1} \tag{8}$$

Furthermore

$$\left.\begin{array}{r} 2^\delta - \bar{t} = 2^{\delta-j} \\ \deg m(x) < n \end{array}\right| \Rightarrow \deg m^{2^{\delta-j}}(x) < 2^{\delta-j}n \tag{9}$$

We finally obtain

$$\theta(x) = (x^{2^{\delta-j}n} + 1)^{2^j-1} m^{2^{\delta-j}}(x) \tag{10}$$

Thanks to (9) and (10) we deduce

$$\begin{aligned} \mathrm{w_H}(\theta(x)) &= \mathrm{w_H}((x^{2^{\delta-j}n} + 1)^{2^j-1})\, \mathrm{w_H}(m^{2^{\delta-j}}(x)) \\ &= \mathrm{w_H}((x^n+1)^{\bar{t}})\, \mathrm{w_H}(m(x)) \\ &= P_{\bar{t}} \cdot d_{\min}(\bar{C}_{\bar{t}}) \end{aligned}$$

The minimum distance d of the code C will be the minimum value reached by $P_{\bar{t}} \cdot d_{\min}(\bar{C}_{\bar{t}})$ for $\bar{t} \geq t$ which proves the theorem. □

COROLLARY 3.2. *Let C be in $\mathcal{C}_q(N)$,* $\quad d_{\min}(C) = \min_{0 \leq j \leq \delta} \{2^j \cdot d_{\min}(\bar{C}_{2^\delta - 2^{\delta-j}})\}$

Proof – if

$$2^{\delta-1} + 2^{\delta-2} + \ldots + 2^{\delta-s} \geq t \geq 2^{\delta-1} + 2^{\delta-2} + \ldots + 2^{\delta-j+1} \tag{11}$$

then

$$2^{\delta-1} + 2^{\delta-2} + \ldots + 2^{\delta-j} = 2^\delta - 2^{\delta-j} \tag{12}$$

is the radix-2 expansion of $\bar{t}$ with $s \leq j \leq \delta$ and $P_{\bar{t}} = 2^j$. We obtain

$$d_{\min}(C) = \min\{2^j \cdot d_{\min}(\bar{C}_{2^\delta - 2^{\delta-j}}) \mid s \leq j \leq \delta\} \tag{13}$$

Moreover when $0 \leq j < s$ the generator polynomial of $\bar{C}_{2^\delta - 2^{\delta-j}}$ is $x^n + 1$. If we state that the minimal distance of such a code is infinite we obtain the final result. □

PROPOSITION 3.3. *Let $(d_i)_{1 \leq i \leq t}$ be the list of the minimal distances reached by the codes of $\mathcal{C}(n)$. Then the minimal distances actually reached by the codes of $\mathcal{C}(2^\delta n)$ are exactly the elements of the set $\mathcal{D}$ defined hereunder*

$$\mathcal{D} = \{2^s d_i \mid 0 \leq s \leq \delta, \quad 1 \leq i \leq t\}$$

Proof – Thanks to Corollary 3.2 we first remark that the minimal distance of any code of $\mathcal{C}(2^\delta n)$ lies necessarily in $\mathcal{D}$. We just have to show that for every value d in $\mathcal{D}$, we are able to construct a code of $\mathcal{C}(2^\delta n)$ whose minimal distance is actually d.
We set $d = 2^s \bar{d}$ where $\bar{d}$ is the minimal distance of the code $\bar{C}(n, \bar{k}, \bar{d}) =< \bar{g}(x) >$. Let $C(2^\delta n, k, d)$ denote the code generated by :

$$(x^n + 1)^{2^\delta - 2^{\delta-s}} \bar{g}^{2^{\delta-s}}(x) \tag{14}$$

We deduce from the formula (13) that

$$d = \min\{2^j \cdot d_{\min}(\bar{C}_{2^\delta - 2^{\delta-j}}) \mid s \leq j \leq \delta\} \tag{15}$$

and from the expression (14) of the generator polynomial of C that

$$\bar{C}_{2^\delta - 2^{\delta-j}} =< \bar{g}(x) >$$

for all $j \in \{s, s+1, \ldots \delta\}$. By definition $\bar{d}$ denotes the minimal distance of $\bar{C}_{2^\delta - 2^{\delta-j}}$ and so

$$\min\{2^j \bar{d} \mid s \leq j \leq \delta\} = 2^s \bar{d} \tag{16}$$

which proves that we can for every value of $\mathcal{D}$ construct a suitable code of $C(2^\delta n)$ □

Problem – The matter is now to determine the minimal degree generator polynomial of any code $C(2^\delta n, k, d)$ for a fixed $d = 2^s \bar{d}$.

4. Characterization of the "best code"

Firstly we recall that the notion of good code is related to the value of the rate $\mathcal{R} = k/N$ where k and N are respectively the dimension and the length of the code : the larger $\mathcal{R}$ is the better the code is considered. Since we have seen that on each length N there exists only a finite number of different repeated-root cyclic codes (by the splitting of $x^N + 1$), we can actually speak of "best code". We are now going to see how to characterize them in order to get their construction in the last part.

DEFINITION 4.1. *Let C be in $C(2^\delta n)$, the splitting of C is the $(\delta+1)$-tuple $(\bar{C}_0, \bar{C}_1, \ldots \bar{C}_\delta)$ satisfying $\bar{C}_i = \bar{C}_{2^\delta - 2^{\delta-i}}$ for all i in $\{0, 1, \ldots \delta\}$.*

Remarks –
a) The $\bar{C}_i$'s belong to $C(n)$.
b) $\bar{C}_0 \subseteq \bar{C}_1 \subseteq \ldots \bar{C}_\delta$.
c) All the codes of $C(2^\delta n)$ which have the same splitting do have equal minimum distances. The reciprocal is false.

THEOREM 4.2. *Let $(\bar{C}_0, \bar{C}_1, \cdots \bar{C}_\delta)$ be the splitting of $C(2^\delta n, k, d)$ satisfying*

$$\bar{C}_i(n, k_i, d_i) =< \bar{g}_i(x) >, \forall i \in \{0, 1, \cdots \delta\}$$

Then the minimal degree generator polynomial of C is

$$\mu(x) = \bar{g}_0(x).\bar{g}_1^{2^{\delta-1}}(x).\bar{g}_2^{2^{\delta-2}}(x) \cdots \bar{g}_{\delta-1}^2(x).\bar{g}_\delta(x)$$

We have thus $\left| \begin{array}{l} k = k_0 + 2^{\delta-1}k_1 + \cdots + 2k_{\delta-1} + k_\delta \\ d = \min\{2^i.d_i \mid 0 \leq i \leq \delta\} \end{array} \right.$

Proof – By Corollary 3.2 we have

$$d = \min_{0\leq j\leq \delta}\{2^j.d_{\min}(\bar{C}_{2^\delta-2^{\delta-j}})\} = \min_{0\leq j\leq \delta}\{2^j.d_{\min}(\bar{C}_i)\} = \min_{0\leq j\leq \delta}\{2^j.d_j\} \tag{1}$$

By simple calculations we obtain

$$\begin{aligned} k &= 2^\delta n - \deg\mu(x) \\ &= 2^\delta n - [(n-k_0) + 2^{\delta-1}(n-k_1) + \dots 2(n-k_{\delta-1}) + (n-k_\delta)] \\ &\dots \\ &= k_0 + 2^{\delta-1}k_1 + \dots + 2k_{\delta-1} + k_\delta \end{aligned} \tag{2}$$

Firstly we are going to show that $(\bar{C}_0, \bar{C}_1, \dots \bar{C}_\delta)$ is actually the splitting of $C =< \mu(x) >$: From Remark b) we deduce that $i < j \Rightarrow \bar{g}_j(x)$ divides $\bar{g}_i(x)$ where $\bar{g}_j(x)$ is never zero. So we can set

$$\bar{\sigma}_i(x) = \frac{\bar{g}_i(x)}{\bar{g}_{i+1}(x)} \qquad \text{with } 0 \leq i \leq \delta - 1 \tag{3}$$

Then for all ℓ in $\{0, 1, \dots \delta - 1\}$

$$\bar{g}_\ell(x) = \bar{g}_\delta(x)\prod_{i=\ell}^{\delta-1}\bar{\sigma}_i(x) \tag{4}$$

Using these new notations, $\mu(x)$ can be rewritten

$$\begin{aligned} \mu(x) &= \bar{g}_0(x).\bar{g}_1^{2^{\delta-1}}(x).\bar{g}_2^{2^{\delta-2}}(x)\cdots\bar{g}_{\delta-1}^2(x).\bar{g}_\delta(x) \\ &= \bar{\sigma}_0(x).\bar{\sigma}_1^{2^{\delta-1}+1}(x).\bar{\sigma}_2^{2^{\delta-1}+2^{\delta-2}+1}(x)\cdots\bar{\sigma}_{\delta-1}^{2^{\delta-1}+\dots 2+1}(x).\bar{g}_\delta^{2^{\delta-1}+\dots 2}(x) \\ &= \bar{\sigma}_0(x).\bar{\sigma}_1^{2^\delta-2^{\delta-1}+1}(x).\bar{\sigma}_2^{2^\delta-2^{\delta-2}+1}(x)\cdots\bar{\sigma}_{\delta-1}^{2^\delta-1}(x).\bar{g}_\delta^{2^\delta}(x) \end{aligned} \tag{5}$$

We have then for all $\bar{t} = 2^\delta - 2^{\delta-i}$

$$\bar{\mu}_{\bar{t}}(x) = \bar{\sigma}_i(x).\bar{\sigma}_{i+1}(x)\cdots\bar{\sigma}_{\delta-1}(x).\bar{g}_\delta(x) = \bar{g}_i(x) \tag{6}$$

We conclude that $\bar{C}_{2^\delta-2^{\delta-i}} = \bar{C}_i$ and it follows that $(\bar{C}_0, \bar{C}_1, \dots \bar{C}_\delta)$ is the splitting of C. Show now that $\mu(x)$ is the lowest degree polynomial satisfying

$$\bar{\mu}_{2^\delta-2^{\delta-i}}(x) = \bar{g}_i(x) \qquad \forall i \in \{0, 1, \dots \delta\} \tag{7}$$

The lowest degree polynomial $g(x)$ satisfying

$$\bar{g}_{2^\delta-1}(x) = \bar{g}_\delta(x) \tag{8}$$

is $\bar{g}_\delta^{2^\delta}(x)$. If we desire further that

$$\bar{g}_{2^\delta-2}(x) = \bar{g}_{\delta-1}(x) \tag{9}$$

with the condition that the degree of $g(x)$ is minimal, it suffices to multiplicate $g(x)$ by a polynomial relatively prime to it, whose degree is as low as possible in such a way that (7) were satisfied. $\bar{g}_{\delta-1}(x) = \bar{\sigma}_{\delta-1}(x)\bar{g}_\delta(x)$ implies that this requested factor is $\bar{\sigma}_{\delta-1}^{2^\delta-1}(x)$ which gives a new expression of $g(x)$:

$$g(x) = \bar{\sigma}_{\delta-1}^{2^\delta-1}(x)\bar{g}_\delta^{2^\delta}(x) \tag{10}$$

We proceed in the same way to construct step by step $g(x)$ and we obtain finally the desired expression of $\mu(x)$:

$$\mu(x) = \bar{g}_0(x).\bar{g}_1^{2^{\delta-1}}(x).\bar{g}_2^{2^{\delta-2}}(x)\cdots\bar{g}_{\delta-1}^2(x).\bar{g}_\delta(x) \tag{11}$$

□

5. Construction of the table - Algorithm

Up to the present we have seen how to determine out of the splitting of any code in $\mathcal{C}(2^\delta n)$ the lowest degree generator polynomial, the minimum distance and the dimension of the code. Virtually the inverse way is much more interesting : we dispose of the table compiled by Chen [5, Appendix D] that gives the characteristics (n, k, d, generator polynomial) of binary cyclic codes of odd length. We have seen how to construct $\mathcal{D}$ the field of the possible minimal distances of the codes in $\mathcal{C}(2^\delta n)$. We are now going to show the way to determine, for any d in $\mathcal{D}$ on any even length, the lowest degree generator polynomial which enables us to set up the table of the best codes.

For this purpose we introduce two new notions :

DEFINITION 5.1. *Let (u_i) and (v_i) be two real sequences. (u_i) makes flush (v_i) if*

$$(u_i) \text{ overvalues } (v_i) \text{ and } \exists \ell \mid u_\ell = v_\ell$$

DEFINITION 5.2. *$\mathcal{C}(N,k,d)$ is $\bar{d}$-maximal if it is the largest dimension code of $\mathcal{C}(N)$ which satisfies $d \geq \bar{d}$.*

We have then

PROPOSITION 5.3. *$C_i(n,k_i,d_i)$ $i \in \{0,1,\cdots t\}$ the codes of $\mathcal{C}(n)$. If $\exists d \in \mathbf{N}$ and $C_{i_0} \subseteq C_{i_1} \subseteq \cdots C_{i_\delta}$ with $(d_{i_j})_{0\leq j\leq\delta}$ making flush $(\frac{d}{2^j})_{0\leq j\leq\delta}$ then $(C_{i_j})_{0\leq j\leq\delta}$ is the splitting of a code $C \in \mathcal{C}(2^\delta n)$ whose minimal distance is d. If furthermore the C_{i_j} are $\frac{d}{2^j}$-maximal then C is d-maximal.*

Proof – a) If $(d_{i_j})_{0\leq j\leq\delta}$ makes flush $(\frac{d}{2^j})_{0\leq j\leq\delta}$ then we have by definition

$$\left\{\begin{array}{ll} \forall j & d_{i_j} \geq \frac{d}{2^j} \\ \exists \ell & d_{i_\ell} = \frac{d}{2^\ell} \end{array}\right| \Rightarrow \left\{\begin{array}{ll} \forall j & 2^j d_{i_j} \geq d \\ \exists \ell & 2^\ell d_{i_\ell} = d \end{array}\right. \tag{1}$$

The minimal distance of C is the minimum of $\{2^j d_{i_j} \mid 0 \leq j \leq \delta\}$. This minimum is reached for $j = \ell$ and so

$$d_{\min}(C) = d \tag{2}$$

We remark that the condition $C_{i_0} \subseteq C_{i_1} \subseteq \cdots C_{i_\delta}$ is necessary to be allowed to consider $(C_{i_0}, C_{i_1} \cdots C_{i_\delta})$ as the splitting of a code C.

b) Suppose that $C(2^\delta n, k, d)$ is not d-maximal. Then by Definition 5.2 there exists a code $C'(2^\delta n, k', d')$ with $d' \geq d$ and $k' > k$. We have by Theorem 4.2 :

$$k = k_{i_0} + 2^{\delta-1} k_{i_1} + \cdots + k_\delta \quad \text{and} \quad k' = k'_{i_0} + 2^{\delta-1} k'_{i_1} + \cdots + k'_\delta \tag{3}$$

$k' > k$ implies that there exists ℓ yielding $k'_{i_\ell} > k_{i_\ell}$ and so $d'_{i_\ell} < \frac{d}{2^\ell}$ since $(C_{i_j})_{0\leq j\leq\delta}$ are $\frac{d}{2^\ell}$-maximal. Thus $2^\ell d'_{i_\ell} < d$ which implies

$$d' = \min\{2^j d'_{i_j} \mid 0 \leq j \leq \delta\} < d \tag{4}$$

This is inconsistent with the hypothesis $d' \geq d$. □

Remark – If C is d-maximal then we cannot conclude that the codes (C_{i_j}) are $\frac{d}{2^j}$-maximal.

ALGORITHM

We can now settle the list of the best binary cyclic codes in $\mathcal{C}(2^\delta n)$ using the following algorithm :

data : the list $C_i(n, k_i, d_i) = < g_i(x) >$ of the codes of $\mathcal{C}(n)$ arranged in k_i decreasing order with $i \in \{0, 1, \cdots t\}$.

1) We construct $\mathcal{D} = \{2^j.d_i,\ 0 \leq j \leq \delta,\ i \in \{0, 1, \cdots t\}\}$.

2) We fix $d \in \mathcal{D}$; we construct $(d, \frac{d}{2}, \cdots \frac{d}{2^\delta})$.

3) We point out $C_{i_v} \in \mathcal{C}(n)$ satisfying $d_{i_v} = \frac{d}{2^v}$ with v as large as possible. So d_{i_v} will be the lowest and generally k_{i_v} the largest.

4) We construct $C_{i_0} \subseteq C_{i_1} \subseteq \cdots C_{i_v} \subseteq \cdots C_{i_\delta}$ with $(d_{i_j})_{0 \leq j \leq \delta}$ overvaluing $(\frac{d}{2^j})_{0 \leq j \leq \delta}$ and making flush at d_{i_v} (hypothesis in 3)

5) If the C_{i_j} are $\frac{d}{2^j}$-maximal then we obtain the best code (Theo 4.2 $\Rightarrow \mu(x)$ and k). Otherwise we try to construct other sequences $(C_{i'_j})_{0 \leq j \leq \delta}$ with at least one $i'_j > i_j$ (i.e. one $k'_{i_j} > k_{i_j}$).

6) We keep the sequence with the largest k.

Example : we are going to construct best codes on the length

$$2^\delta n = 2^2.9 = 36$$

The codes are specified by the roots of the generator polynomial $g(x)$ in the manner that $(3)^2$ indicates that α^3 has multiplicity 2 in $g(x)$ where α is a fixed element of order n in the appropriate extension of $\mathbb{F}_2$. Only one root in each conjugate class is specified.

n	k	D	Roots of gen. pol.	code
9	9	1	(-)	C_0
	8	2	(0)	C_1
	3	3	(1)	C_2
	2	6	(0)(1)	C_3
	0	∞	(0)(1)(3)	C_∞

1) $\mathcal{D} = \{1, 2, 3, 4, 6, 8, 12, 24, \infty\}$

2) $\boxed{d = 4}$ $\longrightarrow$ $(4, \underline{2}, \underline{1})$ made flush in 2 or 1.
3) 4) (C_3, C_1, C_0) maximal sequence $\longrightarrow$ $(6, \underline{2}, \underline{1})$
5) $k = 2 + 2 \times 8 + 9 = 27, \quad \mu(x) \cong (0)(1).(0)^2.(-) = (0)^3(1)$

2) $\boxed{d = 6}$ $\longrightarrow$ $(\underline{6}, \underline{3}, \frac{3}{2})$ made flush in 6 or 3.
3) 4) (C_3, C_2, C_2) $\longrightarrow$ $(6, 3, 3)$
5) $k = 2 + 2 \times 3 + 3 = 11$

3) 4) (C_3, C_3, C_1) $\longrightarrow$ $(6, 6, 2)$
5) $k = 2 + 2x2 + 8 = 14$ it's better !
6) $\mu(x) \cong (0)(1).(0)^2(1)^2.(0) = (0)^4(1)^3$

6. Conclusion

As it can be seen in Table 1, binary cyclic codes of even length may have rather good characteristics which perfectly insert themselves in Chen's table. They are even sometimes better than B.C.H. codes ! For instance we have constructed a (42, 35, 4) code whereas the best cyclic code on neighbouring odd length and minimal distance is (45, 35, 4). Idem for (42, 24, 8) and (45, 22, 8)... In fact the characteristics of repeated-root cyclic codes are closely connected [Corollary 3.2] to the characteristics of the odd length code which is associated to each of them (their rate cannot be better than the rate of the associated code [2, par.5]). It would be interesting to study sequences of repeated-root cyclic codes constructed out of good B.C.H. codes : the codes of length 14, 28 or 42 are good because the codes which have given rise to them (of length 7 and 21) have relative efficient characteristics. Finally we can point out that repeated-root cyclic codes are virtually convenient to use : thanks to rather simple formulas [Theorem 4.2] we can easily construct even length codes corresponding to some required characteristics ; we can fix the minimal distance, the length or the dimension and construct -if possible- the desired code out of some appropriate B.C.H. code.

7. References

[1] F.J. MacWilliams and N.J.A. Sloane, *The theory of error-correcting codes*, Amsterdam, North-Holland, 1977.

[2] G.Castagnoli, J.L.Massey, P.A.Schoeller and N.von Seemann, "On repeated-root cyclic codes", *IEEE Trans. Inform. Theory*, vol.37, no.2, pp.337-342, Mar.1991.

[3] J.H. van Lint, "Repeated-root cyclic codes", *IEEE Trans. Inform. Theory*, vol.37, no.2, pp.343-345, Mar.1991.

[4] H. Hasse, "Theorie der höheren Differentiale in einem algebraischen Funktionenkörper mit vollkommenem Konstantenkörper bei beliebiger Charakteristik", *J. Reine. Ang. Math.*, vol.175, pp.50-54, 1936.

[5] W.W.Peterson and E.J.Weldon, Jr, *Error-Correcting Codes*, Second ed. Cambridge, MA : MIT Press, 1972.

(*) – Codes the minimal distance d of which is equal to 2 do not appear in that list to reduce its volume. As a matter of fact, these codes are always generated by $(1+x)$ and so their characteristics are $(N, N-1, 2)$. For the same reason we have not noted the codes with $k=1$ which are generated by $(1+x^N)/(1+x)$ and the characteristics of which are $(N, 1, N)$.

<u>**Table 1 :**</u> Proceeding as we did in this example, the complete table of binary cyclic codes of even length up to 64 has been constructed, giving for each possible minimal distance the lowest degree generator polynomial. We give here an abridged version in which lie the only very best codes by the following criterium :

- The codes are ordered by increasing length.
- On the same length they are ordered by increasing minimal distance.
- The restriction for a code $C(N, k, d)$ to appear in the list is $d > 2$ and $k > 1$ $^{(*)}$ and no code with better characteristics must figure previously in the list i.e. a previous code must not satisfy in the same time : a larger (or equal) minimal distance and a larger (or equal) dimension.

N	k	d	Roots of generator polynomial
6	2	4	$(0)^2(1)$
7	4	3	(1)
	3	4	$(0)(1)$
9	2	6	$(0)(1)$
10	4	4	$(0)^2(1)$
12	7	4	$(0)^3(1)$
	2	8	$(0)^4(1)^3$
14	9	4	$(0)^2(1)$
	4	6	$(0)(1)^2(3)$
	3	8	$(0)^2(1)^2(3)$
15	11	3	(1)
	10	4	$(0)(1)$
	7	5	$(1)(3)$
	6	6	$(0)(1)(3)$
	5	7	$(1)(3)(5)$
	4	8	$(0)(1)(3)(5)$
	2	10	$(0)(1)(3)(7)$
17	9	5	(1)
	8	6	$(0)(1)$
18	2	12	$(0)^2(1)^2(3)$
20	13	4	$(0)^3(1)$

N	k	d	Roots of generator polynomial
21	16	3	$(3)(7)$
	15	4	$(0)(3)(7)$
	12	5	$(1)(3)$
	11	6	$(0)(1)(3)$
	9	8	$(0)(1)(3)(7)$
	5	10	$(0)(1)(3)(5)$
	3	12	$(0)(1)(3)(5)(9)$
23	12	7	(1)
	11	8	$(0)(1)$
24	17	4	$(0)^5(1)$
	2	16	$(0)^8(1)^7$
27	2	18	$(0)(1)(3)$
28	22	4	$(0)^3(1)$
	19	3	$(1)^3$
	12	8	$(0)^4(1)^3(3)$
	4	12	$(0)^3(1)^4(3)^3$
	3	16	$(0)^4(1)^4(3)^3$
30	24	4	$(0)^2(1)$
	22	3	$(1)^2$
	18	5	$(1)^2(3)$
	17	6	$(0)(1)^2(3)$
	14	8	$(0)^2(1)^2(3)(5)$
	9	10	$(0)(1)^2(3)^2(7)$
	6	12	$(0)^2(1)^2(3)^2(5)(7)$
	5	14	$(0)(1)^2(3)^2(5)(7)^2$
	4	16	$(0)^2(1)^2(3)^2(5)^2(7)$
	2	20	$(0)^2(1)^2(3)^2(5)(7)^2$
31	26	3	(1)
	25	4	$(0)(1)$
	21	5	$(1)(3)$
	20	6	$(0)(1)(3)$
	16	7	$(1)(3)(5)$
	15	8	$(0)(1)(3)(5)$
	11	11	$(1)(3)(5)(7)$
	10	12	$(0)(1)(3)(5)(11)$
	6	15	$(1)(3)(5)(7)(11)$
	5	16	$(0)(1)(3)(5)(7)(11)$
33	22	6	$(0)(1)$
	13	10	$(1)(3)$
	2	22	$(0)(1)(3)(5)$
35	28	4	$(5)(7)$
	3	20	$(0)(1)(3)(5)(7)$
36	29	4	$(0)^3(1)$
	2	24	$(1)^4(1)^4(3)^3$

N	k	d	Roots of gen. polynomial
39	26	6	(0)(1)
	15	10	(1)(3)
	13	12	(1)(3)(13)
	2	26	(0)(1)(3)(7)
40	31	4	$(0)^5(1)$
41	21	9	(1)
	20	10	(0)(1)
42	35	4	$(0)^2(3)(7)$
	24	8	$(0)^2(1)(3)^2(7)^2$
	14	12	$(0)^2(1)^2(3)^2(5)(7)$
	9	16	$(0)^2(1)^2(3)^2(5)(7)^2(9)$
	5	20	$(0)^2(1)^2(3)^2(5)^2(7)(9)$
	3	24	$(0)^2(1)^2(3)^2(5)^2(7)^2(9)$
	2	28	$(0)^2(1)^2(3)^2(5)^2(7)(9)^2$
43	29	6	(1)
	15	13	(1)(3)
	14	14	(0)(1)(3)
45	16	10	(0)(1)(3)(7)
	6	18	(0)(1)(3)(5)(7)(9)
	5	21	(1)(3)(5)(7)(9)(15)
	4	24	(0)(1)(3)(5)(7)(9)(15)
	2	30	(0)(1)(3)(5)(7)(9)(21)
46	11	16	$(0)^2(1)^2(3)$
47	24	11	(1)
	23	12	(0)(1)
48	37	4	$(0)^9(1)$
	2	32	$(0)^1 6(1)^1 5$
49	3	28	(0)(1)(3)(7)
51	43	3	(1)
	42	4	(0)(1)
	35	5	(1)(3)
	34	6	(0)(1)(3)
	27	9	(1)(3)(5)
	26	10	(0)(1)(3)(5)
	19	14	(1)(3)(5)(9)
	17	16	(1)(3)(5)(9)(17)
	11	17	(1)(3)(5)(9)(19)
	10	18	(0)(1)(3)(5)(11)(19)
	9	19	(1)(3)(5)(9)(17)(19)
	8	24	(0)(1)(3)(5)(9)(17)(19)
	2	34	(0)(1)(3)(5)(9)(11)(19)
54	2	36	$(0)^2(1)^2(3)^2(9)$
55	34	8	(0)(1)
	30	10	(0)(1)(11)
	25	11	(1)(5)
	24	12	(0)(1)(5)
	21	15	(1)(5)(11)
	20	16	(0)(1)(5)(11)

N	k	d	Roots of generator polynomial
56	48	4	$(0)^5(1)$
	3	32	$(1)^8(1)^8(3)^7$
57	38	6	(0)(1)
	2	38	(0)(1)(3)(5)
60	53	4	$(0)^3(1)$
	41	5	$(0)^3(1)^3(3)$
	40	6	$(0)^4(1)^3(3)$
	38	8	$(0)^4(1)^2(3)(5)$
	11	20	$(0)^3(1)^4(3)^3(5)^3(7)$
	5	28	$(0)^3(1)^4(3)^4(5)^4(7)^3$
	4	32	$(0)^4(1)^4(3)^4(5)^4(7)^3$
	2	40	$(0)^4(1)^4(3)^4(5)^3(7)^4$
62	55	4	$(0)^2(1)$
	47	5	$(1)^2(3)$
	46	6	$(0)(1)^2(3)$
	40	8	$(0)^2(1)^2(3)(5)$
	37	7	$(1)^2(3)^2(5)$
	32	10	$(1)^2(3)^2(5)(7)$
	22	14	$(1)^2(3)^2(5)^2(7)(11)$
	11	22	$(0)(1)^2(3)^2(5)^2(7)^2(9)(11)$
	10	24	$(0)^2(1)^2(3)^2(5)^2(7)^2(9)(11)$
	6	30	$(0)(1)^2(3)^2(5)^2(7)^2(9)(11)^2$
	5	32	$(0)^2(1)^2(3)^2(5)^2(7)^2(9)(11)^2$
63	57	3	(1)
	56	4	(0)(1)
	51	5	(1)(3)
	50	6	(0)(1)(31)
	45	8	(0)(1)(5)(9)(21)
	39	9	(1)(3)(5)(7)
	38	10	(0)(1)(3)(5)(7)
	36	11	(1)(3)(5)(7)(9)
	35	12	(0)(1)(3)(5)(7)(9)
	30	13	(1)(3)(5)(7)(9)(11)
	29	14	(0)(1)(3)(5)(7)(9)(11)
	28	15	(1)(3)(5)(7)(9)(11)(21)
	27	16	(0)(1)(3)(5)(7)(9)(11)(21)
	21	18	(0)(1)(3)(5)(7)(9)(15)(21)(23)
	19	19	(1)(3)(5)(7)(9)(21)(23)(27)(31)
	18	21	(1)(3)(5)(7)(11)(15)(27)(31)
	17	22	(0)(1)(3)(5)(7)(11)(15)(27)(31)
	16	23	(1)(3)(5)(7)(9)(11)(15)(21)(27)(31)
	15	24	(0)(1)(3)(5)(7)(11)(15)(21)(27)(31)
	11	26	(0)(1)(3)(5)(7)(9)(11)(15)(23)(27)(31)
	10	27	(1)(3)(5)(7)(9)(11)(15)(23)(27)(31)
	9	28	(0)(1)(3)(5)(7)(11)(15)(21)(23)(27)(31)
	7	31	(1)(3)(5)(7)(9)(11)(15)(21)(23)(27)(31)
	6	32	(0)(1)(3)(5)(7)(9)(11)(13)(15)(21)(23)(27)(31)
	3	36	(0)(1)(3)(5)(7)(11)(15)(21)(23)(27)(31)
	2	42	(0)(1)(3)(5)(7)(9)(11)(13)(15)(23)(27)(31)

SELF-DUAL AND DECOMPOSABLE GEOMETRIC GOPPA CODES

C. Munuera
University of Valladolid, Valladolid, Spain

R. Pellikaan
Eindhoven University of Technology, Eindhoven, The Netherlands

Abstract

In this paper, a necessary and sufficient criterion for self-duality of geometric Goppa codes is given.

1 Introduction

The aim of this note is to establish a necessary and sufficient criterion for self-duality of geometric Goppa codes. Self-dual codes play an important role in coding theory, so it is of interest to find conditions under which a code is self-dual.

Here we consider geometric Goppa codes, defined as follows:

Let $\mathcal{X}$ be a projective, nonsingular, absolutely irreducible curve of genus g defined over the finite field $\mathbf{F}_q$. Let $\mathbf{F}_q(\mathcal{X})$ be the function field of $\mathcal{X}$ over $\mathbf{F}_q$ and $\Omega_{\mathcal{X}}$ the vector space of rational differential forms on $\mathcal{X}$ over $\mathbf{F}_q$.

Let $\mathcal{P} = \{P_1, ..., P_n\}$ be a set of n distinct rational points on the curve $\mathcal{X}$. We fix the order of the P_i and denote by D the divisor $P_1 + ... + P_n$. For a rational divisor G on $\mathcal{X}$ of degree m and support disjoint from $\mathcal{P}$ we consider the vector spaces $\mathcal{L}(G) = \{f \in \mathbf{F}_q(\mathcal{X})^* \mid (f) \geq -G\} \cup \{0\}$ and $\Omega(G) = \{\omega \in \Omega_{\mathcal{X}} \setminus \{0\} \mid (\omega) \geq G\} \cup \{0\}$. The algebraic-geometric or geometric Goppa codes asociated to $\mathcal{P}$ and G over $\mathbf{F}_q$ are defined by

$$\begin{aligned} C_L(\mathcal{X}, \mathcal{P}, G, \mathbf{F}_q) &= \{(f(P_1), ..., f(P_n)) \mid f \in L(G)\}, \\ C_\Omega(\mathcal{X}, \mathcal{P}, G, \mathbf{F}_q) &= \{(res_{P_1}(\omega), ..., res_{P_n}(\omega)) \mid \omega \in \Omega(G-D)\}, \end{aligned}$$

see Goppa [5], [6], [7] .

For the properties of geometric Goppa codes we refer to the textbook [11]. In order to be able to say something about the dimension and the minimum distance of these codes we restrict to the case $2g-2 < \deg(G) < n$, in which $C_L(\mathcal{X}, \mathcal{P}, G, \mathbf{F}_q)$ has dimension $\deg(G) + 1 - g$ and at least minimum distance $n - \deg(G)$; and $C_\Omega(\mathcal{X}, \mathcal{P}, G, \mathbf{F}_q)$ has dimension $n - \deg(G) - 1 + g$ and at least minimum distance $\deg(G) - 2g + 2$. From now on we write, for short, $C(\mathcal{P}, G)$ instead of $C_L(\mathcal{X}, \mathcal{P}, G, \mathbf{F}_q)$ whenever it is clear which curve and field are meant.

Let us remember that a code $C(\mathcal{P}, G)$ is self-dual if it coincides with its dual $C(\mathcal{P}, G)^\perp$, that is with its orthogonal for the bilinear form

$$< \mathbf{x}, \mathbf{y} >= \sum x_i y_i.$$

In this paper we obtain a neccesary and sufficient condition for self-duality of a geometric Goppa code $C_L(\mathcal{X}, \mathcal{P}, G, \mathbf{F}_q)$ in terms of G and D when the length n, of the code satifies $n > 2g + 2$. Our main result is the following

Theorem 3.12 *Assume $n > 2g+2$. The code $C(\mathcal{P}, G)$ is self dual if and only if there exists a differential form η with simple poles and residue 1 at every $P_i \in \mathcal{P}$ such that $2G = D + K$, where $K = (\eta)$.*

A sufficient condition for self-duality has been given by H. Stichtenoth without any condition on n (see [9]): let ω be a differential form on $\mathcal{X}$, and $W = (\omega)$. Since there is an isomorphism $\Omega(G - D) \cong \mathcal{L}(D + W - G)$, we have the following

Proposition 1.1 *There exists a differential form ω on $\mathcal{X}$ with simple poles and residue 1 at every $P_i \in \mathcal{P}$ such that $C(\mathcal{P}, G)^\perp = C(\mathcal{P}, D + W - G)$, where W is the divisor of ω.*

A sufficient condition for self-duality follows from Proposition 1.1

Theorem 1.2 *If there exists a differential form η with simple poles and residue 1 at every $P_i \in \mathcal{P}$ such that $2G = D + K$, where $K = (\eta)$, then $C(\mathcal{P}, G)$ is self dual.*

Proof: Let ω be the differential form cited in 1.1. There exists a rational function f such that $\eta = f\omega$; since both η and ω have simple poles at every P_i it is $v_{P_i}(f) = 0$ and $f(P_i) = res_{P_i}(\eta)/res_{P_i}(\omega) = res_{P_i}(\eta) = 1$. Then $K = W + (f)$, so $D + W - G \sim D + K - G$ and $C(\mathcal{P}, G) = C(\mathcal{P}, D + K - G) = C(\mathcal{P}, D + W - G) = C(\mathcal{P}, G)^\perp$; thus $C(\mathcal{P}, G)$ is self dual. □

This result is due to H. Stichtenoth (see [9]). In some cases the condition stated there is also necessary. For example, in the $g > 1$ case, Y. Driencourt and Stichtenoth (see [3]) gave the following

Theorem 1.3 *Let $g > 1$. Suppose G is a divisor with:*
a) $G \geq 0$;
b) $\deg(G) \geq 4g - 1$; and
c) $v_P(G) \geq \deg(G) - 2g + 1$ for all places $P \leq G$;
then, the converse of 1.2 holds. □

Later this result has been generalized by C.P. Xing (see [12]) by assuming $n > 6g-4$ only.

However the converse of 1.2 is not always true. For instance, in the same paper [3] a counterexample with $g = 1$ and $n = 4 = 2g + 2$ is found. Moreover, as shown in [8], it is possible to get a similar example over $\mathbf{F}_q$ where q is an even power of 2, with a code arising from an hyperelliptic curve of genus $g = \frac{1}{2}q - 1$, and of lenght $n = 2g + 2$.

Furthermore, the converse of Theorem 1.2 is not true for so-called *decomposable* codes, which we study in the next section. In section 3 we prove theorem 3.12. Both sections, 2 and 3, are based on the paper [8] by C. Munuera and R. Pellikaan. Self-dual Goppa codes have been studied in [1], [2], [3] and [9] too.

2 Decomposable Codes

Definition 2.1 If C_1 is an $[n_1, k_1]$ code, and C_2 is an $[n_2, k_2]$ code, then we say that C is the *direct sum* of C_1 and C_2 if (up to reordering of coordinates)

$$C = \{(\mathbf{x}, \mathbf{y}) \mid \mathbf{x} \in C_1, \mathbf{y} \in C_2\}.$$

We denote this by $C = C_1 \oplus C_2$. If moreover C_1 and C_2 are nonzero, then we say that C *decomposes into* C_1 and C_2. We call a linear code C *decomposable* if there exist nonzero codes C_1 and C_2 such that C decomposes into C_1 and C_2.

Remark 2.2 1) The code C decomposes into C_1 and C_2 if and only if (up to reordering of coordinates) C has a generator matrix of the form

$$M = \begin{pmatrix} M_1 & 0 \\ 0 & M_2 \end{pmatrix},$$

where M_1 and M_2 are nonempty generator matrices for C_1 and C_2, respectively.
2) If C decomposes into an $[n_1, k_1, d_1]$ and an $[n_2, k_2, d_2]$ code, then obviously C is an $[n_1 + n_2, k_1 + k_2, d]$ code, where $d = \min(d_1, d_2)$. Hence

$$d \le \frac{n-k}{2} + 1$$

by the Singleton bound. Thus there are no MDS decomposable codes.

Example 2.3 Let C be a code of minimum distance one and length greater than one. If $\mathbf{x}$ is a codeword of weight one and $\{\mathbf{x}, \mathbf{y}_1, ..., \mathbf{y}_{k-1}\}$ is a basis of C, then so is $\{\mathbf{x}, \mathbf{y}_1 - \lambda_1\mathbf{x}_1, ..., \mathbf{y}_{k-1} - \lambda_{k-1}\mathbf{x}_1\}$. Thus C has a generator matrix of the form

$$\begin{pmatrix} 1 & 0 \\ 0 & M' \end{pmatrix},$$

where M' is an $(n-1) \times (k-1)$ matrix, so C is decomposable. We say that codes of minimum distance one and length greater than one are *trivial decomposable.*

In the following we will discuss nontrivial decomposable geometric Goppa codes.

Proposition 2.4 *Let G be a divisor such that $\deg(G) < n$. Then the following statements are equivalent:*
1) $C(\mathcal{P}, G)$ is a decomposable code.
2) There are two nonzero effective divisors D_1 and D_2 such that $D_1 + D_2 = D$, both $\mathcal{L}(G - D_1)$ and $\mathcal{L}(G - D_2)$ are not zero, and $\mathcal{L}(G) = \mathcal{L}(G - D_1) \oplus \mathcal{L}(G - D_2)$.

Proof: We have $\mathcal{L}(G - D_1) \cap \mathcal{L}(G - D_2) = \mathcal{L}(G - D) = (0)$, since $\deg(G) < n$. Both $\mathcal{L}(G - D_1)$ and $\mathcal{L}(G - D_2)$ are subspaces of $\mathcal{L}(G)$. If C is decomposable, then there exist two effective divisors D_1 and D_2 such that $D_1 + D_2 = D$, and a basis $\{f_1, ..., f_s, g_1, ..., g_t\}$ with $(f_i) \geq D_2$ for $i = 1, ..., s$ and $(g_i) \geq D_1$ for $i = 1, ..., t$, so $\{f_1, ..., f_s\} \subseteq \mathcal{L}(G - D_2)$ and $\{g_1, ..., g_t\} \subseteq \mathcal{L}(G - D_1)$ and (2) is proved. Conversely, if $\mathcal{L}(G - D_1) = < g_1, ..., g_s >$ and $\mathcal{L}(G - D_2) = < f_1, ..., f_t >$, then $\mathcal{L}(G) = < f_1, ..., f_s, g_1, ..., g_t >$ so $C(\mathcal{P}, G)$ decomposes into $C(\text{supp } (D_1), G - D_2)$ and $C(\text{supp } (D_2), G - D_1)$. □

Corollary 2.5 *If $\deg(G) < n$ and $C(\mathcal{P}, G)$ is a decomposable code of dimension k, then $k \leq \ell(2G - D) + 1$.*

Proof: For two effective divisors E_1, E_2 we have $\ell(E_1) + l(E_2) \leq \ell(E_1 + E_2) + 1$, see [11]. According to Proposition 2.4 there are two effective divisors D_1 and D_2 such that $D = D_1 + D_2$ and $l(G - D_1) > 0$ and $\ell(G - D_2) > 0$, so

$$k = \ell(G - D_1) + \ell(G - D_2) \leq \ell(2G - D) + 1.$$

□

Corollary 2.6 *If $\deg(G) < n$ and $C(\mathcal{P}, G)$ is nontrivial decomposable, then $n \leq 2g+2$.*

Proof: Let $m = \deg(G)$. If $C(\mathcal{P}, G)$ is decomposable then, according to 2.4, there are two nonzero effective divisors D_1 and D_2 such that $D_1 + D_2 = D$, both $\mathcal{L}(G - D_1)$ and $\mathcal{L}(G - D_2)$ are not zero, and $\mathcal{L}(G) = \mathcal{L}(G - D_1) \oplus \mathcal{L}(G - D_2)$. Let $n_1 = \deg(D_1)$ and $n_2 = \deg(D_2)$. We may assume that $n_1 \leq n_2$ and $m - n_1 \geq 0, m - n_2 \geq 0$ (otherwise $\mathcal{L}(G - D_1) = \{0\}$ or $\mathcal{L}(G - D_2) = \{0\}$). Now there are several cases.

Case 1: If $m - n_1 > 2g - 2$ and $0 \leq m - n_2 \leq 2g - 2$, then

$$m + 1 - g \leq m - n_1 + 1 - g + \frac{m - n_2}{2} + 1,$$

by Proposition 2.4, the Riemann-Roch theorem and Clifford's theorem. So $n + n_1 \leq m + 2$. Moreover $\deg(G) < n$, hence $m = n - 1$ and $n_1 = 1$. Thus the code is trivial decomposable, which is a contradiction.

Case 2: If $m - n_1 > 2g - 2$ and $m - n_2 > 2g - 2$ then .

$$m + 1 - g = m - n_1 + 1 - g + m - n_2 + 1 - g,$$

so $n+g=m+1$. Moreover $\deg(G)<n$, hence $m=n-1$ and $g=0$. Thus the code is MDS and decomposable, so it is trivial decomposable, which is a contradiction.

Case 3: If $0 \leq m-n_1 \leq 2g-2$ then $0 \leq m-n_2 \leq 2g-2$ (as $n_1 \leq n_2$); thus

$$m+1-g \leq \frac{m-n_1}{2}+1+\frac{m-n_2}{2}+1,$$

so $n \leq 2g+2$. □

Example 2.7 Let G be a divisor of degree m on a curve $\mathcal{X}$ of genus g. Suppose $m<n=2g+2$.

a) If either

1) G has degree $2g$ and there are two nonzero effective divisors D_1 and D_2 such that $D_1+D_2=D$ and $G-D_1$ is canonical and $G-D_2$ is principal, or

2) G is not special and $\mathcal{X}$ is hyperelliptic and there are two nonzero effective divisors D_1 and D_2 such that $D_1+D_2=D$ and both $G-D_1$ and $G-D_2$ are hyperelliptic divisors (see [11], p. 342),

then $C(D,G)$ is decomposable (this is a direct consequence of Proposition 2.4).

b) Let $C(\mathcal{P},G)$ be a decomposable code of lenght $n=2g+2$ and dimension $k=g+1$; then $m=2g$, so according to 2.5

$$\begin{aligned} \deg(2G-D) &= 2g-2 \\ \ell(2G-D) &= g \end{aligned}$$

so $2G-D$ is a canonical divisor with simple poles at every point in $\mathcal{P}$, $2G-D=(\omega)$. If $C(\mathcal{P},G)$ decomposes about the codes made from the points in $\mathcal{P}_1$, $\mathcal{P}_2$ ($\mathcal{P}_1 \cap \mathcal{P}_2 = \emptyset, \mathcal{P}_1 \cup \mathcal{P}_2 = \mathcal{P}$), in such a way that the residues of ω are equal to λ_1 at every point in $\mathcal{P}_1$, and the residues of ω are equal to λ_2 at every point in $\mathcal{P}_2$, with $\lambda_1 \neq \lambda_2$, then according to 1.2, the code is self-dual. To see this, let f be a rational function such that (fw) has simple poles and residue 1 at every point $P \in \mathcal{P}$ (so $f(P)=\lambda_i^{-1}$ if $P \in \mathcal{P}_i$; such f exists by the independence of valuations, see [11]). Let $K=(f\omega)$. According to 1.2,

$$C(\mathcal{P},G)^{\perp}=C(\mathcal{P},D+K-G)$$

and since $D+K-G=D+(f)+2G-D-G=G+(f)$, we have

$$C(\mathcal{P},G)^{\perp}=C(\mathcal{P},G+(f))=(\lambda_1,...,\lambda_n)C(\mathcal{P},G)=C(\mathcal{P},G)$$

that is to say, the code $C(\mathcal{P},G)$ is self-dual but the converse of 1.2 fails.

In the same way the converse of 1.2 is not true, in general, for self-dual decomposable geometric Goppa codes.

3 Equality of codes and equivalence of divisors

According to the results obtained in the last section, we must restric ourselves to $n > 2g + 2$ (and $\deg(G) < n$), in which case there are no decomposable nontrivial geometric Goppa codes.

The problem we want to solve, can be viewed as a particular case of the following: given two divisors G, H on $\mathcal{X}$ with support disjoint from $\mathcal{P}$, and such that $C(\mathcal{P}, G) = C(\mathcal{P}, H)$, then, which relationship between G and H is there?

This problem has been treated by C.P. Xing in [12]. In his work he assumes $2g - 1 < \deg(G) = \deg(H) < n - 1$ and $n > 2g + 2$. In this case he obtains

Proposition 3.1 *Suppose $n > 2g + 2$. Let G and H be two effective divisors with support disjoint from $\mathcal{P}$ on the curve $\mathcal{X}$. If $2g - 1 < \deg(G), \deg(H) < n - 1$, then $C(\mathcal{P}, G) = C(\mathcal{P}, H)$ if and only if $G = H$.*

However equality is not the most general situation in which two divisors provide the same code as we shall see. Firstly we define a special case of linear equivalence of divisors

Definition 3.2 1) Two divisors G, H on $\mathcal{X}$ are called *equivalent about* $\mathcal{P}$, denoted $G \sim_{\mathcal{P}} H$, if there exists a rational function f such that $f(P_i) = 1$ for every $P_i \in \mathcal{P}$ and $H - G = (f)$.
2) Let C be a linear code in $\mathbf{F}_q^n$ and $\lambda = (\lambda_1, ..., \lambda_n)$ an n-tuple of nonzero elements in $\mathbf{F}_q$. For every $\mathbf{x} = (x_1, ..., x_n) \in C$, define $\lambda\mathbf{x} = (\lambda_1 x_1, ..., \lambda_n x_n)$ and $\lambda C = \{\lambda\, \mathbf{x} \mid \mathbf{x} \in C\}$.

Remark 3.3 We can view λ as a linear map in $\mathbf{F}_q^n$ leaving the Hamming metric invariant. Note that a linear map of $\mathbf{F}_q^n$ leaves the Hamming metric invariant if and only if it is of the form $\lambda\sigma$ for some permutation σ of $\{1, ..., n\}$.

The relation between the notion of equivalence about $\mathcal{P}$ and our problem is given by the following property (see [10])

Proposition 3.4 *Let G, H be two divisors on $\mathcal{X}$ with support disjoint from $\mathcal{P}$. If $G \sim_{\mathcal{P}} H$ then $C(\mathcal{P}, G) = C(\mathcal{P}, H)$.*

Proof: If $H - G = (f)$ then the map

$$\mathcal{L}(H) \longrightarrow \mathcal{L}(G) \;\; ; \;\; h \mapsto fh$$

is an isomorphism of vector spaces. Thus $\mathcal{L}(G) = \{fh \mid h \in \mathcal{L}(H)\}$, so $C(\mathcal{P}, G) = (f(P_1), ..., f(P_n))C(\mathcal{P}, H) = (1, ..., 1)C(\mathcal{P}, H) = C(\mathcal{P}, H)$ □

In the rest of this section we shall prove the converse of 3.4 .

Proposition 3.5 *Assume $n > 2g+2$ and $2g-1 < \deg(G) < n-1$. If there is a word in $C(\mathcal{P},G)$ with weight n then $C(\mathcal{P},G) = C(\mathcal{P},H)$ if and only if $G \sim_{\mathcal{P}} H$*

Proof: After 3.4 we have to prove only one direction of the assertion. Let $(x_1,...,x_n) \in C(\mathcal{P},H)$ be a codeword of weight n and let $f \in \mathcal{L}(G), h \in \mathcal{L}(H)$ be the functions such that $f(P_i) = x_i = h(P_i)$ for every $P_i \in \mathcal{P}$. Let us consider the divisors $G' = G+(f), H' = H+(h)$. It is clear that G' and H' are effective divisors with support disjoint with $\mathcal{P}$. Furthermore, as we have seen in 3.4

$$C(\mathcal{P},G) = (f(P_1),...,f(P_n))C(\mathcal{P},G') = (x_1,...,x_n)C(\mathcal{P},G')$$

and

$$C(\mathcal{P},H) = (h(P_1),...,h(P_n))C(\mathcal{P},H') = (x_1,...,x_n)C(\mathcal{P},H')$$

so

$$(x_1,...,x_n)C(\mathcal{P},G') = (x_1,...,x_n)C(\mathcal{P},H')$$

that is $C(\mathcal{P},G') = C(\mathcal{P},H')$. Now, according to 3.1, $G' = H'$ so $G-H = (h/f)$, and $G \sim_{\mathcal{P}} H$ □

Remark 3.6 The above proposition is not true without the restriction on the degree of the divisors $2g-1 < \deg(G) < n-1$ as we can see as follows: let K be a canonical divisor and P, Q two rational points not in $\mathcal{P}$. Since $\ell(K+P) = \ell(K+Q) = \ell(K)$ we have $\mathcal{L}(K+P) = \mathcal{L}(K+Q)$ so $C(\mathcal{P},K+P) = C(\mathcal{P},K+Q)$ but $K+P \not\sim K+Q$. Futhermore, if $K = (\eta)$ where η is a differential form with simple poles and residue 1 at every point $P \in \mathcal{P}$, then $C(\mathcal{P},D-P) = C(\mathcal{P},K+P)^{\perp} = C(\mathcal{P},K+Q)^{\perp} = C(\mathcal{P},D-Q)$ but $D-P \not\sim D-Q$.

The result stated in 3.5 seems to depend on the existence of codewords with weight large enough but, in fact, this condition is superfluous.

Proposition 3.7 *If C is a code in $\mathbf{F}_q^n$ not contained in a coordinate hyperplane, and $n < q$, then there is a codeword in C with weight n.*

Proof: For every $i = 1,...,n$ let $C_i = C \cap (x_i = 0)$. $C_i \neq C$ for all i since C is not contained in any hyperplane of the form $x_i = 0$. If there is no codeword with weight n then $C \subseteq \cup_{i=1}^n C_i$, so, by taking cardinalities we have $q^k \leq nq^{k-1}$ (where $k = \dim C$), that is $q \leq n$ □

Corollary 3.8 *If $2g-1 < \deg(G) < n-1$ and $n < q$, then there exists in $C(\mathcal{P},G)$ a codeword with weight n.*

Proof: If $C(\mathcal{P},G)$ is contained in a hyperplane $x_i = 0$, then P_i is a base point for G, but since $2g-1 < \deg(G)$, G is base point free □

For a given positive integer r let us consider the vector space over $\mathbf{F}_{q^r}$

$$\mathcal{L}(G,\mathbf{F}_{q^r}) = \{f \in \mathbf{F}_{q^r}(\mathcal{X})^* \mid (f) \geq -G\} \cup \{0\}$$

and the code $C(\mathcal{X},\mathcal{P},G,\mathbf{F}_{q^r})$, image of the map

$$\mathcal{L}(G,\mathbf{F}_{q^r}) \longrightarrow \mathbf{F}_{q^r}^n \ ; \ f \mapsto (f(P_1),\ldots,f(P_n)).$$

Lemma 3.9 *$\dim_{\mathbf{F}_q}(\mathcal{L}(G,\mathbf{F}_q)) = \dim_{\mathbf{F}_{q^r}}(\mathcal{L}(G,\mathbf{F}_{q^r}))$ for every divisor G rational over $\mathbf{F}_q$.*

Proof: See [11], chap. (2.3.1) □

Lemma 3.10 *Let G,H be two divisors on $\mathcal{X}$, rational over $\mathbf{F}_q$. If $C(\mathcal{X},\mathcal{P},G,\mathbf{F}_q) = C(\mathcal{X},\mathcal{P},G,\mathbf{F}_q)$ then $C(\mathcal{X},\mathcal{P},G,\mathbf{F}_{q^r}) = C(\mathcal{X},\mathcal{P},G,\mathbf{F}_{q^r})$.*

Proof: Let $\{f_1,\ldots,f_k\}$ be a basis for $\mathcal{L}(G,\mathbf{F}_q)$. Since $\mathcal{L}(G,\mathbf{F}_q) \subseteq \mathcal{L}(G,\mathbf{F}_{q^r})$, according to 3.9 , it is a basis for $\mathcal{L}(G,\mathbf{F}_{q^r})$ too, so both, $C(\mathcal{X},\mathcal{P},G,\mathbf{F}_q)$ and $C(\mathcal{X},\mathcal{P},G,\mathbf{F}_{q^r})$ have the same generator matrix □

At this moment we are able to prove the following

Theorem 3.11 *Assume $n > 2g+2$. If $2g-1 < \deg(G), \deg(H) < n-1$ then $C(\mathcal{P},G) = C(\mathcal{P},H)$ if and only if $G \sim_{\mathcal{P}} H$.*

Proof: After 3.4 it is enough to prove the 'only if' part. Assume $C(\mathcal{P},G) = C(\mathcal{P},H)$ and take an integer r such that $q^r > n$. According to 3.7 there is a codeword of weight n in $C(\mathcal{X},\mathcal{P},G,\mathbf{F}_{q^r})$ so, from 3.5, $G \sim_{\mathcal{P}} H$ over $\mathbf{F}_{q^r}$. Thus $\ell(G-H,\mathbf{F}_{q^r}) = 1$, and there is a function $f \in \mathcal{L}(G-H,\mathbf{F}_{q^r})$ such that $f(P_1) = \ldots = f(P_n) = 1$. Since G and H are rational over $\mathbf{F}_q$, $\ell(G-H,\mathbf{F}_q) = 1$ according to 3.9 , so there exists a nonzero function $f' \in \mathcal{L}(G-H,\mathbf{F}_q)$, so $f = \lambda f'$ for some $\lambda \in \mathbf{F}_{q^r}^*$, so $f'(P_1) = 1/\lambda$, and (after scaling f' if necessary) we can assume $f'(P_1) = 1$, so $\lambda = 1$, so $f = f'$, and finally $G \sim_{\mathcal{P}} H$ over $\mathbf{F}_q$. □

Let us return now to our problem of settling a necessary condition for the self-duality of geometric Goppa codes.

As we know, the dimension of a code $C(\mathcal{P},G)$ is $k = \ell(G) - \ell(G-D)$, thus,
a) if $\deg(G) \geq n$ then $k > n-g$;
b) if $2g-2 < \deg(G) < n$ then $g \leq k \leq n-g$; and

c) if $\deg(G) \leq 2g-2$ then $k \leq g$;
(these results follow from Riemann-Roch and Clifford's theorems). If $C(\mathcal{P}, G)$ is self-dual then $k = \frac{n}{2}$, so $2g - 1 < \deg(G) < n - 1$ (because $n > 2g + 2$). Furthermore its minimum distance d verifies $d > 1$; thus it is not trivial decomposable and we can apply 3.11 to obtain the following

Theorem 3.12 *Assume $n > 2g+2$. The code $C(\mathcal{P}, G)$ is self dual if and only if there exists a differential form η with simple poles and residue 1 at every $P_i \in \mathcal{P}$ such that $2G = D + K$, where $K = (\eta)$.*

Proof: Since $C^{\perp}(\mathcal{P}, G) = C(\mathcal{P}, D + W - G)$, where $W = (\omega)$, ω with simple poles and residue 1 at every $P_i \in \mathcal{P}$, according to 3.11 there is a rational function f such that $f(P_i) = 1$ for all $P_i \in \mathcal{P}$ and $2G = D+W+(f)$. Then it is enough to take $\eta = f\omega$ □

When $n \leq 2g + 2$ the conclusion of the theorem fails as said in sections 1 and 2.

References

[1] Y. Driencourt and J.F. Michon, Rapport sur les codes géométriques, oct. 1986.

[2] Y. Driencourt and J.F. Michon, Remarques sur les codes géométriques, C.R. Acad. Sc. Paris **301** (1985), 15-17.

[3] Y. Driencourt and H. Stichtenoth, A criterion for self-duality of geometric codes, Communications in Algebra **17**(4) (1989), 885-898.

[4] V.D. Goppa, Codes associated with divisors, Probl. Peredachi Inform. **13**(1) (1977), 33-39. Translation: Probl. Inform. Transmission **13** (1977), 22-26.

[5] V.D. Goppa, Codes on algebraic curves, Dokl. Akad. Nauk SSSR **259** (1981), 1289-1290. Translation: Soviet Math. Dokl. **24** (1981), 170-172.

[6] V.D. Goppa, Algebraico-geometric codes, Izv. Akad. Nauk SSSR **46** (1982), Translation: Math. USSR Izvestija **21** (1983), 75-91.

[7] V.D. Goppa, Codes and information, Russian Math. Surveys **39** (1984), 87-141.

[8] C. Munuera and R. Pellikaan, Equality of geometric Goppa codes and equivalence of divisors, preprint, 1992.

[9] H. Stichtenoth, Self-dual Goppa codes, J. Pure Appl. Algebra **55** (1988), 199-211.

[10] H. Stichtenoth, The automorphisms of geometric Goppa codes, J. Algebra **130** (1990), 113-121.

[11] M. Tsfasman and S. Vladut, Algebraic Geometric Codes, Kluwer Ac. Publ, Dordrecht, 1991.

[12] C.-P. Xing, When are two geometric Goppa codes equal ?, IEEE Trans. Information Theory IT-**38** (1992), 1140-1142.

ON A CONJECTURE OF MACWILLIAMS AND SLOANE

F. Rodier
Laboratoire de Mathématiques Discrètes, Marseille, France

1. Introduction

Let C_m be a primitive binary BCH code, of length $2^m - 1 = q - 1$, and of designed distance $\delta = 2t + 1$. We want to study the dual of this code, which we denote by $C_m^{\perp}$.

From the Carlitz-Uchiyama bound [4], the weight w of a nonzero word of $C_m^{\perp}$ is such that

$$|w - 2^{m-1}| \leq (t-1)2^{m/2}.$$

MacWilliams and Sloane suggest a slightly better result for m odd ([4], research problem 9.5):

$$|w - 2^{m-1}| \leq \frac{1}{\sqrt{2}}(t-1)2^{m/2}. \qquad (A)$$

The purpose of this paper is to give a survey of recent results in [6] and [7] on this problem, where MacWilliams and Sloane's bound is disproved. Actually one will determine precisely the minimum distance of the duals of BCH codes when m goes to infinity for some classes of designed distance.

2. Discussion of the problem

Denote by $(a_i)_{1\leq i\leq q-1}$ the elements of $\mathbf{F}_q^*$ and by Tr the trace of $\mathbf{F}_{2^m}$ into $\mathbf{F}_2$. One shows that the dual of the code C_m is the image of the application

$$f \longmapsto c_f = (\operatorname{Tr} f(a_1), \operatorname{Tr} f(a_2), \ldots, \operatorname{Tr} f(a_{q-1}))$$

from the space of polynomials $f \in \mathbf{F}_q[x]$ of degree $d = \delta - 2$ such that $f(0) = 0$, to the space $(\mathbf{F}_2)^{q-1}$ (cf. [7]). Moreover the weight w of the codeword c associated to f is given by

$$S_m(f) = \sum_{\mathbf{F}_q} (-1)^{\operatorname{Tr} f(x)} = 2^m - 2w. \tag{B}$$

Then the Carlitz-Uchiyama bound is equivalent to the Weil inequality [8]:

$$|S_m(f)| \leq (d-1)2^{m/2},$$

and MacWilliams and Sloane's conjectural bound (A) becomes

$$|S_m(f)| \leq \frac{1}{\sqrt{2}}(d-1)2^{m/2}$$

for m odd.

One knows (cf. Lang and Weil [2]) that for a given f and for m going to infinity the former inequality is the best possible. This means that asymptotically one has

$$\limsup_{m\to\infty} \frac{|S_m(f)|}{2^{m/2}} = d - 1.$$

So the question is to compute

$$\limsup_{m\to\infty} \frac{|S_m(f)|}{2^{m/2}}$$

for m odd.

3. The results

For $d \leq 5$ one has

$$\limsup_{\substack{m\to\infty \\ m \text{ odd}}} \frac{|S_m(f)|}{2^{m/2}} \leq \frac{1}{\sqrt{2}}(d-1).$$

This can be shown easily by embedding the code into a Reed and Muller code (cf. [4], p. 451 for $d = 3$, or [6]). So, for $d \leq 5$, that is for codes dual to BCH codes with designed distance less than or equal to 7, MacWilliams and Sloane's bound (A) is true.

For $d \geq 7$, the situation is different, as shows already the case where $d = 7$ (cf. [5]). Actually, the computations in small degrees lead to formulate the following conjecture.

Conjecture. *For each odd d, there exists a polynomial f of degree d with coefficients in* $\mathbf{F}_2$ *such that*

$$\limsup_{\substack{m\to\infty \\ m \text{ odd}}} \frac{|S_m(f)|}{2^{m/2}} = d - 1.$$

In this paper, this conjecture will be shown to be true for d belonging to two series. These series are conjecturally infinite by Artin [1].

Let d be a prime number. Let us suppose that either

$$(1) \quad \begin{cases} \text{a)} & d \text{ is congruent to } -1 \pmod{8}, \\ \text{b)} & d \text{ does not divide any number } 2^i - 1 \text{ for } 1 \leq i < (d-1)/2, \end{cases}$$

or

$$(2) \quad \begin{cases} \text{a)} & d \neq 3, \\ \text{b)} & d \text{ is congruent to } -1, 3, 19 \pmod{28}, \\ \text{c)} & d \text{ does not divide any number } 2^i - 1 \text{ for } 1 \leq i < d-1. \end{cases}$$

The case (1) is realized for the numbers

$$d = 7, 23, 47, 71, 79, 103, 167, 191, 199 \ldots$$

The case (2) is realized for the numbers

$$d = 19, 59, 83, 131, 139 \ldots$$

One has the following theorem. Let us first define the Dickson polynomial $g_d(x)$ of odd degree d as being the reduction (mod. 2) of

$$\sum_{j=0}^{(d-1)/2} \frac{d}{d-j} \binom{d-j}{j} x^{d-2j}.$$

Theorem. *Let d be a prime number fulfilling one of the two previous conditions (1) or (2). Let us set*

$$\text{– in case 1,} \quad f(x) = x^d$$

$$\text{– in case 2,} \quad f(x) = g_d(x).$$

Then we have

$$\limsup_{\substack{m \to \infty \\ m \text{ odd}}} \frac{|S_m(f)|}{2^{m/2}} = d - 1.$$

By the expression (B) linking $S_m(f)$ and w, we get the following corollary.

Corollary 1. *Let d be a prime number fulfilling one of the two previous conditions (1) or (2). Let C_m be the primitive binary BCH code of length $2^m - 1$ and designed distance $\delta = d + 2 = 2t + 1$. Then there exist for each m some codewords c_m of weight w_m in the dual code $C_m^\perp$ such that*

$$\limsup_{\substack{m \to \infty \\ m \text{ odd}}} \frac{|w_m - 2^{m-1}|}{2^{m/2}} = \frac{d-1}{2} = t - 1.$$

Here is another way to state corollary 1.

Corollary 2. *Let d and C_m be as in the preceding corollary and let $\delta = d+2 = 2t+1$ the designed distance of the code C_m. Then, for every positive ϵ there exists an odd value of m and a codeword c_m in $C_m^\perp$ of weight w_m such that*

$$|w_m - 2^{m-1}| > (1 - \epsilon)(t - 1)2^{m/2}.$$

As a consequence the inequality expected by MacWilliams and Sloane is not true for such d and for infinitely many values of m.

The next corollary gives an inequality on the minimal distance d_m of the code $C_m^\perp$.

Corollary 3. *Let d, C_m and t be as in the preceding corollary. Then for every positive ϵ there exists an odd value of m such that*

$$2^{m-1} - (t-1)2^{m/2} < d_m < 2^{m-1} - (1 - \epsilon)(t - 1)2^{m/2}.$$

Hence for these values of d and m, d_m is approximatively equal to $2^{m-1} - (t-1)2^{m/2}$.

4. Sketch of the proof of the theorem

For a complete proof, one may see [6] and [7].

By Weil [8], one knows that there exist complex numbers π_i for $1 \leq i \leq d-1$ such that

$$-S_m = \sum_{1 \leq i \leq d-1} \pi_i^m$$

The numbers π_i are algebraic integers and pairwise complex conjugate and their absolute value is equal to $\sqrt{2}$.

If $g = (d-1)/2$ does not divide m, then because of the hypothesis made on d, the polynomial f is a permutation polynomial of the field $\mathbf{F}_{2^m}$ and the sum S_m is zero. This is clear if $f = x^d$ and results from theorem 7.16 of [3] for f a Dickson polynomial. This implies that there is a complex number ϕ such that $\pi_i^g = \phi$ or $\overline{\phi}$ and thus one has

$$-S_{gm} = g(\phi^m + \overline{\phi}^m) = 2g(\sqrt{2})^{gm}\,\Re e\,\omega^m.$$

where $\omega = \phi/\sqrt{2}$. Then ω is a complex number of absolute value 1 which is a root of the equation (by letting $m = 1$ in the preceding formula)

$$X^2 + \frac{S_g}{g \cdot 2^{g/2}} X + 1 = 0.$$

One wants to show that the number ω is not a root of unity. From this equation it is enough to show that

$$\frac{S_g}{g \cdot 2^{g/2}} \neq 0 \quad \text{and} \quad \pm\sqrt{2}.$$

This can be deduced from the following lemma. Let us call K_g the Kloostermann sum

$$K_g = \sum_{\mathbf{F}_{2^g}^*} (-1)^{\mathrm{Tr}(x+x^{-1})}.$$

Lemma. *The number S_g is congruent modulo d to*

$$\begin{cases} 1 & \text{if } f = x^d, \\ \dfrac{K_g + d}{2} & \text{if } f \text{ is a Dickson polynomial.} \end{cases}$$

Then a little computation shows the desired result.

From a result of Kronecker, the ω^m for m a positive integer are everywhere dense on the torus $\mathbf{T}$ of complex numbers of absolute value 1. It is easy to show the same for the ω^m for m a positive odd integer. Thus we have

$$\limsup_{\substack{m\to\infty \\ m \text{ odd}}} |\Re e\, \omega^m| = 1.$$

This implies

$$\limsup_{\substack{m\to\infty \\ m \text{ odd}}} \frac{|S_m|}{2^{m/2}} = \limsup_{\substack{m\to\infty \\ m \text{ odd}}} \frac{|S_{gm}|}{2^{gm/2}} = 2g \limsup |\Re e\, \omega^m| = d-1.$$

5. Conclusion

So MacWilliams and Sloane's bound is true for codes dual to BCH codes with designed distance not greater than 7.

For codes with larger designed distance, we have shown that it is not true in general. On the contrary, for many parameters d and m, the Carlitz-Uchiyama bound is almost reached.

The question remains to show that for every odd d greater to 7 there exists a sequence of m such that the codes $C_m^\perp$ almost reach the Carlitz-Uchiyama bound (cf. Conjecture in §3).

6. References

1. Hooley, C.: "On Artin's conjecture", *J. reine angew. Math.*, vol. 225 (1967), 209-20.

2. Lang, S. and Weil, A.: "Number of points of varieties in finite fields", *Amer. J. Math.*, vol. 76 (1954), 819-827.

3. Lidl, R. and Niederreiter, H.: *Finite Fields,* Encyclopedia of mathematics and its applications, vol. 20, Cambridge University Press, Cambridge, 1983.

4. MacWilliams, F.J. and Sloane, N.J.A.: *The Theory of Error-Correcting Codes,* North-Holland, Amsterdam, 1977.

5. Rodier, F.: "On the spectra of the duals of binary BCH codes of designed distance $\delta = 9$", *IEEE transactions on Information Theory,* vol 38, n° 2 (1992), 478-479.

6. Rodier, F.: "Minoration de certaines sommes exponentielles binaires", in *Coding Theory and Algebraic Geometry*(Eds. H. Stichtenoth and M.A. Tsfasman), Lecture Notes in Math. n° 1518, Springer-Verlag, 1992.

7. Rodier, F.: "Sur la distance minimale d'un code BCH", submitted to *Discrete Math.* (1992).

8. Weil, A.: "On some exponential sums", *Proc. Nat. Acad. Sci. U.S.A.*, Vol. 34 (1948), 204-207.

A CHARACTERIZATION OF THE q-ARY IMAGES OF q^m-ARY CYCLIC CODES

G.E. Séguin

Royal Military College of Canada, Kingston, Ontario, Canada

I. Woungang

G.E.C.T., University of Toulon and Var, La Garde, France

† Part of the work reported in this paper was done in the summer of 1991 while the first author was at the Université de Toulon, FRANCE, supported by a research fellowship from the French government. This work was also partly funded by research grant number OGP00G288 from the Natural Sciences and Engineering Research Council of Canada, held by the first author at the Ecole Polytechnique de Montréal.

ABSTRACT

We give a simple characterization of the q-ary images of the q^m-ary cyclic codes of length n in the special case when the order of q module $(q^m - 1)/(n,q^m - 1)$ is m. This is done by introducing an appropriate modulo structure on F_q^{mn}.

1. INTRODUCTION

Let $\underline{\alpha} = (\alpha_0, \alpha_1, ..., \alpha_{m-1})$ be a basis for F_{q^m} over F_q, we then define the function $d_{\underline{\alpha}}:F_{q^m} \to F_q^m$ by setting,

$$d_{\underline{\alpha}}(x) = (x_0,x_1,\dots,x_{m-1}) \tag{1}$$

where $x \in F_{q^m}$ and

$$x = \sum_0^{m-1} x_i\alpha_i, \quad x_i \in F_q. \tag{2}$$

It follows that $d_{\underline{\alpha}}$ is a bijective F_q-linear mapping. We extend $d_{\underline{\alpha}}$ to $F_{q^m}^n$ by setting,

$$d_{\underline{\alpha}}(\underline{x}) = (d_{\underline{\alpha}}(x_0), d_{\underline{\alpha}}(x_1), \ldots, d_{\underline{\alpha}}(x_{n-1})) \tag{3}$$

where $\underline{x} = (x_0, x_1, \cdots, x_{n-1}) \in F^n_{q^m}$.

Let $\beta \in F_{q^m}$, then $x \to \beta x$, $x \in F_{q^m}$ defines a linear transformation on F_{q^m}. If

$$\beta\alpha_i = \sum_{j=0}^{m-1} b_{ij}\alpha_j, \quad b_{ij} \in F_q \tag{4}$$

for i = 0, 1, ..., m - 1; we then denote the mxm matrix with (i,j) entry b_{ij} by (β). If $x = \sum_0^{m-1} x_i\alpha_i$, then

$$\beta x = \sum_0^{m-1} x_i\beta\alpha_i = \sum_{j=0}^{m-1}\left(\sum_{i=0}^{m-1} x_i b_{ij}\right)\alpha_j \tag{5}$$

and so

$$d_{\underline{\alpha}}(\beta x) = \underline{x}(\beta) = d_{\underline{\alpha}}(x)(\beta). \tag{6}$$

This extends naturally to:

$$d_{\underline{\alpha}}(\beta\underline{x}) = d_{\underline{\alpha}}(\underline{x})(\beta)_n \tag{7}$$

where $\beta\underline{x} = (\beta x_0, \beta x_1, \ldots, \beta x_{n-1})$, $\underline{x} \in F^n_{q^m}$ and where $(\beta)_n$ is the super-circulant with first super-row,

$$[(\beta), 0, 0, -, 0]. \tag{8}$$

i.e. $(\beta)_n$ is an nxn circulant matrix whose entries are mxm matrices and a circulant matrix is one whose ith row is obtained by cyclically shifting row i-1 by one position, i=2,3,.... The matrix $(\beta)_n$ is also seen to be the tensor product of the nxn identity matrix and (β). Equation (7) says that diagram (9) is commutative.
The intrinsic cyclic shift operator T is defined by:

$$T: (x_0, \ldots, x_{k-1}) \to (x_{k-1}, x_0, \ldots, x_{k-2}) \tag{10}$$

and T^i, i ≥ 1, is defined recursively. T^m considered as a mapping on F_q^{mn} is an F_q -

$$\begin{array}{ccc} F^n_{q^m} & \overset{\beta}{\rightarrow} & F^n_{q^m} \\ d_{\underline{\alpha}} \downarrow & & \downarrow d_{\underline{\alpha}}. \\ F^{mn}_q & \overset{(\beta)_n}{\rightarrow} & F^{mn}_q \end{array} \tag{9}$$

linear operator and its matrix with respect to the canonical basis $\underline{e}_i=(0, 0, ..., 0, 1, 0, .., 0)$, $0 \leq i < mn$, where in $\underline{e}_i$ the unique 1 occurs in position i, the positions being labelled from 0 to mn-1, is the nxn super-circulant with first super-row:

$$[0, I, 0, -, 0] \tag{11}$$

In (11), I is the mxm identity matrix and 0 is the mxm all-0 matrix.
It is rather obvious that the following diagram commutes;

$$\begin{array}{ccc} F^n_{q^m} & \overset{T}{\rightarrow} & F^n_{q^m} \\ d_{\underline{\alpha}} \downarrow & & \downarrow d_{\underline{\alpha}}. \\ F^{mn}_q & \overset{T^m}{\rightarrow} & F^{mn}_q \end{array} \tag{12}$$

If $V \subset \mathbf{F}^n_{q^m}$ is a cyclic code, i.e. an $\mathbf{F}_{q^m}$ - linear subspace invariant under T, then $W = d_{\underline{\alpha}}(V)$ is an F_q - linear subspace of $\mathbf{F}^{mn}_q$ invariant under T^m. W is also called a quasi-cyclic code [1,2,3] and more precisely a quasi-cyclic (QC) code of index a divisor of m [3]. We call W the q-ary image of V with respect to $\underline{\alpha}$. The collection $d_{\underline{\alpha}}(V)$, $V \subset \mathbf{F}^n_{q^m}$ a cyclic code will define a subclass of the class of all QC-codes of index a divisor of m contained in $\mathbf{F}^{mn}_q$. The purpose of this paper is to give a simple characterization of this subclass in the case when $F_{q^m}=F_q(\beta)=F_q(\beta^n)$ for some $\beta \in F_{q^m}$.

2. A MODULE STRUCTURE ON F^{mn}_q

From the commutative diagrams (9) and (12), we immediately obtain the following lemma:

Lemma 1: An F_q-linear subspace $W \subset F^{mn}_q$ is the image under d_α of a cyclic code $V \subset F^n_{q^m}$ if and only if W is invariant under T^m and $(\gamma)_n$ for every γ in F_{q^m}.

Remark: In lemma 1, W is invariant under $(\gamma)_n$ for every $\gamma \in F_{q^m}$ if and only if it is invariant under $(\beta)_n$ for some β in F_{q^m} such that $F_{q^m} = F_q(\beta)$.

Hence in the sequel $\underline{\alpha}$ is a fixed basis for F_{q^m} over F_q and β is some fixed element such that $F_{q^m} = F_q(\beta)$. From equations (8) and (11) it is apparent that $C = T^m(\beta)_n = (\beta)_n T^m$ and the 1st super-row of C is:

$$[0, (\beta), 0, -, 0]. \tag{13}$$

To simplify the notation, we shall set,

$$\mathbf{B} = (\beta). \tag{14}$$

From Equation (13), it follows that the first super-row of C^2 is,

$$[0, 0, B^2, 0, -, 0]$$

and the first super-row of C^n is,

$$[B^n, 0, 0, -, 0].$$

Consequently, all the matrices C^j, j = 0, 1, ... are "super-circulants", i.e. nxn circulant matrices whose entries are mxm matrices. Whence, if we are only interested in matrices of the form $f(C)$, where $f(X) \in F_q[X]$, then we may limit ourselves to the consideration of the first super-row of $f(C)$. We may conveniently represent the first super-row of $f(C)$ as a polynomial in Y of degree inferior to n and whose coefficients are polynomials in B. More precisely, if $f(X) = \sum_l f_l X^l \in F_q[X]$, and if

$$f_i(x) = \sum_j f_{jn+i} X^j, \quad 0 \le i < n \tag{15}$$

then

$$f(X) = \sum_{i=0}^{n-1} X^i f_i(X^n) \tag{16}$$

and the first super-row of $f(C)$ is:

$$f_0(B^n)+Bf_i(B^n)Y+\cdots+B^{n-1}f_{n-1}(B^n)Y^{n-1}. \tag{17}$$

Definition: We make F_q^{mn} into an $F_q[X]$ -module by setting $f(X)\underline{a}=\underline{a}f(C), f(X)\in F_q[X]$ *and* $\underline{a}\in F_q^{mn}$.

Remark: The sub-modules of F_q^{mn} are then simply the subspaces of F_q^{mn} which are invariant under C.

Thereom 1: If there exists a polynomial $f(X)\in F_q[X]$ such that $T^m = f(C)$ (*or* $(\beta)_n = f(C)$) , then the q-ary images (with respect to $\underline{\alpha}$) of the cyclic codes in $F_{q^m}^n$ are the sub-modules of F_q^{mn} .

Proof: If $T^m = f(C)$, then $(\beta)_n = g(C)$ for some polynomial $g(X)\in F_q[X]$ because $(\beta)_n = C(T^m)^{-1}$ and $(T^m)^{-1}$ is a polynomial in C. Hence, if W is a sub-module of F_q^{mn} , then it is a subspace and it is invariant under T^m and $(\beta)_n$, hence by lemma 1, it is the q-ary image of a cyclic code $V\subset F_{q^m}^n$. Conversely, if $W = d_{\underline{\alpha}}(V)$, $V\subset F_{q^m}^n$ a cyclic code, then by lemma 1 it is invariant under $(\beta)_n$ and T^m, hence invariant under C, hence a submodule.■

The next thereom tells us when the condition of thereom 1 is met.

Theorem 2: There exists a polynomial $f(X)\in F_q[X]$ such that $T^m = f(C)$ if and only if $F_{q^m} = F_q(\beta^n)$, i.e. if and only if the order of q module $(q^n-1)/(n,q^m-1)$ is m; where (,) denotes the greatest common divisor.

Proof: From (11) and (17), there exists a polynomial f(x) in $F_q[X]$ such that $T^m = f(C)$ if and only if there exists n polynomials $f_i(X)\in F_q[X]$, $0\le i<n$, such that,

$$I = Bf_i(B^n),\ 0 = B^i f_i(B^n),\quad 0\le i<n,\quad i\ne 1.$$

Hence, if and only if,

$$I = Bf_i(B^n),\quad 0=f_i(B^n),\quad 0\le i<n,\quad i\ne 1.$$

This is equivalent to

$$1 = \beta f_i(\beta^n),\quad 0=f_i(\beta^n),\quad 0\le i<n,\quad i\ne 1.$$

Hence, if and only if there exists a polynomial $f_i(X)$ such that,

$$1 = \beta f_i(\beta^n)$$

and this will occur if and only if $\beta \in F_q(\beta^n)$, i.e. if and only if $F_{q^m} = F_q(\beta) = F_q(\beta^n)$. In turn, this will be the case if and only if the order of q module $(q^m-1)/(n,q^m-1)$ is m.■

To determine the module structure of F_q^{mn}, we must determine the minimal polynomial of the matrix C. This is done in the next lemma.

<u>Lemma 2</u>: The minimal polynomial of C over F_q is $f(X^n)$ where $f(X) = irr(\beta^n, F_q)$ is the irreducible polynomial of β^n over F_q.

<u>Proof</u>: The minimal polynomial of C is the polynomial $f(X) \neq 0$ of least degree such that $f(C) = 0$. Let $f(X) \neq 0$ be a polynomial such that $f(C) = 0$, then $f(X) = \sum_{i=0}^{n-1} X^i f_i(X^n)$, $f_i(X) = \sum_j f_{jn+i} X^j$ and according to (17), $f(C) = 0$ if and only if:

$$B^i f_i(B^n) = 0, \quad 0 \le i < n$$

which is equivalent to

$$f_i(\beta^n) = 0, \quad 0 \le i < n.$$

It is therefore rather evident that the degree of f(x) will be minimized by the choice

$$f_0(X) = irr(\beta^n, F_q)$$

and $f_i(X) = 0, \quad 1 \le i < n.$ Hence,

$$f(X) = f_0(X^n).■$$

By Theorem 1, if $T^m = g(C)$ for some polynomial $g(X) \in F_q[X]$ then deg $f_0(X) = m$, where $f_0(X) = irr(\beta^n, F_q)$, and so the degree of the minimal polynomial $f(X) = f_0(x^n)$ is mn which is the dimension of F_q^{mn}. Hence, F_q^{mn} is a cyclic module [4] and so there exists a vector $a \in F_q^{mn}$ such that,

$$\underline{a},\ \underline{a}C,\ \underline{a}C^2,\ \cdots,\ \underline{a}C^{mn-1} \tag{18}$$

forms a basis for F_q^{mn} . If we define,

$$\psi:\ F_q[X]/(f_0(X^n)) \rightarrow F_q^{mn}$$

by

$$\psi(g(X)) = g(X)\underline{a}, \tag{19}$$

where in (19), we write g(x) for the residue class $g(X) + (f_0(x^n))$, then ψ is a bijective module isomorphism ($F_q[X]/(f_0(X^n)$ is made into an $F_q[X]$ -module in a natural fashion). But the sub-modules of $F_q[X]/(f_0(X^n)$ are nothing else but its ideals. As is well known, there is a 1-1 correspondence between the ideals of $F_q[X]/(f_0(X^n))$ and the monic divisors of $f_0((X^n))$.

Hence, the sub-modules of F_q^{mn} (which are precisely the q-ary images, with respect to $\underline{\alpha}$, of the cyclic codes in F_{q^m}) are given as:

$$F_q[X]g(x)\underline{a} = \{b(x)g(x)\underline{a} \mid b(x) \in F_q[X]\} \tag{20}$$

where g(x) runs over the monic divisors of $f_0(X^n)$. Clearly in (20) we may restrict the degree of b(x) to be less than mn - deg g(x). Finally, we remark that it is easy to see that we may choose for $\underline{a}$ the vector:

$$\underline{a} = (1\ 0\ 0 \cdot \cdot \cdot \cdot 0). \tag{21}$$

As a generator matrix for the QC-code of (20) we may choose the matrix,

$$\begin{bmatrix} g(X)\underline{a} \\ Xg(X)\underline{a} \\ \cdot \\ \cdot \\ \cdot \\ X^{k-1}g(X)\underline{a} \end{bmatrix} \tag{22}$$

where,

$$k = mn - deg\ g(X). \tag{23}$$

As an interesting corollary to the above development we have:

Corollary: If m and n are such that the order of q module $(q^m - 1)/(n, q^m - 1)$ is m, then the number of monic divisors of $f_0(X^n)$, $f_0(X) = irr(\beta, F_q)$, $F_{q^m} = F_q(\beta^n)$, over F_q is equal to the number of monic divisors of X^n-1 over F_{q^m}.
We now illustrate the above theory by means of an example.

Example: Let q=m=2 and n=5; then $(q^m - 1)/(n, q^m - 1) = 3/(5,3) = 3$ and the order of q=2 module 3 is 2=m, hence the above theory applies to this case. The field $\mathbf{F}_4$ is generated by the recursion $\alpha^2 = 1 + \alpha$ and we may choose $\beta = \alpha$ so that $f_o(x) = irr(\beta^n, F_2) = irr(a^5, F_2) = 1 + x + x^2$ and so $f(x) = f_0(X^n) = 1 + X^5 + X^{10}$. .
We choose $\underline{\alpha} = (1, \alpha)$ so that the matrix B is:

$$B = \begin{bmatrix} 0 & 1 \\ 1 & 1 \end{bmatrix}.$$

The prime factorization of $X^5 + 1$ over $\mathbf{F}_4$ is:

$$X^5 + 1 = (1 + x)\ (1 + \alpha x + x^2)\ (1 + \alpha^2 x + x^2),$$

and the prime factorization of $1 + x^5 + x^{10}$ over $\mathbf{F}_2$ is:

$$1 + x^5 + x^{10} = (1 + x + x^2)\ (1 + x + x^4)\ (1 + x^3 + x^4).$$

Hence there are 8 4-ary cyclic codes of length 5 and 8 sub-modules of F_2^{10}. The generators of these 8 sub-modules are:

$$1 \cdot \underline{a} = (10\ 00\ 00\ 00\ 00) = \underline{b_1}$$

$$(1 + x + x^2) \cdot \underline{a} = (10\ 01\ 11\ 00\ 00) = \underline{b_2}$$

$$(1 + x + x^4) \cdot \underline{a} = (10\ 01\ 00\ 00\ 01) = \underline{b_3}$$

$$(1 + x^3 + x^4) \cdot \underline{a} = (10\ 00\ 00\ 10\ 01) = \underline{b_4}$$

$$(1 + x^3 + x^4 + x^5 + x^6) \cdot \underline{a} = (01\ 10\ 00\ 10\ 01) = \underline{b_5}$$

$$(1 + x + x^2 + x^3 + x^6) \cdot \underline{a} = (10\ 11\ 11\ 10\ 00) = \underline{b_6}$$

$$(1 + x + x^3 + x^4 + x^5 + x^7 + x^8) \cdot \underline{a} = (01\ 01\ 01\ 01\ 01) = \underline{b_7}$$

$$(1 + x^5 + x^{10}) \cdot \underline{a} = (00\ 00\ 00\ 00\ 00) = \underline{b_8}$$

For example, the sub-module generated by $\underline{b_3}$ has generator matrix:

$$\begin{bmatrix} \underline{b_3} \\ x\underline{b_3} \\ x^2\underline{b_3} \\ x^3\underline{b_3} \\ x^4\underline{b_3} \\ x^5\underline{b_3} \end{bmatrix} = \begin{bmatrix} 10 & 01 & 00 & 00 & 01 \\ 11 & 01 & 11 & 00 & 00 \\ 00 & 10 & 11 & 10 & 00 \\ 00 & 00 & 01 & 10 & 01 \\ 11 & 00 & 00 & 11 & 01 \\ 11 & 10 & 00 & 00 & 10 \end{bmatrix}.$$

W, the row-space of this latter matrix, must therefore be (from its dimension 6) the binary image of the 4-ary cyclic code generated by $1 + \alpha x + x^2$ or $1 + \alpha^2 x + x^2$. The binary images of these two polynomials are, respectively, (10 01 10 00 00) and (10 11 10 00 00). Since the latter is obtained by shifting the last row of the above matrix by 2, we conclude that W is the binary image of the cyclic code generated by $1 + \alpha^2 x + x^2$.■

The following result is an immediate consequence of the above theory:
Theorem 3: If m and n are such that the order of q module $(q^m - 1)/(n, q^m - 1)$ is m, then the q-ary images of the q^m-ary cyclic codes of length n with respect to the basis $\underline{\alpha}$ for F_{q^m} over F_q all have a generator matrix of the following form:

$$\begin{bmatrix} \underline{a}_0 & \underline{a}_1 & \underline{a}_2 & \cdots & \underline{a}_{n-1} \\ \underline{a}_{n-1}B & \underline{a}_0 B & \underline{a}_1 B & \cdots & \underline{a}_{n-2}B \\ \underline{a}_{n-2}B^2 & \underline{a}_{n-1}B^2 & \underline{a}_0 B^2 & \cdots & \underline{a}_{n-3}B^2 \\ & & & & \\ & & & & \end{bmatrix}$$

where $\underline{a}_i$, $0 \leq i < n$, are q-ary m-tuples, B is the matrix of $x \to \beta x,\ x \in F_{q^m} = F_q(\beta) = F_q(\beta^n)$, with respect to $\underline{\alpha}$.

3. CONCLUSIONS

In this paper we have provided a simple characterization of the q-ary images of the q^m-ary cyclic codes of length n in a special case. These results can probably be generalized to characterize the q-ary images of q^m-ary quasi-cyclic codes. This problem is presently under investigation by the authors.

REFERENCES

[1] C.L. Chen, W.W. Peterson and E.J. Weldon, Jr., "Some Results on Quasi-Cyclic Codes", *Information and Control*, Vol. 15, Nov. 1969, pp. 407-423.

[2] G.E. Séguin and G. Drolet, "The Trace Description of Irreducible Quasi-Cyclic Codes". *IEEE Trans. on Information Theory*, Vol. 36, No. 6, Nov. 1990, pp. 1463-1466.

[3] G.E. Séguin and H.I. Huynh, "Quasi-Cyclic codes: A Study", report published by the Laboratoire de Radiocommunications et de Traitement du Signal, Université Laval, Québec, Canada, 1985.

[4] N. Jacobson, *Lectures in Abstract Algebra*. Vol. II. D. Van Nostrand Co., 1953.

WEIGHTS OF PRIMITIVE BINARY CYCLIC CODES FROM NON-PRIMITIVE CODES

J. Wolfmann

G.E.C.T., University of Toulon and Var, La Garde, France

Abstract

Let s, k, integers such that s is a divisor of $2^k - 1$. Let g(x) be a primitive divisor of $x^s - 1$ over $\mathbf{F}_2$, and let $\pi(x)$ be a primitive polynomial of degree k over $\mathbf{F}_2$. We consider the binary cyclic code C of length $N = 2^k - 1$, generated by $\frac{x^N - 1}{g(x)\pi(x)}$. For special cases, we determine the weights of C by using the weights of the irreducible cyclic code of length s, generated by $\frac{x^s - 1}{g(x)}$.

1.INTRODUCTION

The classical definitions and results can be found in [5] for Coding Theory , and in [4] for Finite Fields.
The words of a cyclic code are described by the polynomial representation.
A binary cyclic code is said to be primitive if its length is $N = 2^k - 1$ for any integer k.
In this paper we consider the case where the check polynomial of such a code is the product of two special irreducible divisors of $x^N - 1$ over $\mathbf{F}_2$.
Let s, k, integers such that s is a divisor of $2^k - 1$. Let g(x) be a primitive divisor of $x^s - 1$ over $\mathbf{F}_2$, and let $\pi(x)$ be a primitive polynomial of degree k over $\mathbf{F}_2$. We consider the binary cyclic code C of length $N = 2^k - 1$, generated by $\dfrac{x^N - 1}{g(x)\pi(x)}$, and the binary irreducible cyclic code Γ of length s, generated by $\dfrac{x^s - 1}{g(x)}$. The code words of C and Γ are described in section 2 by using trace functions. In section 3, a general result is given on the weights of C, and the weight set is explicitely determined for special cases. Finaly, a table gives numerical results for $4 \le k \le 12$.

2.PRESENTATION OF THE CODES

We first specify notations and recall classical definitions and properties (see [4] and [5]).

2.0 Preliminaries

The G.C.D. of the integers i and j is denoted by (i,j).
In the following we consider a fixed integer k, $k \ge 1$.
$\mathbf{F} = \mathbf{F}_{2^k}$ is the finite field of order 2^k, and $\mathbf{F}^* = \mathbf{F}_{2^k} \setminus \{0\}$.

2.01 Trace

If $\mathbf{F}_q$ is a finite field and h any integer, $h \ge 1$, we recall that the trace of $\mathbf{F}_{q^h}$ over $\mathbf{F}_q$ is the $\mathbf{F}_q$-linear form defined by $\mathrm{Tr}(z) = z + z^q + \ldots + z^{q^i} + \ldots + z^{q^{h-1}}$.
If i divides j, then the trace of $\mathbf{F}_{2^j}$ over $\mathbf{F}_{2^i}$ is denoted by tr^j_i and (transitivity of traces) :

(T) $\qquad tr^j_1(z) = tr^i_1\,(tr^j_i(z))$

For our fixed k,the trace tr^k_1 of $\mathbf{F}_{2^k}$ over $\mathbf{F}_2$ is denoted by tr (absolute trace of $\mathbf{F}_{2^k}$).

2.02 Support and weight

The support of $a(x) = \sum_{i=0}^{r} a_i x^i$, is $\mathrm{supp}(a(x)) = \{ i \in \mathbf{N} : 0 \le i \le r \text{ and } a_i \ne 0 \}$.

The weight of $a(x) = \sum_{i=0}^{r} a_i x^i$, is $w(a(x)) = |\mathrm{supp}(a(x))|$.

2.03 Minimal polynomial, primitive divisor, primitive polynomial

If θ belongs to an extension field of F_2, then its the minimal polynomial over F_2 is denoted by $m_\theta(x)$. If L is any non-negative integer, then every irreducible divisor g(x) of x^L-1 over F_2 is the minimal polynomial of a L-th root of unity over F_2. If this root is a primitive one (generator of the cyclic group of L-th roots of unity over F_2) then g(x) is called a primitive divisor. A primitive polynomial of degree k over F_2 is the minimal polynomial over F_2 of a primitive root of F_{2^k}. Obviously, it also is a primitive divisor of x^N-1 where $N=2^k-1$.

2.1 Irreducible cyclic codes

An irreducible cyclic code C of length L over F_2, is a cyclic code generated by $\frac{x^L-1}{g(x)}$ where g(x) is an irreducible divisor of x^L-1 over F_2. (We only consider the non-degenerate case, that is g(x) is a primitive divisor). We know that g(x) is the minimal polynomial of a primitive L-th root of unity over F_2.

The following result is well-known (see [1] and [8]).

Proposition 2.1

Let I be an irreducible cyclic code of length L over F_2, with $(L,2)=1$.

Assume that I is generated by $\frac{x^L-1}{m_\theta(x)}$, with $\theta=\beta^{-1}$ where β is a primitive L-th root of unity over F_2, and $m_\theta(x)$ is the minimal polynomial of θ over F_2.

Let F_{2^R} be the splitting field of x^L-1 over F_2. Then I is the set of words:

$$c_a(x)=\sum_{i=0}^{L-1}Tr(a\beta^i)x^i,$$ where $a\in F_{2^R}$ and Tr denotes the trace of F_{2^R} over F_2.

2.2 Description of the codes

Let n, s, k, integers such that $ns=2^k-1$. Let $F=F_{2^k}$, and $F^*=F_{2^k}\setminus\{0\}$.

Let g(x) be a primitive divisor of x^s-1 over F_2, and let $\pi(x)$ be a primitive polynomial of degree k over F_2. Let α be in F such that $\pi(x)$ is the minimal polynomial of α^{-1}.

Define $\delta=\alpha^s$, and $\gamma=\alpha^n$.

We consider the following binary codes:

The cyclic code C of length $N=2^k-1$, generated by $\frac{x^N-1}{g(x)\pi(x)}$.

The irreducible cyclic code Γ of length s, generated by $\frac{x^s-1}{g(x)}$.

We also consider the following useful intermediate code :

The irreducible cyclic code Δ of length n, generated by $\frac{x^n-1}{m_\varepsilon(x)}$ with $\varepsilon=\delta^{-1}$.

Let F_{2^r} be the splitting field of x^s-1 over F_2, and let F_{2^m} be the splitting field of x^n-1 over F_2. From proposition 2.1, we deduce the following description of Γ, Δ, C, and their weights.

2.21 Description of Γ

The polynomial g(x) is a primitive divisor of x^s-1 over F_2, then g(x) is the minimal polynomial of a primitive s-root of unity, which can be defined as γ^{-d}, whith (d, s) = 1. It follows that :

(1) $$\Gamma = \{\mu_u(x) = \sum_{i=0}^{s-1}[tr_1^r(u\gamma^{di})]x^i) : u \in F_{2^r}\}$$

and :

(2) $$w(\mu_u(x)) = \#\{\ 0 \le i \le s-1 : tr_1^r(u\gamma^{di}) = 1\}.$$

2.22 Description of Δ

It comes directly from proposition 2.1.

(3) $$\Delta = \{v_b(x) = \sum_{i=0}^{n-1}[tr_1^m(b\delta^i)]x^i : b \in F_{2^m}\}.$$

(4) $$w(v_b(x)) = \#\{\ 0 \le i \le n-1 : tr_1^m(b\delta^i) = 1\}.$$

2.23 Description of C

Code C is the direct sum (see [5] or [8]) of the two irreducible cyclic codes of length $N = 2^k-1$ generated by $\frac{x^N-1}{\pi(x)}$ and $\frac{x^N-1}{g(x)}$ (remark that the first one is the simplex code)

Therefore, from proposition 2.1 :

(5) $$C = \{m_{(a,v)}(x) = \sum_{i=0}^{2k-2}[tr_1^k(a\alpha^i+v\alpha^{din})]x^i : a \in F \text{ and } v \in F\}.$$

(6) $$w(m_{(a,v)}(x)) = \#\{\ 0 \le i \le 2^k-2 : tr_1^k(a\alpha^i)+tr_1^k(v\alpha^{din}) = 1\}$$
$$= \#\{\ x \in F^* : tr_1^k(ax)+tr_1^k(v(x^{dn})) = 1\}$$

3.THE WEIGHTS OF C

The following theorems give the weight set of C for special cases.
The definitions and notations are as in section 2.

Theorem 3.1

If $k = 2t$, $s = 2^t+1$, then the weight set of C is the set of all integers

$0,\ 2^{2t-1},\ (2^t-1)w,\ 2^{2t-1}-w,\ 2^{2t-1}+2^t-w$

such that w is even and $\left| w - \frac{(2^t+1)}{2} \right| \le 2^{t/2}$.

Theorem 3.2

If : a) $\mathbf{F}_{2^k}$ is the splitting field of x^n-1 over $\mathbf{F}_2$

b) $k = 2t$, and there exists a divisor e of t such that $2^e \equiv -1 \pmod s$

then the weight set of C is the set of all the integers

$$0,\ 2^{2t-1},\ (\frac{2^{2t}-1}{s})w,\ 2^{2t-1}+(\frac{\varepsilon 2^t-1}{s})w,\ 2^{2t-1}+(\frac{\varepsilon 2^t-1}{s})w-\varepsilon 2^t$$

where $\varepsilon = (-1)^{t/e}$, and such that w is a non-zero weight of Γ.

3.1 Preliminary result

In order to prove the two theorems, we first establish a preliminary result giving a description of the weights of C, and which is useful in both cases.

Denote the weight $w(m_{(a,v)}(x))$ in (6), by $w(a,v)$.

Let us consider the following subsets of $\mathbf{F}^*$:

$\bar{H}_a = \{x \in \mathbf{F}^* : tr_1^k(ax) = 1\}$ with $a \in \mathbf{F}$ (affine hyperplane)

$E(v) = \{x \in \mathbf{F}^* : tr_1^k(vx^{dn}) = 1\}$ with $v \in \mathbf{F}$.

In other words, the characteristic function in $\mathbf{F}^*$ of $\bar{H}_a$ is $h_a(x) = tr(ax)$, and the characteristic function in $\mathbf{F}^*$ of E(v) is $tr_1^k(vx^n)$.

It comes from (6) that $w(a,v)$ is the number of solutions in $\mathbf{F}^*$ of $tr_1^k(ax) + tr_1^k(vx^{dn}) = 1$, which also is the cardinality of the symetric difference $\bar{H}_a \,\Delta\, E(v)$. Then :

(7) $$w(a,v) = |\bar{H}_a \,\Delta\, E(v)| = |\bar{H}_a| + |E(v)| - 2|\bar{H}_a \cap E(v)|.$$

Now, we need to calculate the cardinalities of the sets : $\bar{H}_a$, $E(v)$, $\bar{H}_a \cap E(v)$.

The case $v = 0$ is a trivial one and we now assume $v \neq 0$.

Obviously:

(8) if $a = 0$, then $|\bar{H}_a| = 0$; if $a \neq 0$, then $|\bar{H}_a| = 2^{k-1}$.

Calculation of $|E(v)|$.

From the definitions of α and γ, we know that $\gamma^i = (\alpha^i)^n$. Let us consider the multiplicative subgroup G_n of the n-th root of unity in $\mathbf{F}^*$: $G_n = \{1, \delta, \ldots, \delta^i, \ldots \delta^{n-1}\}$, with $\delta = \alpha^s$. Because of $ns = 2^k-1$, the distinct classes modulo G_n are the subsets $\alpha^j G_n$ with $0 \le j \le s-1$. From the definition , if z belongs to E(v) so do the members of the class of z modulo G_n. Then, E(v) is a disjoint union of such classes. If $x = \alpha^j g$, with $g \in G_n$ and $0 \le j \le s-1$, then $g^n = 1$ and:

$tr_1^k(vx^{dn}) = tr_1^k(v\alpha^{jdn}(g^n)^d) = tr_1^k(v(\alpha^n)^{jd}) = tr_1^k(v\gamma^{jd})$.

Using the transitivity property (T) of 2.01, and because γ belongs to $\mathbf{F}_{2^r}$:

$tr_1^k(v\gamma^{jd}) = tr_1^r(\gamma^{jd} tr_r^k(v)) = tr_1^r(u(\gamma^d)^j)$, where $u = tr_r^k(v)$.

Finally, $tr_1^k(vx^{dn}) = tr_1^r(u(\gamma^d)^j)$, where $u = tr_r^k(v)$.

We conclude that $x = \alpha^j g \in E(v)$, if and only if $tr_1^r(u(\gamma^d)^j) = 1$, which means from (1) that $j \in supp(\mu_u(x))$. Therefore :

(9) $$E(v) = \bigcup_{t \in S} (\alpha^j G_n), \text{ where } S = supp(\mu_u(x)) \text{ and } u = tr_r^k(v) .$$

It follows that $| E(v) | = n| supp(\mu_u(x)) |$, that is :

(10) $$| E(v) | = nw(\mu_u(x)) , \text{ where } S = supp(\mu_u(x)) \text{ and } u = tr_r^k(v) .$$

Calculation of $| \bar{H}_a \cap E(v) |$.

a) if $a = 0$: $\bar{H}_a = \emptyset$, and $| \bar{H}_a \cap E(v) | = 0$.

b) if $a \neq 0$: using (9),

$\bar{H}_a \cap E(v) = \bar{H}_a \cap (\bigcup_{j \in S} \alpha^j G_n) = \bigcup_{j \in S} (\bar{H}_a \cap \alpha^j G_n)$, therefore:

(11) $$| \bar{H}_a \cap E(v) | = \sum_{j \in S} |\bar{H}_a \cap \alpha^j G_n|$$

Now, because $G_n = \{1, \delta, \ldots\ldots, \delta^i \ldots\ldots \delta^{n-1}\}$ with $\delta = \alpha^s$:

$| \bar{H}_a \cap \alpha^j G_n | = \#\{ 0 \leq i \leq n-1 : tr_1^k(a\alpha^j \delta^i) = 1\}$.

If $b_j = tr_m^k(a\alpha^j)$ then $tr_1^k(a\alpha^j\delta^i) = tr_1^m(\delta^i tr_m^k(a\alpha^j)) = tr_1^m(b_j\delta^i)$. So we obtain from (3) :

(12) $$| \bar{H}_a \cap \alpha^j G_n | = w(v_{b_j}(x)) \text{ with } b_j = tr_m^k(a\alpha^j).$$

Finally, from (11) :

(13) $$| \bar{H}_a \cap E(v) | = \sum_{j \in S} w(v_{b_j}(x)) \quad , \text{ where } S = supp(\mu_u(x)).$$

Conclusion :

Starting from (7), and using (8), (10), (13), we are now able to calculate w(a,v).

(14) If $a \neq 0$ and $v = 0$: $w(a,0) = 2^{k-1}$.

(15) If $a = 0$ and $v \neq 0$: $w(0,v) = nw(\mu_u(x))$ with $u = tr_r^k(v)$.

(16) If $a \neq 0$ and $v \neq 0$: $w(a,v) = 2^{k-1} + nw(\mu_u(x)) - 2\sum_{j \in S} w(v_{b_j}(x))$,

with $u = tr_r^k(v)$, $S = supp(\mu_u(x))$, $b_j = tr_m^k(a\alpha^j)$. ♦

3.2 Proof of theorem 3.1:

If $k = 2t$ and $s = 2^t + 1$, then $n = 2^t - 1$ and $G_n = \mathbf{F}_{2^t} \setminus \{0\}$.

Consider the two following subsets of **F** :

$H_a = \{x \in \mathbf{F} : tr_1^k(ax) = 0\} = \mathbf{F} \setminus \bar{H}_a$ (vector hyperplane),

$E_j = \alpha^j \mathbf{F}_{2^t} = \alpha^j G_n \cup \{0\}$ with $j \in supp(\mu_u(x))$.

Both are vector subspaces respectively with dimension $2t - 1$ and t.

If H_a contains E_j then $|\bar{H}_a \cap \alpha^j G_n| = 0$. In this case, (12) implies that $w(v_{b_j}(x)) = 0$.

Otherwise, $H_a \cap E_j$ is a $(t-1)$-dimensional subspace and $w(v_{b_j}(x)) = |\bar{H}_a \cap \alpha^j G_n| = 2^{t-1}$. Because of the dimensions and the fact that the $\alpha^j G_n$ are two by two disjoint, H_a contains at most one subspace E_j with $j \in supp(\mu_u(x))$.

For a given v, $v \neq 0$, this gives rise from (16), to the following possible values for $w(a,v)$ when $a \neq 0$:

Case 1 :

If H_a contains one subspace E_j , then $w(a,v) = 2^{k-1} + nw(\mu_u(x)) - 2(2^{t-1})(|S|-1)$.

The number $N_1(v)$ of such non-zero element a in F, is the product of the number of non-zero elements in the orthogonal subspace of every E_j by the number of E_j .

We find : $N_1(v) = (2^t - 1)w$.

Case 2 :

If no subspace E_j is contained in H_a , then $w(a,v) = 2^{k-1} + nw(\mu_u(x)) - 2(2^{t-1})|S)|$, and the number $N_2(v)$ of such non-zero element a in F is : $N_2(v) = 2^{2t} - 1 - N_1(v)$.

Summarizing the two cases with $|S| = w(\mu_u(x))$ and $n = 2^t - 1$, the two possible weights are :

(17) $2^{2t-1} - w$ (Case 1) and $2^{2\tau-1} + 2^t - w$ (Case 2), where w is a weight of Γ.

To complete the proof by using the conclusion of 3.1, we need the weight set of Γ .

It is given by the following proposition (see [3]).

Proposition 3.3

Let Γ be the binary cyclic code of length $s = 2^t + 1$ generated by $\frac{x^s - 1}{m_\beta(x)}$,

where β is a primitive s-th root of 1 over F_2.

The weights of the non-zero words of Γ , are the integers w such that :

$$\left| w - \frac{2^t + 1}{2} \right| \leq 2^{t/2} \quad \text{and w is even.}$$

To complete the proof of theorem 3.1, we remark that $N_1(v)$ is never equal to zero, and that $N_2(v)$ is zero if and only if $2^{2t} - 1 = N_1(v)$, which means $w = 2^t + 1$. From the above propsition, $2^t + 1$ is not a weight of Γ, and thus case 1 and case 2 both occur. From (17) and Proposition 3.3, the proof is now completed.

3.3 Proof of theorem 3.2:

In order to use the conclusion of 3.1, we need the weights of $v_{b_j}(x)$ when $a \neq 0$, $v \neq 0$, $u = tr^k_r(v)$, $j \in S = supp(\mu_u(x))$, $u = tr^k_r(v)$, $b_j = tr^k_m(a\alpha^j)$.

Because F_{2k} is the splitting field of $x^n - 1$ over F_2 , then $m = k$ and $b_j = a\alpha^j$.

Under assumptions of theorem 3.2, the weights of Δ are given by the following proposition (see [1],[2],[7]).

Proposition 3.4

Let F_{2^m} be the splitting field of x^n-1 over F_2, $ns = 2^m-1$.

Let Δ be the cyclic code of length n, generated by $\frac{x^n-1}{m_\varepsilon(x)}$,with $\varepsilon = \delta^{-1}$ and δ a primitive n-th root of 1 over F_2 .

Let $v_b(x)$ be a word of Δ, $v_b(x) = \sum_{i=0}^{n-1} [tr_1^m(b\delta^i)]x^i$, with $b \in F_{2^m}$.

If m = 2t, and if there exists a divisor e of t such that $2^e \equiv -1 \pmod{s}$, then :

If $b^n = 1$: $\quad w(v_b(x)) = 2^{t-1}(\frac{2^t+\eta(s-1)}{s})$

If $b^n \neq 1$: $\quad w(v_b(x)) = 2^{t-1}(\frac{2^t-\eta}{s})$

where $\eta = (-1)^{t/e}$.

In order to apply the above proposition to w(a,v) in (16) when $a \neq 0$ and $v \neq 0$, we need to find the weight of $v_{b_j}(x)$ with $b_j = a\alpha^j$. Now remark that $(b_j)^n = 1$, is equivalent to $a^{-1} \in \alpha^j G_n$. Consequently, for a given a, and because of the fact that the $\alpha^j G_n$ are two by two disjoint, the number of j such that $(b_j)^n = 1$, is 1 or 0.

Case 1 :

If this number is 1 then :

$$w(a,v) = 2^{k-1} + nw(\mu_u(x)) - 2[\, 2^{t-1}(\frac{2^t+\eta(s-1)}{s}) + 2^{t-1}(\frac{2^t-\eta}{s})\,(|S|-1)].$$

In this case, the number $M_1(v)$ of such non-zero element a in F is the product of the cardinality of $E(v) = \bigcup_{i \in S} (\alpha^i G_n)$. That is, $M_1(v) = nw(\mu_u(x))$

Case 2 :

Otherwise ,

$$w(a,v) = 2^{k-1} + nw(\mu_u(x)) - 2[2^{t-1}(\frac{2^t-\eta}{s})\,|S|].$$

The number $M_2(v)$ of such non-zero element a in F, is $M_2(v) = 2^{2t} - 1 - M_1(v)$.

Now using $|S| = w(\mu_u(x))$, $ns = 2^{2t}-1$, and proposition 3.4, we see that both $M_1(v)$ and $M_2(v)$ are never zero, and we obtain the expected result. ♦

4.EXAMPLES

In [6] we can find the weight distribution of Δ for $6 \leq m \leq 26$.

From this result, table 1gives the weights obtained for C by Theorem 3.1 and Theorem 3.2 for $4 \leq k \leq 12$, and k even. Of course, it is also possible to obtain the weights of C for any k and s satistying to the conditions of Theorem 3.1, and also to deduce from [6] the weights of C for $14 \leq k \leq 26$ and k even , under the assumptions of Theorem 3.2.

5.CONCLUSION

In the present work we found the weights of the binary cyclic code C of length $N = 2^k - 1$, generated by $\frac{x^N - 1}{g(x)\pi(x)}$ when $\pi(x)$ be a primitive polynomial of degree k over F_2 , g(x) is a primitive irreducible divisor of $x^s - 1$, and when s is a special integer. An open problem is to generalized the results for other divisors of $x^s - 1$. This can be done when g(x) is any divisor of $x^s - 1$, and the weight distribution of C can be determined. This will be the purpose of a forthcoming paper.

REFERENCES

[1] Baumert, L.D., McEliece, R.J.,*Weights of Irreducible Cyclic Codes,* Information and control 20, (1972), 158-175.

[2] Delsarte, P., Goethals, J.M., *Irreducible Binary Cyclic Codes of Even Dimension,* U. North Carolina Dept. Stat. Mineo, Series n° 600, 27 (1970).

[3] Lachaud, G., Wolfmann, J., *The Weights of the Orthogonals of the Extended Quadratic Binary Goppa Codes,* IEEE Trans. Info. Theory 36 (1990), 686-692.

[4] Lidl, R., Niederreiter, H., *Finite Fields,* Encyclopedia of Mathematics and its applications, 20, Addison-Wesley, Reading (1983).

[5] McWilliams, F.J., Sloane, N.J.A.,*The Theory of Error-correcting Codes,* North-Holland, Amsterdam, (1977).

[6] McWilliams, F.J., Seery, J.,*The Weight Distribution of some Minimal Cyclic Codes,* IEEE Trans. Info. Theory 27 (1981), 796-806.

[7] Wolfmann, J., *Formes Quadratiques et Codes à Deux Poids,* C.R. Acad. Sc. Paris, t. 281 (1975), 533-535.

[8] Wolfmann, J., *New Bounds on Cyclic Codes from Algebraic Curves,* Lecture Notes in Computer Science, Springer-Verlag, 388 (1989), 2055-2060.

TABLE 1

s is a divisor of $N = 2^k - 1$.

g(x) is a primitive divisor of $x^s - 1$ over F_2.

$\pi(x)$ is a primitive polynomial of degree k over F_2 .

C is the the binary cyclic code of length $N = 2^k - 1$, generated by $\frac{x^N - 1}{g(x)\pi(x)}$.

For $4 \le k \le 12$, and k even, this table gives the length N of C, the dimension K of C, an example of $\pi(x)$, an example of g(x), and the weights of C (Note that these weights do not depend on the choices of $\pi(x)$ and g(x)).

<u>k = 4</u> N = 15, $\pi(x) = x^4+x+1$ or x^4+x^3+1.

s = 3 : K = 6, $g(x) = x^2+x+1$ (unique).
Weights : 6-8-10.

s = 5 : K = 8, $g(x) = x^4+x^3+x^2+x+1$ (unique).
Weights : 4-6-8-10-12.

<u>k = 6</u> N = 63, $\pi(x) = x^6+x+1$ (example).

s = 3 : K = 8, $g(x) = x^2+x+1$ (unique).
Weights : 26-32-34-42.

s = 9 : K = 12, $g(x) = x^6+x^3+1$ (unique).
Weights : 14-26-28-30-32-36-38-40-42.

<u>k = 8</u> N = 255, $\pi(x) = x^8+x^6+x^5+x^4+1$ (example).

s = 3 : K = 10, $g(x) = x^2+x+1$ (unique)
Weights : 122-128-138-170.

s = 5 : K = 12, $g(x) = x^4+x^3+x^2+x+1$ (unique).
Weights : 102-118-124-128-134-140-204.

s = 17 : K = 16, $g(x) = x^8+x^5+x^4+x^3+1$ or $x^8+x^7+x^6+x^4+x^2+x+1$.
Weights : 90-116-118-120-122-128-132-134-136-138-150-180.

<u>k = 10</u> N = 1023, $\pi(x) = x^{10}+x^3+1$ (example).

s = 3 : K = 12, $g(x) = x^2+x+1$ (unique)
Weights : 490-512-522-682.

s = 11 : K = 20, $g(x) = x^{10}+x^9+x^8+x^7+x^6+x^5+x^4+x^3+x^2+x+1$ (unique).
Weights : 186-372-482-488-494-500-506-512-514-520-526
532-538-558-744-930.

s = 33 : K = 20, $g(x) = x^{10}+x^7+x^5+x^3+1$ or $x^{10}+x^9+x^5+x+1$.
Weights : 372-434-490-492-494-496-498-500-512-522-524
526-528-530-532-558-620-682.

<u>k = 12</u> N = 4095, $\pi(x) = x^{12}+x^7+x^4+x^3+1$ (example).

s = 3 : K = 14, $g(x) = x^2+x+1$ (unique).
Weights : 1942-2006-2048-2730

s = 5 : K = 16, $g(x) = x^4+x^3+x^2+x+1$ (unique).
Weights : 1638-1996-2022-2048-2060-2086-3276

s = 9 : K = 18, $g(x) = x^6+x^3+1$ (unique).
Weights : 910-1820-1998-2012-2026-2048-2062-2076-2090-2730

$s = 13 : K = 24$, $g(x) = x^{12}+x^{11}+x^{10}+x^9+x^8+x^7+x^6+x^5+ x^4+x^3+x^2+x+1$ (unique).
Weights : 630-1260-1890-1988-1998-2008-2018-2028-2038-2048-2052
2062-2072-2082-2092-2102-2520-3150-3780

$s = 65 : K = 24$, $g(x) = x^{12}+x^8+x^7+x^6+x^5+ x^4+1$ (example).
Weights : 1638-1764-1890-2008-2010-2012-2014-2016-2018-2020-2022
2048-2072-2074-2076-2078-2080-2082-2084-2086-2142-2268
2394-2520

****

2

MATHEMATICAL TOOLS FOR CODING

BOOLEAN FUNCTIONS ON FINITE FIELDS OF CHARACTERISTIC 2

C. Carlet
INRIA, Rocquencourt, Le Chesnay, France
and
University of Picardie, France

Abstract

We study the polynomial representations (in one variable) of the boolean functions on a Galois field of characteristic 2. We investigate the different ways to obtain them, and characterize them as the solutions of a differential equation. We prove that their derivatives are the squares of those polynomials which are invariant under a certain transformation that we introduce, and deduce a result on BCH codes. Eventually, we use them to solve some differential equations.

1 Introduction

Let m be a positive integer and G the Galois field of order 2^m. The boolean functions on G are the functions from G to its prime subfield $F = GF(2)$. These functions play an important role in algebraic coding theory and in cryptography. Usually, they are expressed by means of the coordinate functions $x_1, ..., x_m$ relative to a base of the F-space G. That is not our purpose here. We know that any function f from G to itself admits a unique polynomial representation over G of degree at most $2^m - 1$:

$$a_0 + a_1 x + \ldots + a_{2^m-1} x^{2^m-1}$$

(indeed, to any such polynomial corresponds a unique function from G to itself, and this correspondence is one to one; the number of functions from G to itself being equal to that of such polynomials, this correspondence is a bijection).

We are interested in the polynomial representations of the boolean functions. These representations appear in a natural way when we consider a cyclic binary code of length $2^m - 1$. Such a code is, by definition, an ideal of the algebra $F[x]/(x^{2^m-1} - 1)$ (cf.[5] ch. 7). To each element of this algebra, we may apply the Fourier transform (also called Mattson-Solomon polynomial, cf. [5] ch.8 §6) which is the polynomial representation of a boolean function. The Mattson-Solomon polynomial leads to a very simple definition of the primitive narrow-sense BCH codes (cf. [5] ch.8, §6, remark 3) , and to the simplest proof of the BCH bound in that case (cf. [5] ch.8, §6, theorem 28). In section 2, we investigate the different ways to obtain the polynomial representations of all boolean functions, using the trace function, the Mattson-Solomon polynomials and the boolean function of support $\{0\} : x \to 1 + x^{2^m-1}$. For any subset E of G, we express the polynomial representation of the boolean function of support E by means of the polynomial $P(x) = \prod_{u \in E} (x + u)$ and its derivative. We prove in section 3 that the polynomial representations of the boolean functions are the solutions of the linear differential equation over G :

$$y + y^2 = (x^{2^m} + x)y'.$$

This property leads to an algebraic expression of the polynomial representations of the boolean functions by means of their derivatives. The question of characterizing these derivatives arises. We introduce in section 4 two representations of polynomials and some related transformations. We characterize in section 5 the derivatives of the polynomial representations of the boolean functions as the squares of those polynomials which are invariant under one of these transformations. We deduce a result on the non-existence of some BCH codes (cf. corollary 7). In section 6, we characterize the even parts of the boolean functions of degrees at most $2^{m-1} + 2^{m-2}$ as the solutions of some linear differential equations.

Notations : If $P(x)$ is any polynomial over G, we will denote by $P(x) \bmod (x^{2^m} + x)$ the unique polynomial of degree smaller than 2^m which is congruent with $P(x)$ modulo $(x^{2^m} + x)$. We denote the cardinality of a subset E of G by $| E |$.

2 The polynomial representations of the boolean functions

Definition 1 *We call polynomial representation of a function f from G to G the unique polynomial $B_f(x)$ over G, of degree smaller than 2^m, such that, for any u in G, $f(u)$ is equal to $B_f(u)$. We denote by $\mathcal{B}$ the set of all the polynomial representations of boolean functions on G.*

There are several ways to obtain all the elements of $\mathcal{B}$:

1. The most usual one consists of using either the trace function tr from G to F or the boolean function $x \rightarrow x^{2^m-1}$: choose any polynomial $P(x)$ over G and compute : $tr(P(x)) \bmod (x^{2^m} + x) = \sum_{i=o}^{m-1} (P(x))^{2^i} \bmod (x^{2^m} + x)$ (respectively $P(x)^{2^m-1} \bmod (x^{2^m} + x)$).

2. The Mattson-Solomon polynomials lead to a second construction, more effective than the previous one : choose any subset E of $G\setminus\{0\}$, consider the following polynomial (where α denotes a primitive element of G) :

$$P(x) = \sum_{i=0}^{2^m-2} \epsilon_i x^i, \quad \text{where} \quad \epsilon_i = 1 \quad \text{if} \quad \alpha^i \in E, \quad \text{and} \quad \epsilon_i = \quad 0 \quad \text{otherwise}$$

and compute the Mattson-Solomon polynomial (cf[5], ch.8, §6) :

$$B(x) = \sum_{i=1}^{2^m-1} P(\alpha^i) x^{2^m-1-i}.$$

According to the inverse formula of the Fourier transform, we have :

$$P(x) = \sum_{i=0}^{2^m-2} B(\alpha^i) x^i.$$

If the cardinality of E is even, then $B(x)$ is the polynomial representation of the boolean function f whose support $\{u \in G / f(u) = 1\}$ is equal to E, according to the inverse formula of the Fourier transform and since $B(0) = P(1)$ is equal to $|E| \bmod 2 = 0$. Otherwise, $B(x)$ is the polynomial representation of the boolean function whose support is equal to $E \cup \{0\}$. We so obtain all the polynomial representations of the boolean functions of even weights (ie of supports of even cardinalities). We deduce all those of the boolean functions of odd weights by adding to the previous polynomials the polynomial representation of the function of support $\{0\}$: $x \rightarrow 1 + x^{2^m-1}$.

Remark
The boolean functions of odd weights are the boolean functions of degree $2^m - 1$.

3. the third construction is as constructive as the second one, and much simpler. It uses the boolean function of support $\{0\}$: $x \rightarrow 1 + x^{2^m-1}$ that we shall denote by γ :
let E be any subset of G, then the boolean function of support E admits as polynomial representation the polynomial :

$$B_E(x) = \sum_{u \in E} \gamma(x + u) = \sum_{u \in E} (x + u)^{2^m-1} + |E| mod 2.$$

Notice that $\gamma(x)$ is also equal to : $\prod_{v\in G\setminus\{0\}} (x+v)$.

Therefore, $B_E(x)$ is equal to : $\sum_{u\in E} \prod_{v\in G\setminus\{u\}} (x+v)$.

We shall deduce a relation between the polynomial $\prod_{u\in E} (x+u)$ and $B_E(x)$:

Proposition 1 *Let E be any subset of G. The polynomial $P_E(x) = \prod_{u\in E}(x+u)$ and the polynomial representation $B_E(x)$ of the boolean function of support E satisfy the relation :*

$$B_E(x) = \frac{P'_E(x)(x^{2^m}+x)}{P_E(x)} = P'_E(x)P_{G\setminus E}(x)$$

where $P'_E(x)$ denotes the derivative of the polynomial $P_E(x)$.

Proof :
According to the classical formula on the differentiation of a product of polynomials, the derivative $P'_E(x)$ of $P_E(x)$ is the polynomial : $\sum_{u\in E} \prod_{v\in E\setminus\{u\}} (x+v)$.

We have : $\dfrac{x^{2^m}+x}{P_E(x)} = \prod_{v\in G\setminus E} (x+v)$.

Therefore : $\dfrac{P'_E(x)(x^{2^m}+x)}{P_E(x)} = \sum_{u\in E} \prod_{v\in G\setminus\{u\}} (x+v) = B_E(x)$. □

3 Characterization of the polynomial representations of the boolean functions

We characterize now the elements of $\mathcal{B}$ as the solutions of a differential equation over G. The following theorem completes theorem 29 [5, p.249] in the primitive case :

Theorem 1 *$\mathcal{B}$ is the set of all the solutions of the differential equation :*

$$y^2 + y = (x^{2^m} + x)y'. \tag{1}$$

Proof :
If f is a boolean function and B is its polynomial representation, then since f^2+f is the zero function, the polynomial $x^{2^m}+x$ divides the polynomial $B^2(x)+B(x)$. There exists a polynomial $R(x)$ of degree at most 2^m-1 such that :

$$B^2(x) + B(x) = (x^{2^m} + x)R(x).$$

By differentiating, we obtain : $B'(x) = R(x) + (x^{2^m} + x)R'(x)$.
The degrees of $R(x)$ and $B'(x)$ being smaller than 2^m, $R'(x)$ must be the zero polynomial, $R(x)$ is equal to $B'(x)$ and :

$$B^2(x) + B(x) = (x^{2^m} + x)B'(x).$$

Conversely, if a polynomial $B(x)$ satisfies this relation, then it has degree at most $2^m - 1$. Let f be the function from G to G whose polynomial representation is $B(x)$, then since $x^{2^m} + x$ divides $B^2(x) + B(x)$, the function $f^2 + f$ is the zero function, and f is a boolean function. □

Corollary 1 *Let $B(x)$ be the polynomial representation of a function f from G to G. Let $R(x)$ and $Q(x)$ be the polynomials of degrees smaller than 2^{m-1} such that $B(x) = R(x) + x^{2^{m-1}}Q(x)$. Then, f is a boolean function if and only if :*

$$B(x) = R^2(x) + xQ^2(x).$$

Proof :

$B(x)$ is a solution of equation (1) if and only if :

$$R(x) + R^2(x) + x^{2^{m-1}}Q(x) + x^{2^m}Q^2(x) = (x^{2^m} + x)B'(x).$$

The polynomials $R(x) + R^2(x) + x^{2^{m-1}}Q(x)$ and $xB'(x)$ have degrees less than 2^m, so that polynomial equation implies : $Q^2(x) = B'(x)$ and f is therefore a boolean function if and only if $R(x)$ and $Q(x)$ satisfy :

$$R(x) + x^{2^{m-1}}Q(x) = R^2(x) + xQ^2(x).$$

□

Corollary 2 *Let $B(x)$ be any element of $\mathcal{B}$, then any root of $B(x)$ in the splitting field of $B(x)$ has an odd order if and only if it belongs to G.*

Proof :
The proof is straightforward : according to theorem 1, we have :

$$B(x)(B(x) + 1) = (x^{2^m} + x)B'(x).$$

The roots of $B'(x)$ are of even orders, and the polynomials $B(x)$ and $B(x) + 1$ are prime each other. □

Corollary 3 *Two elements of $\mathcal{B}$ have same derivative if and only if their difference is 0 or 1.*

Proof :
We may without loss of generality suppose that one of the polynomials is 0.
Let $B(x)$ be any element of $\mathcal{B}$. According to theorem 1, $B'(x)$is the zero polynomial if and only if $B^2(x) = B(x)$ which is equivalent to : $B(x) = 0$ or $B(x) = 1$. □

Corollary 4 *Any element of the narrow-sense primitive BCH code of designed distance δ has weight exactly δ if and only if its Mattson-Solomon polynomial $B(x)$ is such that $B'(x)$ divides $B(x) + 1$.*

Proof :
We have seen (cf. section 2.2) that the Mattson Solomon polynomial associated with such an element belongs to $\mathcal{B}$.
Any element $B(x)$ of $\mathcal{B}$ divides $x^{2^m}+x$ if and only if its derivative $B'(x)$ divides $B(x)+1$, according to theorem 1. □

Let $B(x)$ be any element of $\mathcal{B}$. By iterating equation (1), we can recover $B(x)$ from its derivative $B'(x)$:

$$\begin{aligned}
&B(x) = \\
&(x^{2^m} + x)B'(x) + B^2(x) = \\
&(x^{2^m} + x)B'(x) + (x^{2^{m+1}} + x^2)B'^2(x) + B^4(x) = \ldots = \\
&(x^{2^m} + x)B'(x) + (x^{2^{m+1}} + x^2)B'^2(x) + \ldots + (x^{2^{2m-1}} + x^{2^{m-1}})B'^{2^{m-1}}(x) + B^{2^m}(x)
\end{aligned}$$

and since $d^\circ B(x)$ is less than 2^m, $B(x)$ is the part of the polynomial : $xB'(x)+x^2B'^2(x)+\ldots+x^{2^{m-1}}B'^{2^{m-1}}(x)+B(0)$ consisting of the monomials of degrees less than 2^m.

This equality shows how the knowledge of $B'(x)$ permits to recover the polynomial $B(x)$; it is an effective form of corollary 3.
The derivative $B'(x)$ of an element of $\mathcal{B}$ may have few terms, while the polynomial $B(x)$ has in general a great number of terms. So it seems interesting to study the properties of these derivatives. We shall characterize them in section 5 and deduce some corollaries. Such a characterization will be obtained via the use of two representations of polynomials over G, that we study in next section.

4 Representations of polynomials and deduced transformations

a . the valuation-representation

The representation which is usually used on polynomials over finite fields is the component-representation (cf[6] p. 38):

$$B(x) = \sum_{j=0}^{p-1} x^j (B_j(x))^p \quad \text{where } p \text{ is the characteristic of the field.}$$

Since the characteristic here is 2, this representation reduces to : $B(x) = B_0^2(x) + xB_1^2(x)$, which does not give much help. We will use a generalized version of the component-representation that we call the valuation-representation. Its principle is the following : gather all those monomials involved in $B(x)$ whose degrees admit the same 2-valution (let us recall that the 2-valuation of an integer r is the greatest integer j such that 2^j divides r); for any integer j such that $0 \leq j \leq m-1$, the sum of those monomials of $B(x)$ whose degrees admit j as 2-valuation can be written as $(xB_j^2(x))^{2^j} = x^{2^j} B_j^{2^{j+1}}(x)$:

Definition 2 *We call valuation-representation of any polynomial $B(x)$ of degree at most $2^m - 1$ the following (unique) expansion of $B(x)$:*

$$B(x) = B(0) + \sum_{j=0}^{m-1} (xB_j^2(x))^{2^j} = B(0) + \sum_{j=0}^{m-1} (x^{2^j} B_j^{2^{j+1}}(x)).$$

For example, if $B(x) = 1 + x + x^2 + x^4 + x^5 + x^6$ (we choose a polynomial over F for convenience) then : $B_0(x) = 1 + x^2, B_1(x) = 1 + x$ and $B_2(x) = 1$ and $B(x)$ is equal to : $1 + x(1 + x^2)^2 + x^2(1 + x)^4 + x^4$.

Remark :
We have seen that any element $B(x)$ of $\mathcal{B}$ is equal to the part of the polynomial $xB'(x) + x^2B'^2(x) + \ldots + x^{2^{m-1}} B'^{2^{m-1}}(x) + B(0)$ consisting of the monomials of degrees less than 2^m. For any $j = 0, \ldots, m-1$, all the exponents in the polynomial $x^{2^j} B'^{2^j}(x)$ admit j as 2-valuation. So, this description of $B(x)$ corresponds to its valuation-representation.

Notice that for any $i \geq 1$, the assertion $(d^\circ B(x) < 2^i)$ is equivalent to : $(\forall j < i, d^\circ B_j(x) < 2^{i-j-1})$ and $(\forall j \geq i, B_j(x) = 0)$.

This representation is useful to characterize those polynomials which can be written as the sum of a polynomial and of its square :

Proposition 2 *Let* $B(x) = \sum_{j=0}^{m-1} (xB_j^2(x))^{2^j}$ *be a polynomial over* G *such that* $B(0) = 0$. *There exists a polynomial* $Q(x)$ *such that* $B(x) = Q(x) + Q^2(x)$ *if and only if the sum* $\sum_{j=0}^{m-1} (B_j(x))$ *is equal to zero.*

Proof :
Suppose there exists $Q(x)$ such that $B(x) = Q(x)+Q^2(x)$, then consider the valuation-representation of $Q(x)$: $\sum_{j=0}^{m-2} (xQ_j^2(x))^{2^j}$ (we may suppose : $Q(0) = 0$).
We have : $B_0(x) = Q_0(x)$; $\forall j = 1, \ldots, m-2, B_j(x) = Q_j(x)+Q_{j-1}(x)$ and $B_{m-1}(x) = Q_{m-2}(x)$. So $B_0(x) + \ldots + B_{m-1}(x) = 0$.
Conversely, if $B_0(x) + \ldots + B_{m-1}(x) = 0$, then let us define the polynomials $Q_0(x) = B_0(x)$, and $Q_j(x) = B_j(x) + Q_{j-1}(x), \forall j = 1, \ldots, m-2$.

The polynomial $Q(x) = \sum_{j=0}^{m-2} (xQ_j^2(x))^{2^j}$ satisfies then : $B(x) = Q(x) + Q^2(x)$. □

b . The partitionned representation

Let $B(x)$ be a polynomial over G of degree less than 2^m. For any integer j such that $0 \leq j \leq m-1$, let us gather all those monomials involved in $B(x)$ whose degrees are at least 2^j and less than 2^{j+1} :

Definition 3 *We call partitionned-representation of any polynomial* $B(x)$ *of degree at most* $2^m - 1$ *the following (unique) expansion :*

$$B(x) = B(0) + \sum_{j=0}^{m-1} (x^{2^j} B_j^\natural(x))$$

where, for any j, $B_j^\natural(x)$ *is a polynomial of degree less than* 2^j.

c . Deduced transformations

Suppose that $B(x)$ is a polynomial of degree less than 2^i. Consider its valuation-representation : $B(x) = B(0) + \sum_{j=0}^{i-1} (xB_j^2(x))^{2^j}$. We have $d^\circ B_j(x) < 2^{i-j-1}$, and so, we may consider the polynomial whose partitionned representation is :

$$B(0) + \sum_{j=0}^{i-1} x^{2^{i-j-1}} B_j(x) = B(0) + \sum_{j=0}^{i-1} x^{2^j} B_{i-j-1}(x).$$

Definition 4 *Let $B(x)$ be any polynomial of degree less than 2^i and $B(0)+\sum_{j=0}^{i-1}(xB_j^2(x))^{2^j}$ its valuation-representation, we denote by $B^{<i>}(x)$ the polynomial whose partitionned-representation is equal to : $B(0)+\sum_{j=0}^{i-1} x^{2^{i-j-1}} B_j(x)$.*

We clearly obtain an automorphism $B(x) \to B^{<i>}(x)$ of the F-space of all the polynomials of degrees less than 2^i.

5 Characterization of the derivatives of the polynomial representations of boolean functions

We are now able, by using the automorphims previously defined, to characterize those polynomials of degrees less than 2^{m-1} whose squares are the derivatives of the elements of $\mathcal{B}$:

Proposition 3 *Let $R(x)$ be any polynomial of degree less than 2^{m-1}, then $R^2(x)$ is the derivative of an element of $\mathcal{B}$ if and only if $R(x)$ is invariant under the transformation : $R \to R^{<m-1>}$.*

If R satisfies that property, then the elements of $\mathcal{B}$ which admit $R^2(x)$ as derivative are the polynomials :

$$B(x) = \varepsilon + \sum_{i=0}^{m-1}(xB_i^2)^{2^i} \quad \textit{where } \varepsilon \in \{0,1\}, \quad B_0(x) = R(x) \textit{ and}$$

$$B_i(x) = R(0) + \sum_{j=i}^{m-2} x^{2^{m-j-2}} R_j(x) \; i = 1, \ldots, m-1.$$

Proof :
According to theorem 1 and corollary 1, $R^2(x)$ is the derivative of an element of $\mathcal{B}$ if and only if there exists a polynomial $B(x)$ of degree less than 2^m such that $B^2(x) + B(x) = (x^{2^m} + x)R^2(x)$.
We may suppose that $R(0) = 0$ since any constant a is the derivative of the element of $\mathcal{B}$ equal to : $tr(ax)$.

In order to apply proposition 2, let us calculate the valuation-representation of $P(x) = (x^{2^m} + x)R^2(x)$ by means of the valuation-representation of $R(x)$. We have :

$$R(x) = \sum_{j=0}^{m-2}(xR_j^2(x))^{2^j}, R^2(x) = \sum_{j=1}^{m-1}(xR_{j-1}^2(x))^{2^j}, x^{2^m}R^2(x) = \sum_{j=1}^{m-1}(x.x^{2^{m-j}}R_{j-1}^2(x))^{2^j},$$

and therefore : $P_0(x) = R(x); \forall j = 1, \ldots, m-1, P_j(x) = x^{2^{m-j-1}} R_{j-1}(x)$.

So, according to proposition 2, $R^2(x)$ is the derivative of an element of $\mathcal{B}$ if and only if :

$$R(x)+\sum_{j=1}^{m-1} x^{2^{m-j-1}} R_{j-1}(x)=0, \quad \text{that is} \quad R(x)=\sum_{j=0}^{m-2} x^{2^{m-j-2}} R_j(x)=R^{<m-1>}(x).$$

The expression of $B_i(x)$ when $R(0)=0$ comes from : $B_0(x)=P_0(x)$ and $\forall i\geq 1, B_i(x)=B_{i-1}(x)+P_i(x)$. It is a simple matter to check it by induction. □

Corollary 5 *Let $B(x)$ be any element of $\mathcal{B}$ and $i\in\{1,\ldots,m-1\}$. Then the following assertions are equivalent :*

1) $d^\circ B < 2^{m-1}+2^{m-1-i}$

2) $d^\circ B' < 2^{m-i}$

3) *$B'(x)$ is the 2^{i+1}-th power of a polynomial*

Proof :

According to theorem 1, $d^\circ B'$ is equal to $2d^\circ B-2^m$ and **1)** and **2)** are equivalent. Let $R(x)$ be the polynomial such that $R=R^{<m-1>}$ and $B'(x)=R^2(x)$. We have :
$d^\circ B'(x)<2^{m-i} \Leftrightarrow d^\circ R^{<m-1>}(x)<2^{m-i-1} \Rightarrow$
$(\forall j,(m-j-2\geq m-i-1)\Rightarrow(R_j(x)=0))\Leftrightarrow(\forall j\leq i-1,(R_j(x)=0))\Leftrightarrow$
$(\exists k(x)/R(x)=k^{2^i}(x))$.
So, **2)** implies **3)**.
The converse is similar. □

We shall use now the preceding properties to deduce a result on BCH codes :

Corollary 6 *Let i be a positive integer greater than 1 and smaller than $m-1$. If B is any element of $\mathcal{B}$ of degree less than $2^{m-1}+2^{m-i}$ and such that 2^i does not divide $d^\circ B(x)$, then i divides m and :*

$$d^\circ B(x)=2^{m-1}+(2^{m-1}-2^{i-1})/(2^i-1).$$

If i divides m, then there does exist an element of $\mathcal{B}$ of degree :

$$2^{m-1}+(2^{m-1}-2^{i-1})/(2^i-1)$$

which divides $x^{2^m}+x$.

Proof :

According to corollary 5, $B'(x)$ is the 2^i th power of a polynomial.
By iterating equality $B(x) + B^2(x) = (x^{2^m} + x)B'(x)$, we obtain :

$$B(x) = (x^{2^m} + x)B'(x) + \ldots + (x^{2^{m+i-1}} + x^{2^{i-1}})B'^{2^{i-1}}(x) + B^{2^i}(x).$$

So, there exists a polynomial $M(x)$ such that :

$$B(x) = xB'(x) + \ldots + x^{2^{i-1}}B'^{2^{i-1}}(x) + M^{2^i}(x).$$

The relation : $B(x) + B^2(x) = (x^{2^m} + x)B'(x)$ becomes :

$$x^{2^i}B'^{2^i}(x) + M^{2^i}(x) + M^{2^{i+1}}(x) = x^{2^m}B'(x).$$

So, $B'(x)$ divides $M^{2^i}(x) + M^{2^{i+1}}(x)$ and :

$$x^{2^i}B'^{2^i-1}(x) + \frac{M^{2^i}(x) + M^{2^{i+1}}(x)}{B'(x)} = x^{2^m}.$$

The degree of the polynomial $\frac{M^{2^i}(x)+M^{2^{i+1}}(x)}{B'(x)}$ is equal to $2^{i+1}d^\circ M(x) - d^\circ B'(x)$, and so is not divisible by 2^{i+1} ($d^\circ B'$ being equal to $2d^\circ B - 2^m$, it is not divisible by 2^{i+1}). The degree of $x^{2^i}B'^{2^i-1}(x)$ is divisible by 2^{i+1}, so, the degree of $\frac{M^{2^i}(x)+M^{2^{i+1}}(x)}{B'(x)}$ must be smaller than 2^m, and x^{2^m} is equal to the term of higher degree in $x^{2^i}B'^{2^i-1}(x)$. Therefore: $(2^i - 1)d^\circ B' = 2^m - 2^i$.
Since $2^i - 1$ divides $2^m - 2^i$, it divides $2^{m-i} - 1$, and so i divides $m - i$ and therefore i divides m.
The degree of $B(x)$ is equal to $\frac{1}{2}d^\circ B'(x) + 2^{m-1} = 2^{m-1} + (2^{m-1} - 2^{i-1})/(2^i - 1)$.
Suppose now that i divides m, and $j = \frac{2^m - 2^i}{2^i - 1}$. We shall prove that x^j is the derivative of an element of $\mathcal{B}$ which divides $x^{2^m} + x$:
let us consider the polynomial : $B(x) = 1 + x^{j+1} + \ldots + x^{(j+1)2^{i-1}}$. Its derivative is x^j. We have : $B(x) + B^2(x) = x^{j+1} + x^{(j+1)2^i} = x^j(x + x^{2^m}) = B'(x)(x + x^{2^m})$. So, $B(x)$ is the polynomial representation of a boolean function, and since $B'(x)$ divides $B(x) + 1$, $B(x)$ divides $(x + x^{2^m})$. □

Corollary 6 stated in terms of primitive BCH codes gives :

Corollary 7 *Let i be an integer, $1 < i < m - 1$, and δ an integer greater than $2^{m-1} - 2^{m-i} - 1$ and such that the BCH code of length $2^m - 1$ and designed distance δ is different from the BCH code of designed distance $\delta + 1$. Suppose 2^i does not divide $\delta + 1$, then i divides m, and δ is equal to :*

$$\frac{(2^m - 1)(2^{i-1} - 1)}{2^i - 1}.$$

There exists then an element in the BCH code of length $2^m - 1$ and designed distance δ whose weight is δ.

Proof :
If the BCH code of length $2^m - 1$ and designed distance δ is different from the BCH code of designed distance $\delta + 1$, then there exists an element of $\mathcal{B}$ which has degree $2^m - 1 - \delta$ (cf.[5] ch. 8). So, the proof is straightforward (according to corollary 6), since :

$$2^m - 1 - (2^{m-1} + (2^{m-1} - 2^{i-1})/(2^i - 1)) = \frac{(2^m - 1)(2^{i-1} - 1)}{2^i - 1}.$$

□

Remark :
The last sentence of the statement of corollary 7 is not a new result : it can be deduced from theorem 9.4 in [4] p.279-280 (where q is 2, m has value i and h has value $i - 1$) and corollary 9.3.(where n_1 has value $\frac{2^m-1}{2^i-1}$, n_2 has value $2^i - 1$ and d is $2^{i-1} - 1$).

6 The even parts of the representations of the boolean functions

Let $B(x) = R(x) + x^{2^{m-1}}Q(x) = R^2(x) + xQ^2(x)$ (cf. corollary 1) be any element of $\mathcal{B}$. By derivating this double equality , we have :

$$B'(x) = R'(x) + x^{2^{m-1}}Q'(x) = Q^2(x).$$

Let us assume that $B(x)$ has degree at most $2^{m-1}+2^{m-2}$. The polynomial $Q(x)$ has then degree at most 2^{m-2}. So, according to the previous equality, $Q'(x)$ must be a constant, say a, and $Q^2(x) = R'(x) + ax^{2^{m-1}}$. By squaring the equality $R(x) + x^{2^{m-1}}Q(x) = R^2(x) + xQ^2(x)$ and replacing $Q^2(x)$ by this value, we obtain :

$$R^4(x) + R^2(x) + x^{2^m}R'(x) + x^2R'^2(x) = ax^{2^m+2^{m-1}} + a^2x^{2^m+2}.$$

Conversely, if $R(x)$ satisfies this relation, then the polynomial $B(x) = R^2(x)+ax^{2^{m-1}+1}+ xR'(x)$ is an element of $\mathcal{B}$ and its degree is at most $2^{m-1} + 2^{m-2}$. So :

Proposition 4 *The even parts of the elements of $\mathcal{B}$ of degrees at most $2^{m-1} + 2^{m-2}$ are the solutions of the equations:*

$$y^4 + y^2 + x^{2^m}y' + x^2y'^2 = ax^{2^m+2^{m-1}} + a^2x^{2^m+2}(a \in G).$$

Acknowledgement
We thank Pascale Charpin for helpful comments.

References

[1] D.AUGOT, P.CHARPIN AND N.SENDRIER *Sur une Classe de Polynômes Scindés de l'Algèbre $F_{2^m}[Z]$*, C.R.Acad.Sci.Paris,t.312, Série I, (1991), pp.649-651.

[2] E. R. BERLEKAMP *Algebraic Coding Theory*, McGraw-Hill Book Company New York, St. Louis, San Francisco (1968).

[3] P. CAMION *Factorisation des Polynômes de $F_q[x]$*, rapport de recherche n° 93, INRIA, Sept 1981, Domaine de Voluceau, BP 105, 78153 Le Chesnay Cedex, France.

[4] W.W.PETERSON and E.J.WELDON *Error-correcting codes*, The MIT Press Cambridge, Massachusetts and London, England (1986)

[5] F. J. MAC WILLIAMS AND N. J. SLOANE *The theory of error- correcting codes*, Amsterdam, North Holland (1977).

[6] L. REDEI *Lacunary Polynomials over Finite Fields*, Akadémiai Kiado, Budapest (1973).

BOUND FOR TRACE-EQUATION AND APPLICATION TO CODING THEORY

V. Gillot
University of Toulon and Var, La Garde, France

Abstract – **Using the vector space structure of finite field extension of characteristic two, we transform a one-variable polynomial into a two-variable one. Applying the Deligne bound about exponential sums [1] to the two-variable polynomial, we improve in several cases the bound on the number of solutions for trace-equation. In this way, we obtain different results in Coding Theory. Using the trace-description of the codewords for cyclic codes introduced by Wolfmann [6], we improve the bound on weights for these codes and we also generalize, in different cases, the results of Moreno and Kumar in [2].**

1. Introduction

The main subject of the present work is to evaluate exponential sums over finite field of characteristic two for one-variable polynomial. The problem is to obtain sharp estimates for them. Many authors have been interesting in this subject.

The idea, described in the second section, is to transform a one-variable polynomial into a two-variable one. In the third section, we apply exponential sums bound corresponding to multivariable polynomial, we use the famous Deligne's bound [1]. We improve in several cases the classical corresponding Weil's bound, the section five asserts the improvement with numerical tests. In the section four, we link our results on exponential sums (or equivalently on trace-equations) to bound on the weight of codewords defined by the trace operator. Now, we give several definitions used in this paper and for more general mathematical tools we can refer to [4]. Considering a quadratic finite field extension of $\mathbf{F}_q$ of characteristic two, we note that $\mathbf{F}_{q^2}$ is a two-dimensional $\mathbf{F}_q$-vector space, let $\{u_1, u_2\}$ be a $\mathbf{F}_q$-basis of $\mathbf{F}_{q^2}$. For all element x in $\mathbf{F}_{q^2}$, we consider its coordinates (x_1, x_2) lying in $\mathbf{F}_q$ and satisfying

$$x = x_1 u_1 + x_2 u_2 \ .$$

We use $\{\lambda_1, \lambda_2\}$ the dual basis of $\{u_1, u_2\}$ defined as

$$\text{For all } i, j \in \{1, 2\}, \quad \text{Tr}_{(q^2,q)}(\lambda_i u_j) = \begin{cases} 1 & \text{if } i = j, \\ 0 & \text{else.} \end{cases}$$

If $\mathrm{Tr}_{(q^2,q)}$ denotes the usual trace operator from $\mathbf{F}_{q^2}$ into $\mathbf{F}_q$, for all x in $\mathbf{F}_{q^2}$, we have

$$\mathrm{Tr}_{(q^2,q)}(x) = x + x^q.$$

DEFINITIONS 1.1. *The **support** of a polynomial $f(x) = \sum_{e\in E} a_e x^e$ is defined as*

$$\mathrm{supp}(f) = \{e \in E \mid a_e \neq 0\} \quad \textit{where } E \subset \mathbf{N}.$$

*The **q-ary expansion** of an interger d is the following sum*

$$d = \sum_{i\in I} d_i q^i, \quad I \subset \mathbf{N},$$

*where for each i, the coefficient d_i lie in $0 \leqslant d_i \leqslant q-1$. The **q-ary weight** of an integer d, which has the previous q-ary expansion, is defined by*

$$w_q(d) = \sum_{i\in I} d_i,$$

where all the d_i also lie in $0 \leqslant d_i \leqslant q-1$.

Remark – In the particular case where d is an integer less than q^2, d has a single q-ary expansion

$$d = d_0 + q d_1, \quad \text{where } d_0, d_1 \leqslant q-1.$$

And its q-ary weight is

$$w_q(d) = d_0 + d_1$$

2. Transformation over polynomial

2.1. Construction

Using the two-dimensional vector space structure of $\mathbf{F}_{q^2}$, we transform a one-variable polynomial $f(x)$ over $\mathbf{F}_{q^2}$ of degree less than q^2 into a two-variable one over $\mathbf{F}_q$:

$$F(x_1, x_2) = \mathrm{Tr}_{(q^2,q)}\, f(x_1 u_1 + x_2 u_2).$$

We reduce $F(x_1, x_2)$ modulo $(x_1^q - x_1, x_2^q - x_2)$ to obtain a new two-variable polynomial $F_R(x_1, x_2)$ and we denote [(1)]

$$F_R(x_1, x_2) = F(x_1, x_2) \bmod (x_1^q - x_1, x_2^q - x_2). \tag{1}$$

Let $f(x)$ be the polynomial over $\mathbf{F}_{q^2}$

$$f(x) = \sum_{d\in \mathrm{supp}(f)} a_d\, x^d,$$

we compute the q-ary expansion of every degree d in the support of f as $d = d_0 + q d_1$, and applying the last formulas, we obtain

(1) We write on the left hand side the remainder of the division.

$$F_R(x_1,x_2) = \sum_{d\in\mathrm{supp}(f)} \left(a_d\,(x_1u_1+x_2u_2)^{d_0}(x_1u_1^q+x_2u_2^q)^{d_1}\right.$$
$$\left. +a_d^q\,(x_1u_1+x_2u_2)^{d_1}(x_1u_1^q+x_2u_2^q)^{d_0}\right) \quad (2)$$

If we set $y_1 = x_1u_1+x_2u_2$ and $y_2 = x_1u_1^q+x_2u_2^q$, we obtain an other two-variable polynomial $\Phi(y_1,y_2)$ over $\mathbf{F}_{q^2}$ satisfying

$$\Phi(y_1,y_2) = \sum_{d\in\mathrm{supp}(f)} \left(a_d\,y_1^{d_0}y_2^{d_1} + a_d^q\,y_1^{d_1}y_2^{d_0}\right). \quad (3)$$

We note that

$$F_R(x_1,x_2) = \Phi(x_1u_1+x_2u_2, x_1u_1^q+x_2u_2^q). \quad (4)$$

Using the dual basis $\{\lambda_1,\lambda_2\}$ of $\{u_1,u_2\}$, it is easy to prove that

$$x_1 = \mathrm{Tr}_{(q^2,q)}(\lambda_1 x) \quad \text{and} \quad x_2 = \mathrm{Tr}_{(q^2,q)}(\lambda_2 x). \quad (5)$$

So, we can reconstruct a one-variable polynomial $\phi(x)$ over $\mathbf{F}_{q^2}$ from $F_R(x_1,x_2)$ as

$$\phi(x) = F_R(\mathrm{Tr}_{(q^2,q)}(\lambda_1 x), \mathrm{Tr}_{(q^2,q)}(\lambda_2 x)). \quad (6)$$

Remark – the polynomial $\phi(x)$ is the remainder of the division by $x^{q^2}-x$ of the polynomial $\mathrm{Tr}_{(q^2,q)}\,f(x)$.

Henceforward, we need to resolve a partial derivative system of the polynomial $F_R(x_1,x_2)$ defined as above, the following proposition gives us an easier way to prove it.

PROPOSITION 2.1. *Let $F_R(x_1,x_2)$ and $\Phi(y_1,y_2)$ be the two polynomials described above in (3) and (4). Then we have*

$$\left.\begin{array}{l} \frac{\partial}{\partial x_1}\, F_R(x_1,x_2)=0 \\ \frac{\partial}{\partial x_2}\, F_R(x_1,x_2)=0 \end{array}\right\} \text{ if and only if } \left\{\begin{array}{l} \frac{\partial}{\partial y_1}\,\Phi(y_1,y_2)=0 \\ \frac{\partial}{\partial y_2}\Phi(y_1,y_2)=0 \end{array}\right.$$

Proof – Setting $l_1(x_1,x_2) = x_1u_1+x_2u_2$ and $l_2(x_1,x_2) = x_1u_1^q+x_2u_2^q$, we calculate the partial derivative of $F_R(x_1,x_2)$ in the expression (4) and we obtain

$$\frac{\partial}{\partial x_i}\,F_R(x_1,x_2) = \frac{\partial}{\partial y_1}\,\Phi(l_1(x_1,x_2),l_2(x_1,x_2)) \times \frac{\partial}{\partial x_i}\,l_1(x_1,x_2)$$
$$+\frac{\partial}{\partial y_2}\,\Phi(l_1(x_1,x_2),l_2(x_1,x_2)) \times \frac{\partial}{\partial x_i}\,l_2(x_1,x_2) \quad (7)$$

Computing the partial derivative of each $l_j(x_1,x_2)$ for $j=1,2$, we obtain according to this following notation

$$\begin{pmatrix} \frac{\partial}{\partial x_1}\,F_R \\ \frac{\partial}{\partial x_2}\,F_R \end{pmatrix} = \begin{pmatrix} u_1 & u_1^q \\ u_2 & u_2^q \end{pmatrix} \begin{pmatrix} \frac{\partial}{\partial y_1}\,\Phi \\ \frac{\partial}{\partial y_2}\,\Phi \end{pmatrix}$$

It is easy to check using the dual basis that the above 2×2-matrix is invertible and its inverse is

$$\begin{pmatrix} \lambda_1 & \lambda_2 \\ \lambda_1^q & \lambda_2^q \end{pmatrix}.$$

Then resolving a partial derivative system of F_R is equivalent to resolving a similar one for Φ, we have the result anounced in the proposition. □

2.2. Degree and q-ary order of polynomial

Now, we give some results on the polynomial that we constructed previously. We link the degree of the polynomial $F_R(x_1, x_2)$ defined in (2) to the q-ary order of the polynomial $f(x)$ defined hereunder.

DEFINITION 2.2. *The q-ary order of a polynomial $f(x)$ is*

$$O_q(f) = \max_{e \in \mathrm{supp}(f)} \{w_q(e)\},$$

where w_q denotes the q-ary weight described in the previous part.

Example – Compute the 4-ary order of the polynomial $f(x) = x^{13} + x^8 + 1$ over $\mathbf{F}_{4^2}$. We have $\mathrm{supp}(f) = \{13, 8, 0\}$, we compute the 4-ary weight of each integer in the support of f :

$$\left.\begin{array}{lll} d = 13 & 13 = 3 \times 4 + 1 & w_4(13) = 3 + 1 = 4 \\ d = 8 & 8 = 2 \times 4 + 0 & w_4(8) = 2 \\ d = 0 & & w_4(0) = 0 \end{array}\right\} \quad O_4(f) = \max_{d \in \mathrm{supp}(f)} \{w_4(d)\} = 4$$

PROPOSITION 2.3. *Consider $f(x) = a_d x^d$ a monomial of degree d, less than $q^2 - 1$, over $\mathbf{F}_{q^2}$ and let $F_R(x_1, x_2)$ be the polynomial obtained by the transformation described previously in (1) and (2). Consider the two cases*

(i) $a_d \notin \mathbf{F}_q$.

(ii) $a_d \in \mathbf{F}_q$ and $d_0 \neq d_1$.

If we are in one of the two cases (i) or (ii), then we have

$$\deg(F_R) = w_q(d).$$

Proof – Let $f(x) = a_d x^d$ polynomial over $\mathbf{F}_{q^2}$, we define its transformed polynomial F_R as

$$F_R(x_1, x_2) = a_d(x_1 u_1 + x_2 u_2)^{d_0}(x_1 u_1^q + x_2 u_2^q)^{d_0} + a_d^q(x_1 u_1^q + x_2 u_2^q)^{d_0}(x_1 u_1 + x_2 u_2)^{d_1}$$

The polynomial $F_R(x_1, x_2)$ is the zero polynomial for all $(x_1, x_2) \in \mathbf{F}_q{}^2$ if and only if the polynomial $\mathrm{Tr}_{(q^2,q)}(f(x)) = 0$ for all $x \in \mathbf{F}_{q^2}$, namely

$$a_d x^d + a_d^q x^{dq} = 0,$$

for all $x \in \mathbf{F}_{q^2}$. If we write the q-ary expansion of the degree d, we obtain

$$\frac{a_d}{a_d^q} = x^{(d_0 - d_1)q + (d_1 - d_0)},$$

for all $x \in \mathbf{F}_{q^2}$. The définition of the q-ary expansion of an integer yields that d_0 and d_1 lie in $0 \leqslant d_i \leqslant q - 1$, then we have

$$|(d_1 - d_0)(1 - q)| \leqslant q^2 - 1\,.$$

So, the monomial $x^{(d_1 - d_0)(1-q)}$ is constant if and only if

$$d_0 = d_1$$

Moreover this constant is 1, we obtain

$$a_d = a_d^q.$$

Thus, F_R is the zero polynomial if and only if $a_d \in \mathbf{F}_q$ and $d_0 = d_1$. In all the other cases, we can see with the expression of F_R that the degree of this polynomial is $w_q(d) = d_0 + d_1$. □

COROLLARY 2.4. *Let $f(x)$ and $F_R(x_1, x_2)$ be the polynomials described previously in (1) and (2). Suppose that the q-ary order of f, $O_q(f)$, is reached for a single integer d in the support of f. Let a_d be the non zero coefficient corresponding to the monomial term of degree d of f. Consider the two cases*

(i) $a_d \notin \mathbf{F}_q$.

(ii) $a_d \in \mathbf{F}_q$ and $d_0 \neq d_1$.

If we are in one of the two cases (i) or (ii), then we have

$$\deg(F_R) = O_q(d).$$

Proof – If the q-ary order, $O_q(f)$, of the polynomial f is reached for a single integer d in the support of f, the degree of F_R only depends on the q-ary weight of the degree d and its corresponding coefficient a_d. We are again in the case of the monomial term of the previous proposition. □

3. A new bound for trace-equation

The main subject of this paper is to evaluate, by a new way, the number N of solution of the following set

$$\{x \in \mathbf{F}_{q^2} \mid \mathrm{Tr}_{(q^2,2)} f(x) = 0\}\ . \tag{1}$$

We name this equation : trace-equation.

LEMMA 3.1. *Consider $f(x)$ and $F_R(x_1, x_2)$ the polynomials described in the previous section satisfying (1) and (2). We have*[(2)]

$$\{x \in \mathbf{F}_{q^2} \mid \mathrm{Tr}_{(q^2,2)} f(x) = 0\} = \{(x_1, x_2) \in (\mathbf{F}_q)^2 \mid \mathrm{Tr}_{(q,2)} F_R(x_1, x_2) = 0\} \tag{2}$$

Proof – Using the construction of F_R of the last section and applying the transitivity of the trace operator

$$\mathrm{Tr}_{(q^2,2)}(x) = \mathrm{Tr}_{(q,2)}\left(\mathrm{Tr}_{(q^2,q)}(x)\right),$$

we obtain the result of the lemma. □

Let Ψ an additive character over $\mathbf{F}_q$, we consider the following exponential sums

$$S(f) = \sum_{x \in \mathbf{F}_{q^2}} \Psi\big(\mathrm{Tr}_{(q^2,q)}(f(x))\big) \quad \text{and} \quad S(F_R) = \sum_{x_1, x_2 \in \mathbf{F}_q} \Psi(F_R(x_1, x_2)).$$

(2) $\mathrm{Tr}_{(q,2)}$ denotes the usual operator trace from $\mathbf{F}_q$ to $\mathbf{F}_2$ and $\mathrm{Tr}_{(q^2,2)}$ the usual one from $\mathbf{F}_{q^2}$ to $\mathbf{F}_2$.

LEMMA 3.2. *We have*

$$S(f) = S(F_R) \quad and \quad S(f) = 2N - q^2. \tag{3}$$

Proof – The previous lemma gives us the first result of this lemma. On the other hand, an additive character Ψ over $\mathbf{F}_q$ is defined as

$$\Psi(x) = (-1)^{\mathrm{Tr}_{(q,2)}(x)}.$$

In our case the characteristic of the field is two, so the trace operator takes only two values 0 or 1. If we name N the number of x in $\mathbf{F}_{q^2}$ such that the trace of f is zero, we have

$$S(f) = N - (q^2 - N).$$

□

Applying the two above lemmas, to evaluate exponential sum with polynomial argument $f(x)$ is equivalent to evaluate the exponential one for the polynomial F_R stemed from $f(x)$ as in the previous section. Now, we can apply bound on exponential sum with polynomial in several variables argument. We use the following Deligne's bound.

THEOREM. *(Deligne [1]) Let Q be a n-variable polynomial with degree d over $\mathbf{F}_q$. Let Q_d be the homogeneous part of degree d of Q. Let $\Psi : \mathbf{F}_q \mapsto \mathbf{C}^*$ be a non-trivial additive character over $\mathbf{F}_q$. We suppose :*
(i) d is prime to p (field characteristic)
(ii) The homogeneous part Q_d defines a smooth hypersurface H_0 in $\mathbf{P}^{n-1}(\mathbf{F}_q)$.
Then

$$\Big| \sum_{x_1,\ldots,x_n \in \mathbf{F}_q} \Psi(Q(x_1,\ldots,x_n)) \Big| \leqslant (d-1)^n \, q^{\frac{n}{2}}.$$

If we apply this theorem to the polynomial $F_R(x_1, x_2)$ over $\mathbf{F}_q$, we obtain the following theorem over $f(x)$.

THEOREM 3.3. *Let $f(x)$ be a polynomial over $\mathbf{F}_{q^2}$. If $O_q(f)$ is reached for a single odd integer in the support of f, and if $O_q(f)$ and $\deg(f)$ are both odd then*

$$| 2N - q^2 | \leqslant (O_q(f) - 1)^2 \, q. \tag{4}$$

Proof – The corollary 2.4 gives us that the degree of F_R is the q-ary order of f, and we saw in its proof that the part of highest degree of F_R is constructed from a monomial of f. Let $d = d_0 + qd_1$ be the integer of the support of f which gives the q-ary order of f. The oddness of d and $w_q(d) = d_0 + d_1$ for an even q yields an odd d_0 and an even d_1.
Let $F_R^d(x_1, x_2)$ be the homogeneous polynomial stemed from the monomial of f of degree d, the corollary 2.4 in the case $d_0 \neq d_1$ for any coefficient a_d yields

$$\deg(F_R^d) = d_0 + d_1 = w_q(d) = O_q(f).$$

We saw previously that $d_0 + d_1$ is odd, so the first hypothesis of the Deligne theorem is satisfied.

To prove the nonsingularity of the hypersurface H_0 which defines F_R^d, we need to resolve a partial derivative system, we apply the proposition 2.2 of the second section, it is equivalent to resolve a partial system for the corresponding $\Phi^d(y_1, y_2)$ where

$$\Phi^d(y_1, y_2) = a_d\, y_1^{d_0} y_2^{d_1} + a_d^q\, y_1^{d_1} y_2^{d_0}.$$

In a field of characteristic two, an odd d_0 and an even d_1 give us

$$\begin{cases} \frac{\partial}{\partial y_1}\Phi(y_1, y_2) = d_0 a_d\, y_1^{d_0-1} y_2^{d_1} = 0 \\ \frac{\partial}{\partial y_2}\Phi(y_1, y_2) = d_0 a_d^q\, y_1^{d_1} y_2^{d_0-1} = 0 \end{cases}$$

the solution of this system is $(y_1, y_2) = (0, 0)$, in our particular case, we have

$$y_i = l_i(x_1, x_2) = x_1 u_1^{q^{i-1}} + x_2 u_2^{q^{i-1}}.$$

The family $\{u_1, u_2\}$ is a $\mathbb{F}_q$-basis, then the system has no non-zero solution. The second hypothesis of the Deligne's theorem is satisfied. Apply the lemma 3.2 to obtain the result anounced. □

Remark – Take a good note of the transparency of the register of the polynomial F_R in the statement of our theorem 3.3. Actually, we only have conditions over the polynomial f.

COROLLARY 3.4. *If $f(x)$ is a monomial over $\mathbb{F}_{q^2}$ of odd degree d and if $w_q(d)$ is odd, then*

$$| 2N - q^2 | \leqslant (w_q(d) - 1)^2\, q.$$

Proof – We have already seen that the case of a monomial is a particular case of the previous theorem 3.3. □

3.1. Comparison with the Weil's bound and the trivial one

Let N be the number of solutions of the following trace-equation

$$\mathrm{Tr}_{(q^k, p)}\, f(x) = 0, \qquad \text{for } x \in \mathbb{F}_{q^k}$$

The famous Weil's bound, translated in term of number of solution of trace-equation for a one-variable polynomial of degree d over $\mathbb{F}_{q^k}$, is defined by

$$|pN - q^k| \leqslant \frac{(p-1)(d-1)}{2} 2q^{\frac{k}{2}},$$

In particular, we consider a finite field of characteristic two, and N denotes the number of solution of the trace-equation (8). Thus, we have the hereunder corollary.

Corollary. *Let $f(x)$ be a polynomial over $\mathbf{F}_{q^2}$ of odd degree d and N the number of solution of the trace-equation (8). Then*

$$|2N - q^2| \leqslant (d-1)\, q.$$

Our theorem 3.3 give results for degrees d lying in $q+1 \leqslant d \leqslant q^2-1$, so

$$(d-1) \geqslant q$$

In fact, the Weil bound is always trivial in this case. We just have to compare our bound with the trivial bound for exponential sum gived by the cardinality of the fields q^2. Virtually, we improve the trivial bound when the following relation is satisfied

$$(O_q(f)-1)^2 \leqslant q.$$

In the section 5, we give some numerical test of this improvement.

4. Application to Coding Theory

J. Wolfmann gives in [6] a new bound on cyclic codes using algebraic curves. Namely, he describes cyclic codewords with the trace operator, and he applies the Weil's bound, defined in the last section, about number of solutions of trace-equation to estimate the weight of these codewords. Thus, he improves the classical bounds on weights of these codes. In the previous section, we improve this Weil's bound and the trivial one for several polynomials. Virtually, we express in this section our bound of the theorem 3.3 in term of weight of codewords. First, we recall the trace code described in [6] in our particular case of quadratic finite field extension of characteristic two. The elementary notions for Coding Theory are in [5].

4.1. Trace code description

Let n be an integer prime to q such that $\mathbf{F}_{q^2}$ is the splitting field of x^n-1 over $\mathbf{F}_2$, and let β in $\mathbf{F}_{q^2}$ be a primitive n-th root of unity. For any integer i, the cyclotomic class of i is defined to be

$$\Gamma(i) = \{0 \leqslant t \leqslant n-1 \mid t \equiv i\, 2^j (\operatorname{mod} n) \quad \text{for some } j\},$$

and the minimal polynomial of β^i over $\mathbf{F}_2$ is

$$m_{\beta^i}(x) = \prod_{t\in\Gamma(i)} (x-\beta^t).$$

Now, we consider an arbitrary cyclic code C of length n over $\mathbf{F}_2$. Let $x^n - 1 = g(x)h(x)$ when $g(x)$ is the generator polynomial, we denote the reciprocal polynomial of $h(x)$ by $h^{\perp}(x)$. A subset J of $\{0,\ldots,n-1\}$ such that

$$h^{\perp}(x) = \prod_{j\in J} m_{\beta^j}(x)$$

is called a β-check set of C. According to the proposition 2.1 of [6], the code C is the set of words as

$$c_{\underline{a}}(x) = \sum_{i=0}^{n-1} [\mathrm{Tr}_{(q^2,2)}(f_{\underline{a}}(\beta^i)]\, x^i, \tag{1}$$

where

$$f_{\underline{a}}(x) = \sum_{j \in J} a_j x^j \ , \ \underline{a} = (a_j)_{j \in J} \in (\mathbf{F}_{q^2})^{|J|}. \tag{2}$$

Let r be an integer such that $nr = q^2 - 1$. Then the set of n-th roots of unity in $\mathbf{F}_{q^2}$ is also the set of r-th powers in $\mathbf{F}_{q^2}^{\times}$. Consider the trace-equation $E_{\underline{a}}$ defined by

$$\mathrm{Tr}_{(q^2,2)}\ f_{\underline{a}}(x^r) = 0, \tag{3}$$

and observe that

$$N_{\underline{a}} = \sharp\{x \in \mathbf{F}_{q^2}^{\times} \mid \mathrm{Tr}_{(q^2,2)}\ f_{\underline{a}}(x^r) = 0\} = r(n - w(\underline{a})),$$

where $w(\underline{a})$ is the Hamming weight of the codeword $c_{\underline{a}}(x)$ corresponding to the polynomial $f_{\underline{a}}(x)$.

THEOREM 4.1. *Let C be a binary cyclic code as above. Let rJ be the set $\{rj \mid j \in J\}$ and θ an odd integer such that $\theta = \max_{i \in rJ}\{w_q(i)\}$* [(3)].
Suppose that
(i) Each integer in rJ is odd.
(ii) All the q-ary weight of each integer in rJ are odd and distinct.
Thus, for $w(\underline{a}) \neq 0$ we have

$$|q^2 - 2rw(\underline{a}) - 1| \leqslant (\theta - 1)^2 q + 1.$$

Proof – According to the theorem [6, 4.3] and to the statement of our theorem 3.3, we can translate in terms of weight of codewords, the bound of the theorem 3.3. First, if the q-ary weight of each odd integer in rJ is different, we are in the case where the q-ary order of the polynomial $f_{\underline{a}}(x^r)$ is reached for a single integer in rJ. Moreover, the oddness of the elements of rJ and the oddness of their corresponding q-ary weight gives us the hypothesis of the theorem 3.3. We also have to treat the case where zero is in J or not. First, suppose that $0 \in J$, and note N the number of solutions corresponding to the theorem 3.3, two different cases appear :

if $\mathrm{Tr}_{(q^2,2)}\, f_{\underline{a}}(0) = 0$, then
$$N = N_{\underline{a}} + 1 = q^2 - rw(\underline{a})$$
if $\mathrm{Tr}_{(q^2,2)}\, f_{\underline{a}}(0) = 1$, then
$$N = N_{\underline{a}} = q^2 - 1 - rw(\underline{a})$$

If we suppose that $0 \notin J$, we have

$$N = N_{\underline{a}} + 1 = q^2 - rw(\underline{a})$$

Now, we can replace N by $N_{\underline{a}}$ or $N_{\underline{a}} + 1$ and $O_q(f)$ by θ in the bound of the theorem 3.3 to obtain the result of the theorem. □

(3) $w_q(i)$ denotes the q-ary weight of the integer i defined in the first section.

Remark – Note that this theorem is a generalization of the works of O. Moreno and P. Kumar in their paper [2], when the authors use the Deligne bound. Actually, they apply the Deligne's bound for cyclic codes after the same construction, but they have to verify for every case the statement of the Deligne 's theorem. Moreover, they study only four families of cyclic codes, and most of the codewords in them satisfy our theorem 4.1.

To illustrate the bound of the theorem 4.1, we give an example of cyclic codes described by the trace operator and its associated polynomial.

Example – Let $n = q^2 - 1$ and q a power of two. We consider the primitive cyclic code C of length n over $\mathbf{F}_2$ and the following polynomial according to the previous notations

$$f_{\underline{a}}(x) = ax^{2q+3} + bx^{2q+1} + cx,$$

where the coefficients a, b, c lie in $\mathbf{F}_{q^2}$ and the β-check set is

$$J = \{2q+3, 2q+1, 1\}$$

According to the statement of the theorem 4.1, we have $r = 1$, $0 \notin J$ and furthermore

$$\theta = \max_{i \in J}\{w_q(i)\} = 5.$$

The integer θ is odd, then

$$\frac{q^2 - 16q}{2} \leqslant w(\underline{a}) \leqslant \frac{q^2 + 16q}{2}.$$

The theorem [6, 4.3] and its corollary gives

$$\frac{q^2 - (2q+3)q}{2} \leqslant w(\underline{a}) \leqslant \frac{q^2 + (2q+3)q}{2}.$$

We can note that the last bound is greater than the trivial one, namely more than q^2. Thus, in this case our bound is better than the bound gived by J. Wolfmann in [6]. For $q^2 = 2^{10}$ (resp. $q^2 = 2^{12}$), we also obtain in this example an improvement of the BCH bound on the minimum distance that is 240 (resp. 992) and our bound give $256 \leqslant w(\underline{a}) \leqslant 816$ (resp. $1536 \leqslant w(\underline{a}) \leqslant 2560$).

5. Sharp estimates

In the following table we give some numerical tests to assert the improvement of the Weil's bound anounced at the end of the section 4. We assume that we are not in a trivial case. The polynomial f is of the form

$$f = x^d + f'$$

where the q-ary order of f' satisfies

$$O_q(f') < w_q(d).$$

Namely, the q-ary order of f is reached for a single integer in the support of f, and it is exactly $w_q(d)$. We test only the odd degree d that has an odd q-ary weight satisfying

$$w_q(d) \leqslant [\sqrt{q} + 1]$$

to belong less than the trivial bound. We give in the figure 1 the bound on the number N of solutions corresponding to our theorem 3.3 when q is a power of two. We write T when the bound is trivial (i.e. : more than q^2).

$\|\mathbb{F}_{q^2}\|$	d	$w_q(d)$	New bound
2^6	$2q+1$	3	$16 \leqslant N \leqslant 48$
2^8	$2q+1$	3	$96 \leqslant N \leqslant 160$
2^{10}	$2q+1$	3	$448 \leqslant N \leqslant 576$
	$2q+3$	5	$256 \leqslant N \leqslant 816$
	$4q+1$	5	$256 \leqslant N \leqslant 816$
2^{12}	$2q+1$	3	$1920 \leqslant N \leqslant 2176$
	$2q+3$	5	$1536 \leqslant N \leqslant 2560$
	$2q+5$	7	$896 \leqslant N \leqslant 3200$
	$4q+1$	5	$1536 \leqslant N \leqslant 2560$
	$4q+3$	7	$896 \leqslant N \leqslant 3200$
	$6q+1$	7	$896 \leqslant N \leqslant 3200$
2^{14}	$2q+1$	3	$7936 \leqslant N \leqslant 8448$
	$2q+3$	5	$7168 \leqslant N \leqslant 9216$
	$2q+5$	7	$5888 \leqslant N \leqslant 10496$
	$2q+7$	9	$4096 \leqslant N \leqslant 12288$
	$2q+9$	11	$1792 \leqslant N \leqslant 14592$
	$4q+1$	5	$7168 \leqslant N \leqslant 9216$
	$4q+3$	7	$5888 \leqslant N \leqslant 10496$
	$4q+5$	9	$4096 \leqslant N \leqslant 12288$
	$4q+7$	11	$1792 \leqslant N \leqslant 14592$
	$6q+1$	7	$5888 \leqslant N \leqslant 10496$
	$6q+3$	9	$4096 \leqslant N \leqslant 12288$
	$6q+5$	11	$1792 \leqslant N \leqslant 14592$
	$8q+1$	9	$4096 \leqslant N \leqslant 12288$
	$8q+3$	11	$1792 \leqslant N \leqslant 14592$
	$10q+1$	11	$1792 \leqslant N \leqslant 14592$

Fig. 1. Computing table of the new bound.

6. Conclusion

We can also apply our improvement of the Weil's bound to other code described with the trace operator, for example the code described by J. Wolfmann in his paper [7].
The construction of the second section and the application of the Deligne bound to trace-equation can be generalized, today we can extend the construction to several variables, namely for any extension of finite field, and furthermore we can take any-one prime integer for the characteristic of the field, but other hypothesis appear over the degree of the polynomial considered.

References

[1] P. Deligne, "La conjecture de Weil I", *Institut Hautes Etudes Sci. Publ. Math.*, N° 43, (1974).

[2] P. Kumar and O. Moreno, "Minimum distance bounds for cyclic codes and Deligne's theorem", preprint.

[3] P. Kumar and O. Moreno, "Prime-phase Sequences with Periodic Correlation Properties Better than Binary Sequences", *IEEE Trans. Inform. Theory*, vol. 37, 3, May 1991.

[4] R. Lidl and H. Niederreiter, *Finite Fields*, Encyclopædia of Mathematics and its applications 20, Addison-Wesley, Reading (1983).

[5] F.J. MacWilliams and N.J.A. Sloane, *The theory of Error-correcting codes*, North-Holland, Amsterdam (1977).

[6] J. Wolfmann, " New bounds on cyclic codes from algebraic curves", *Lectures Notes in Computer Science*, -388- pp 47-62, (1989).

[7] J. Wolfmann, "Polynomial description of binary linear codes and related properties", *A.A.E.C.C.*, 2, pp119-138 (1991).

ON GENERALIZED BENT FUNCTIONS
Sur les fonctions courbes généralisées

P. Langevin
University of Toulon and Var, La Garde, France

Abstract In this paper, we compare binary bent functions and the generalized bent functions from the metric and degree point of view. We give an upper bound on the covering radius of the affine Reed-Muller codes defined over a finite and commutative ring. The paper also gives a bound on the degree of a generalized bent function.

Résumé Dans cet article, nous comparons les fonctions courbes binaires et les fonctions courbes généralisées des points de vue métrique et du degré. Nous donnons une majoration du rayon de recouvrement des codes de Reed-Muller affines construits sur des anneaux finis et commutatifs. Une majoration du degré des fonctions courbes généralisées termine l'article.

1 Introduction

In his paper "On bent functions" [7], O. Rothaus studied functions from $\mathbb{F}_2{}^m$ onto $\mathbb{F}_2$ whose the Fourier transform have constant magnitude. In fact, such functions exist if and only if m is an even integer. In this case, these functions are at maximal distance from the first order Reed-Muller code and this may be the reason why he named them "bent functions". Moreover, he proved that for any integer t within the range $[2, \frac{m}{2}]$, there are bent functions of degree exactly t.

Later, the paper [4] of P. V. Kumar, R. A. Scholtz and L. R. Welch defined generalized bent functions as the maps from $(\mathbb{Z}/q\mathbb{Z})^m$ into $\mathbb{Z}/q\mathbb{Z}$ such that their Fourier transform with respect to an one-to-one character have also constant magnitude. As in the binary case, the non-degenerate quadratic forms provide a family of generalized bent functions. The authors gave precise existence criterions and built explicitely some of them in all the cases for which they exist.

Here, we want to compare the bent functions and the generalized bent functions both from the metric point of view and the algebraic point of view.

- Are the generalized bent functions really bent functions ?

- What can be said about the degree of generalized bent functions ?

2 Affine Reed-Muller Codes

Let A be a commutative and finite ring. The full space of maps from A^m into A is a metric space when equipped with the distance function

$$\operatorname{dist}(f,g) = \operatorname{wt}(f-g),$$

where $\operatorname{wt}(h) = \sharp\{x \in A^m \mid f(x) \neq 0\}$.

Here, the affine or first order Reed-Muller code with m variables, denoted by $\mathrm{RM}_A(m)$, is identified with the A-module of the affine maps from A^m into A. The covering radius of $\mathrm{RM}_A(m)$ is the smallest integer r such that the union of spheres centered at the points of $\mathrm{RM}_A(m)$ cover the full space of functions, let us denote it by $\rho_A(m)$:

$$\rho_A(m) = \sup_f \operatorname{dist}(f, \mathrm{RM}_A(m)) = \sup_f \inf_{h\in \mathrm{RM}_A(m)} \operatorname{dist}(f,h).$$

3 Generalized Bent Functions

For each commutative group G, the set of homomorphisms from G into the group $\mathbb{C}^*$ is also a group. This group, denoted by $\hat{G}$, is said to be the character group (or dual group) of G. When A is a commutative ring, one distinguishes between two types of characters : the additive and multiplicative characters depending on whether reference is made to the additive or multiplicative structure of the ring A.

Let χ be an additive character of the ring $\mathbb{Z}/q\mathbb{Z}$. We define the Fourier transform of a map f from $(\mathbb{Z}/q\mathbb{Z})^m$ into $\mathbb{Z}/q\mathbb{Z}$ at the point z with respect to χ by :

$$\hat{f}_\chi(z) = \sum_{x\in(\mathbb{Z}/q\mathbb{Z})^m} \chi(f(x) - z.x),$$

where $z.x$ is the usual dot product over $(\mathbb{Z}/q\mathbb{Z})^m$. We can notice that the map $\chi_z : x \mapsto \chi(z.x)$ is a character of the module $(\mathbb{Z}/q\mathbb{Z})^m$, and that the element $\hat{f}_\chi(z)$ is the usual Fourier transform of the function $\chi \circ f$ (denoted by f_χ) at the character χ_z.

A generalized bent function of dimension m is a map f from $(\mathbb{Z}/q\mathbb{Z})^m$ to $\mathbb{Z}/q\mathbb{Z}$ such that there exists a one-to-one character χ of $(\mathbb{Z}/q\mathbb{Z}, +)$ which satisfies :

$$\|\hat{f}_\chi(z)\| = \sqrt{q^m}, \quad \forall z \in (\mathbb{Z}/q\mathbb{Z})^m.$$

One can prove (see [4]) that this definition does not depend on the choice of the one-to-one character χ. If f is a boolean bent function then it is a generalized bent function and there exists an another boolean function g such that :

$$\hat{f}_\chi(z) = \chi(g(z))\sqrt{2^m}$$

But in the general case, this is not true and we say that a generalized bent function f is regular at the point z if there exists a q-root of unity $\lambda_z \in \mathbb{C}$ such that

$$\hat{f}_\chi(z) = \lambda_z \sqrt{q^m}.$$

A generalized bent function which is regular at every point of $(\mathbb{Z}/q\mathbb{Z})^m$ is said to be a regular generalized bent function.

4 Upper bound on $\rho_A(m)$

Let us suppose A is a finite and commutative ring with q elements. An additive character χ of A is said non-degenerate if the greatest ideal contained in its kernel is (0). For example, if A is a field then all the non-trivial characters are non-degenerate. In the case where A is the ring $\mathbb{Z}/q\mathbb{Z}$, there is a one-to-one correspondance between ideals of A and subgroups of A. Thus, a character is non-degenerate if and only if it is a one-to-one homomorphism and so, they are $\phi(q)$ non-degenerate characters, where ϕ is the Euler function. In what follows $n(A)$ denotes the number of non-degenerate characters of A. These characters satisfy :

LEMMA 1. *If χ is a non-degenerate character of A, then*

$$\sum_{a \in A^m} |\hat{f}_\chi(a)|^2 = q^{2m}.$$

Proof. Since χ is a non-degenerate character, the map $a \mapsto \chi_a$ decribes the full set of characters of A^m when a runs over A^m. Hence the lemma follows from the Parseval identity :

$$\sum_{\psi \in \widehat{A^m}} |\hat{f}(\psi)|^2 = q^m \sum_{x \in A^m} |f(x)|^2 = q^{2m},$$

where $\hat{f}(\psi)$ is the usual Fourier transform of the map f at the character ψ. □

For any map f from A^m into A, let $S(f,a,b)$ denote the sums :

$$S(f,a,b) = \sum_{\chi \neq 1} \hat{f}_\chi(a)\bar{\chi}(b).$$

It is easy to prove that these exponentials sums have the properties :

$$\sum_{b \in A} S(f,a,b) = 0 \tag{1}$$

and the distance between f and the affine map $a.x + b$ is equal to

$$(q-1)q^{m-1} - \frac{1}{q}S(f,a,b). \tag{2}$$

This can be used to show :

THEOREM 2. *The covering radius of the affine Reed-Muller code with m variables over the ring A satisfies the inequality*

$$\boxed{\rho_A(m) \leq (q-1)q^{m-1} - \frac{\sqrt{n(A)}}{q(q-1)}q^{m/2}}$$

Where $n(A)$ is the number of non-degenerated characters of A.

Here we want to point out that this upper bound depend on with the ring structure of A. On one hand, the additive characters reflect the additive structure and on the other hand the notion of non degenerate character reflects the multiplicative structure.

Proof. Let us consider the sum $\Sigma(f)$ defined by :

$$\Sigma(f) = \sum_{b \in A, a \in A^m} |S(f,a,b)|^2.$$

One has :

$$\Sigma(f) = \sum_{a \in A^m} \sum_{\chi \neq 1, \psi \neq 1} \widehat{f_\chi}(a)\widehat{f_{\bar{\psi}}}(a) \sum_{b \in A} \chi(b)\bar{\psi}(b)$$

$$= q \sum_{\chi \neq 1} \sum_{a \in A^m} |\widehat{f_\chi}(a)|^2$$

and by Lemma 1 we get :

$$\Sigma(f) \geq q\,n(A)\,q^{2m}.$$

Looking at the average of the terms of this sum, we deduce that there exists an ordered pair $(a, b) \in A^m \times A$ such that :

$$|S(f, a, c)| \geq \sqrt{n(A)}q^{m/2}.$$

i. If $S(f, a, b) \geq 0$ then

$$d(f, ax + b) \leq (q-1)q^{m-1} - \frac{\sqrt{n(A)}}{q}q^{m/2}.$$

ii. Otherwise, we use the property (1) to get

$$\sum_{c \neq b} S(f, a, b) \geq \sqrt{n(A)}q^{m/2}.$$

Once more, there is a term of the LHS which is greater or equal to the average. So there exists an element $c \in A$ such that $S(f, a, c) \geq \frac{\sqrt{n(A)}}{(q-1)}q^{m/2}$. Hence :

$$d(f, ax + b) \leq (q-1)q^{m-1} - \frac{\sqrt{n(A)}}{q(q-1)}q^{m/2}.$$

The two inequalities (i) and (ii) lead to the result. □

PROPOSITION 3. *Let f be a perfect non-linear function, that is :*

$$f_\chi \times f_\chi(z) = \sum_{x \in A^m} f_\chi(x) f_{\bar{\chi}}(x + z) = 0, \quad \forall z \in A^m - \{0\}.$$

Then :

$$d(f, \mathrm{RM}_A(m)) \leq (q-1)q^{m-1} - \frac{1}{q\sqrt{q-1}}q^{m/2}.$$

Proof. We have

$$\Sigma(f) = q(q-1)q^{2m}$$

and the proof follows as in the preceding theorem. □

5 Distance between a bent function and $RM_{\mathbb{F}_p}(m)$

From now on, we suppose that p is prime number, $p \neq 2$. In her paper [5], K. Nyberg calculates the distance from a generalized bent function to the affine Reed-Muller code over $\mathbb{F}_p$. Here, we propose to show the usefulness of property (2) to calculate the distance between a generalized bent function and any hyperplane. For the the different properties that we use in sections 5 and 6, one can refer to [2] for the Gaussian quadratic sums and to [3] for the arithmetic of the cyclotomic fields.

Let f be a generalized bent function over $\mathbb{F}_p$. First, property (2) states that the distance from f to the affine map $x \mapsto x.u + b$ is

$$d(f, u.x + b) = (p-1)p^{m-1} - \frac{1}{p}\mathrm{S}(f, u, b)$$

but here, $\mathrm{S}(f, u, b)$ takes the form of a trace :

$$\mathrm{S}(f, u, b) = \sum_{\psi \neq 1} \hat{f}_\psi(u)\bar{\psi}(b) = \mathrm{tr}_{\mathbb{Q}(\zeta_p)/\mathbb{Q}}(\hat{f}_\chi(u)\bar{\chi}(b)) \tag{3}$$

where χ is a non-trivial character and ζ_p is a primitive $(p-1)^{\text{th}}$ root of unity. Because the automorphism $z \mapsto \bar{z}$ is in the decomposition group of p in $\mathbb{Q}(\zeta_p)$, it follows, see e.g. [4], that :

$$\hat{f}_\chi(u) = \lambda(u)\sqrt{p^m}$$

where $\lambda(u)$ is a root of unity. When m is even, $\lambda(u)$ is an element of $\mathbb{Q}(\zeta_p)$, and when m is odd, the Gaussian quadratic sum

$$\sum_{k=0}^{p-1} \zeta_p^{k^2} = \theta\sqrt{p}$$

where $\theta = \begin{cases} 1, & \text{if } p = 1[4]\,; \\ i, & \text{if } p = 3[4]\,; \end{cases}$

shows that $\frac{\lambda(u)}{\theta}$ is a root of unity in $\mathbb{Q}(\zeta_p)$.

Now, we are able to evaluate the trace (3), and deduce the complete distance distribution between a generalized bent function and the set of the hyperplanes.

Proposition 4. *Suppose that m is even. When b runs over $\mathbb{F}_p$ then*

$$\mathrm{S}(f, u, b) = \begin{cases} -p^{\frac{m}{2}} & , p-1 \text{ times} \\ (p-1)p^{\frac{m}{2}}, & 1 \text{ time} \end{cases}$$

if f is regular at u, and

$$\mathrm{S}(f, u, b) = \begin{cases} p^{\frac{m}{2}} & , p-1 \text{ times} \\ -(p-1)p^{\frac{m}{2}}, & 1 \text{ time} \end{cases}$$

if f is not regular at u.

Proof. We have to evaluate (3) :

$$S(f,u,b) = \mathrm{tr}_{\mathbb{Q}(\zeta_p)/\mathbb{Q}}(\hat{f}_\chi(u)\bar{\chi}(b)) \\ = p^{\frac{m}{2}}\,\mathrm{tr}_{\mathbb{Q}(\zeta_p)/\mathbb{Q}}(\lambda(u)\zeta_p^{-b})$$

- If f is regular at u, there exists one and only one b such that

$$\lambda(u) = \zeta_p^b$$

and we get

$$\mathrm{tr}_{\mathbb{Q}(\zeta_p)/\mathbb{Q}}(\lambda(u)\zeta_p^{-b}) = \begin{cases} p-1, & \text{if } \lambda(u) = \zeta_p^b; \\ -1, & \text{otherwise.} \end{cases}$$

- If f is not regular at u there exists one and only one b such that

$$-\lambda(u) = \zeta_p^b$$

and we get

$$\mathrm{tr}_{\mathbb{Q}(\zeta_p)/\mathbb{Q}}(\lambda(u)\zeta_p^{-b}) = \begin{cases} 1-p, & \text{if } \lambda(u) = -\zeta_p^b; \\ 1, & \text{otherwise.} \end{cases}$$

□

PROPOSITION 5. *Suppose that m is odd, then for each b in $\mathbb{F}_p$, we have :*

$$S(f,u,b) = \begin{cases} -p^{\frac{m+1}{2}}, & \frac{p-1}{2} \text{ times}; \\ 0, & 1 \text{ time}; \\ p^{\frac{m+1}{2}}, & \frac{p-1}{2} \text{ times}; \end{cases}$$

Proof. Denote by K the field $\mathbb{Q}(\theta\sqrt{p})$ and by L the field $\mathbb{Q}(\zeta_p)$. We know the following field inclusions :

$$\mathbb{Q} \subseteq K \subseteq L$$

Let $\mathcal{R}$ and $\mathcal{N}$ be respectively the sets of squares and non-squares in $\mathbb{F}_p^*$, and let us set :

$$R = \sum_{s\in\mathcal{R}} \zeta_p^s \qquad N = \sum_{s\in\mathcal{N}} \zeta_p^s$$

From the Gaussian quadratic sum we deduce that :

$$R = \frac{\theta\sqrt{p}-1}{2} \qquad N = \frac{-\theta\sqrt{p}-1}{2}$$

Moreover the Galois group of L/K is the set of automorphisms σ_s which send ζ_p to ζ_p^s, where $s \in \mathcal{R}$. And so :

$$\mathrm{tr}_{L/K}(\zeta_p^s) = \begin{cases} R, & \text{if } s \in \mathcal{R}; \\ \frac{p-1}{2}, & \text{if } s = 0; \\ N, & \text{if } s \in \mathcal{N}. \end{cases}$$

Now, we are able to calculate $\mathrm{S}(f,u,b)$:

$$\begin{aligned} \mathrm{S}(f,u,b) &= p^{[\frac{m}{2}]}\mathrm{tr}_{L/\mathbb{Q}}(\lambda(u)\sqrt{p}\zeta_p^{-b}) \\ &= p^{[\frac{m}{2}]}\mathrm{tr}_{K/\mathbb{Q}}(\theta\sqrt{p}\mathrm{tr}_{L/K}(\frac{\lambda(u)}{\theta}\zeta_p^{-b})) \end{aligned}$$

Let $\epsilon = 1$ or -1, according to whether $\frac{\lambda(u)}{\theta}$ is or is not a p-root of unity.

- There are exactly $\frac{p-1}{2}$ elements b such that

$$\epsilon\frac{\lambda(u)}{\theta}\zeta_p^{-b} \in \zeta_p^{\mathcal{R}}$$

  For these elements, we have

$$\mathrm{S}(f,u,b) = \epsilon\theta^2 p^{[\frac{m}{2}]+1}.$$

- There are exactely $\frac{p-1}{2}$ elements b such that

$$\epsilon\frac{\lambda(u)}{\theta}\zeta_p^{-b} \in \zeta_p^{\mathcal{N}}$$

  For these elements, we have

$$\mathrm{S}(f,u,b) = -\epsilon\theta^2 p^{[\frac{m}{2}]+1}.$$

- There is one and only one b such that

$$\epsilon\frac{\lambda(u)}{\theta}\zeta_p^{-b} = 1$$

  In this case,

$$\mathrm{S}(f,u,b) = 0.$$

□

THEOREM 6. *The covering radius of the affine Reed-Muller code with m variables over the prime field $\mathbb{F}_p$ satisfies the inequalities*

$$\boxed{(p-1)p^{m-1} - p^{\lceil\frac{m-1}{2}\rceil} \leq \rho_{\mathbb{F}_p}(m) \leq (p-1)p^{m-1} - \frac{1}{\sqrt{p-1}}p^{\frac{m}{2}-1}}$$

The case where $p = 2$ leads to the Norse bound.

Proof. The RHS is a straightforward application of the section 4. The lower bound is obtained by building generalized bent functions that are not regular, for example :

$$f(x_1, x_2, \ldots, x_m) = sx_1^2 + x_2^2 + \ldots + x_m^2,$$

where s is not a square of $\mathbb{F}_p$. □

This permits us to answer partly the first question. A generalized bent function over a prime field is at maximal distance from the affine Reed-Muller codes $\mathrm{RM}_{\mathbb{F}_p}(m)$ when $m = 1$ and when $m = 2$ provided it is not regular.

6 Degree of generalized bent functions

Once again, we restrict to the prime case. The reason for is that we do not know if the generalized bent functions over $\mathbb{Z}/q\mathbb{Z}$ are polynomial or not.

Let S be a subset of $\mathbb{F}_p{}^m$. We denote by 1_S the caracteristic function of S, i.e. the map with codomain $\mathbb{F}_p$ defined by :

$$1_S(x) = \begin{cases} 1, & \text{if } x \in P; \\ 0, & \text{otherwise.} \end{cases}$$

PROPOSITION 7. *Let f be a generalized bent function and let S be an affine subspace of dimension k, $k \geq \frac{m}{2} + 2$. Then f is orthogonal to 1_S.*

Proof. Let $b \in \mathbb{F}_p$, and denote by $n(b)$ the number of elements s in S which satisfy $f(x) = b$.

as usual, we calculate $n(b)$ by means of characters :

$$pn(b) = \sum_{s \in S} \sum_{\chi \in \widehat{\mathbb{F}_p}} \chi(f(s) - b)$$

so

$$n(b) = p^{k-1} - \frac{1}{p} \sum_{s \in S} \sum_{\chi \neq 1} \chi(f(s) - b)$$

and the Poisson formula gives :

$$\sum_{s\in S}\chi(f(s)) = p^{k-m}\sum_{t\in S^\perp}\hat{f}_\chi(t)$$

The ideal decomposition of (p) in $\mathbb{Z}[\zeta_p]$ is $(p) = P^{(p-1)}$ where P is the prime principal ideal $(1-\zeta_p)$. Since the automorphism $z \mapsto \bar{z}$ is in the decomposition group of P, the P-adic valuation of $\hat{f}_\chi(s)$ is :

$$\nu_P(\hat{f}_\chi(s)) = \frac{(p-1)m}{2}$$

and

$$\nu_P(p^{k-m-1}\sum_{t\in S^\perp}\hat{f}_\chi(t)\bar{\chi}(b)) \geq (k-m-1)(p-1) + \frac{(p-1)m}{2}$$

$$\geq (p-1)(k-\frac{m}{2}-1)$$

and because the dimension of S is great enough :

$$\nu_P(\hat{f}_\chi(s)) > (p-1)$$

Therefore, the integer $n(b)$ has a positive p-adic valuation. On an other hand,

$$f.1_S = \sum_{b\in\mathbb{F}_p} n(b) = 0 \pmod{p}$$

and so, the vectors f and 1_S are orthogonal to each other. □

This combined with the proposition of P. Delsarte [1] which states that the caracteristic functions of affine spaces of dimension k span the Reed-Muller code of order $(p-1)(m-k)$, give :

COROLLARY 8. *If f is a generalized bent function then*

$$\deg(f) \leq \begin{cases} (p-1)(\frac{m+4}{2})-1, & \text{if } m \text{ is even ;} \\ (p-1)(\frac{m+3}{2})-1, & \text{if } m \text{ is odd.} \end{cases}$$

Proof. r be the smallest integer than $\frac{m}{2}+1$. Assume $m = 2t+1$ or $m = 2t$, then $r = t+1$.The function f lies in the dual of the space generated by the caracteristic functions of the affine spaces of dimensions r :

$$f \in \mathrm{RM}_{\mathbb{F}_p}((p-1)(m-r), m)^\perp$$

But the dual of $\mathrm{RM}_{\mathbb{F}_p}((p-1)r, m)$ is $\mathrm{RM}_{\mathbb{F}_p}((p-1)r-1, m)$ (See [6]) and thus

$$\deg(f) \leq (p-1)t + 2p - 3.$$

□

7 References

[1] P. Delsarte, J. M. Goethals and F. J. Mac Williams
"On generalized Reed-Muller Codes and their relatives",
Info. and Control, vol. **16**,
pp 403–443, 1970.

[2] L. K. Hua
"Introduction to number theory",
Springer-Verlag, 1982.

[3] K. Ireland, M. Rosen
"A Classical Introduction to Modern Number Theory",
Graduate Texts in Mathematics, vol. **84**,
Springer-Verlag, 1990.

[4] P. V. Kumar, R. A. Scholtz and L. R. Welch
"Generalized bent functions and their properties",
J. C. T, Series A **40**,
pp 90–107, 1985.

[5] K. Nyberg
"Constructions of bent functions and differerence sets",
Eurocrypt, 1989.

[6] F. J. Macwilliams, N. J. A. Sloane
"The Theory of Error-Correcting Codes",
North-Holland Mathematical Library,
Amsterdam, 1977.

[7] O. S. Rothaus
"On bent functions",
J. C. T. **20**,
pp 300–305, 1976.

BOUNDS FOR SELF-COMPLEMENTARY CODES AND THEIR APPLICATIONS

V.I. Levenshtein
Russian Academy of Sciences, Moscow, Russia

Abstract

Self-complementary codes in the Hamming space, i.e., binary codes that together with any vector contain its complement, are considered. The bound on the size of a self-complementary code with a given minimum distance d presented here is in general better than the corresponding bound for arbitrary binary codes in the Hamming space. Some applications of this bound for estimating the minimum distance of self-dual binary codes, the cross-correlation of arbitrary binary codes, the modulus of sums of Legendre symbols of polynomials, and some parameters of randomness properties of binary codes, are given.

1 Bounds for codes in the Hamming space

The Hamming space F_2^n is referred to as the set of binary vectors $a = (a_1, \ldots, a_n)$ where a_i is 1 or -1, $i = 1, \ldots, n$, with the distance $d(a,b)$ between vectors $a = (a_1, \ldots, b_n)$ and $b = (b_1, \ldots, b_n)$ being equal to the number of places where they differ. An arbitrary code $C \subseteq F_2^n$ is characterized by the two basic parameters : the minimum distance

$$d(C) = \min_{a,b \in C, a \neq b} d(a,b)$$

and the dual distance $d'(C)$ which is defined $[D]$ in general case using the Krawtchouk polynomials

$$K_k^n(x) = \sum_{j=0}^{k}(-1)^j \binom{x}{j}\binom{n-x}{k-j}, \quad k = 0,1,\dots,n. \tag{1}$$

If $\mathcal{B}(C) = (\mathcal{B}_0(C), \mathcal{B}_1(C), \dots, \mathcal{B}_n(C))$ with $\mathcal{B}_i(C)$ being equal to the number of ordered pairs of points of C at Hamming distance i from one another then the minimum distance $d(C)$ of C is the lowest $i, 1 \le i \le n$, such that $\mathcal{B}_i(C) \ne 0$. Let $\mathcal{B}'(C) = (\mathcal{B}'_0(C), \mathcal{B}'_i(C), \dots, \mathcal{B}'_n(C))$ be the MacWilliams transform of $\mathcal{B}(C)$, i.e.

$$\mathcal{B}'_i(C) = \frac{1}{|C|^2}\sum_{j=0}^{n}\mathcal{B}_j(C)K_i^n(j), \quad i = 0,1,\dots,n.$$

The dual distance $d'(C)$ of C is the lowest $i, 1 \le i \le n$, such that $\mathcal{B}'_i(C) \ne 0$. In the case of a linear code C the dual distance $d'(C)$ equals (see [MS]) the minimum distance of the dual code

$$C' = \left\{ a \in F_2^n : (a,b) = \sum_{i=1}^{n} a_i b_i = 0 \quad \text{for any } b \in C \right\}$$

The well-known Rao and Hamming bounds can be written in the following form :

$$H_n(d'(C)) \le |C| \le \frac{2^n}{H_n(d(C))}. \tag{2}$$

where

$$H_n(d) = \begin{cases} \sum_{i=0}^{k}\binom{n}{i} & \text{if } d = 2k+1, \\ 2\sum_{i=0}^{k}\binom{n-1}{i} & \text{if } d = 2k+2. \end{cases}$$

It turns out that a second pair of similar bounds is valid. Let $d_k(n)$ be the lowest root of the equation $K_k^n(x) = 0$ (it is assumed that $d_0(n) = \infty$ and $d_{n+1}(n) = 0$). It can be shown that $d_k(n-2) < d_k(n-1) < d_{k-1}(n-2)$ for $1 \le k \le n-1$. Let

$$A_{n,k}(x) = \sum_{i=0}^{k-1}\binom{n}{i} - \binom{n}{k}\frac{K_{k-1}^{n-1}(x-1)}{K_k^n(x)}$$

and

$$A_n(x) = \begin{cases} A_{n,k}(x) & \text{if } d_k(n-1)+1 \le x \le d_{k-1}(n-2)+1, \\ 2A_{n-1,k}(x) & \text{if } d_k(n-2)+1 \le x \le d_k(n-1)+1. \end{cases}$$

Then for any code $C \subseteq F_2^n$, the following inequalities hold :

$$\frac{2^n}{A_n(d'(C))} \le |C| \le A_n(d(C)). \tag{3}$$

The upper bound in (3) was published by the author [L1] in 1978 (in Russian). It was obtained by using certain polynomials in the Delsarte inequality for codes [D]. In 1980 Sidelnikov [S2] proved that these polynomials are in certain sense optimal. The lower bound in (3) is a new one will be submitted to the IEEE Transactions on Information Theory. It can be obtained by using the dual polynomials in the Delsarte inequality for designs [D]. There are many cases when these bounds are tight (see L3,L4]). The upper bound in (3) is better than the McEliece-Rodemich-Rumsey-Welch bound [MRRW] for any n and any $C \subseteq F_2^n$ with $d(C) \leq n/2$ and also in some asymptotical processes. However in the case when the minimum distance grows linearly with n these bounds coincide asymptotically.

2 Bounds for self-complementary codes

A code $C \subseteq F_2^n$ is called self-complementary if for any $a \in C$ the complementary vector $-a$ belongs to C as well. A question is how it is possible to improve the bounds (2) and (3) using the specific properties of self-complementary codes. The problem can be solved by considering the folded (or projective) Hamming space PF_2^n whose elements are pairs of complementary vectors of F_2^n and whose distance is the minimum Hamming distance between representatives of these pairs. After identifying complementary vectors, an arbitrary code $C \subseteq F_2^n$ can be considered as the code $\bar{C}$ in PF_2^n. Obviously $|C| = 2|\bar{C}|$ if C is a self-complementary code and $|C| = |\bar{C}|$ if C does not contain complementary vectors. We will also characterize an arbitrary code $C \subseteq F_2^n$ by the value

$$\rho(C) = \max_{a,b\in C, a\neq b} \left|\sum_{i=1}^{n} a_i b_i\right| = \max_{a,b\in C, a\neq b} |n - 2d(a,b)|.$$

Since

$$d(-a,b) = n - d(a,b) \quad \text{for any } a,b \in F_2^n \tag{4}$$

(and hence $n - 2d(-a,b) = -(n-2d(a,b))$) for any code $C \subseteq F_2^n$ with $\rho(C) < n$, the minimum distance of the code $\bar{C}$ in PF_2^n is equal to $\frac{1}{2}(n-\rho(C))$. The folded Hamming space PF_2^n, as the Hamming space F_2^n, are examples of so-called polynomial metric spaces (P- and Q-polynomial association schemes [D, BI, BCN]). For such spaces the author published in Russian [L1,L3] a general upper bound on the maximum size of a code with a given minimum distance. The special case of the bound for PF_2^n was also published without proof in English in [L2]. However afterwards some weaker results were published by other authors. For this reason I have decided to publish a new direct proof of the following bound for a self-complementary code C :

$$|C| \leq 2B_n(d(C)) \tag{5}$$

where

$$B_n(x) = \begin{cases} \sum_{i=0}^{k-1} \binom{n}{2i} - \binom{n}{2k} \frac{K^{n-2}_{2k}(x-1)}{K^n_{2k}(x)} & \text{if } d_{2k}(n-2) \le x-1 < d_{2k-1}(n-2), \\ \sum_{i=0}^{k-1} \binom{n}{2i+1} \frac{K^{n-2}_{2k-1}(x-1)}{K^n_{2k+1}(x)} - \binom{n}{2k+1} & \text{if } d_{2k+1}(n-2) \le x-1 < d_{2k}(n-2). \end{cases} \tag{6}$$

It is possible to prove (see [L4]) that for the intervals $d_{2k+1}(n-2)+1 < d(C) < d_{2k}(n-2)+1, k \ge 2$, the bound (5) is slightly better than the one in [L2], and that for any $d, 1 \le d \le n/2$, $2B_n(d) \le A_n(d)$. In conclusion some applications of the bound are presented.

It is very useful to apply the functional approach to explain some properties of the Krauwchouk polynomials. Consider $a = (a_1, \ldots, a_n) \in F_2^n$ as an variable vector and determine the following sets of functions being factorizations of coordinates of the vector :

$$V_i = \{a_{j_1} a_{j_2} \ldots a_{j_i} : 1 \le j_1 < \ldots < j_i \le n\}, i = 1, \ldots, n$$

and $V_0 = \{1\}$. It is clear that $|V_i| = \binom{n}{i}$. All functions $v \in \bigcup_{i=0}^n V_i$ form an orthonormal system with respect to the inner product

$$< v, u >= 2^{-n} \sum_{a \in F_2^n} v(a)u(a).$$

It is easy to verify that for any $a \in F_2^n$ and $b \in F_2^n$

$$K_i^n(d(a,b)) = \sum_{v \in V_i} v(a)v(b). \tag{7}$$

This gives rise to the orthogonality conditions

$$\sum_{d=0}^{n} K_i^n(d) K_j^n(d) \binom{n}{d} = 2^n \binom{n}{i} \delta_{ij} \tag{8}$$

and to the property

$$K_i^n(x) K_j^n(x) \equiv \sum_{l=0}^{n} p_{i,j}^l K_l^n(x) \; (mod K_{n+1}^n(x)) \quad \text{with } p_{i,j}^l \ge 0 \tag{9}$$

($K_{n+1}^n(x)$ is assumed to be $x(x-1)\ldots(x-n)$). Since

$$v(-a) = (-1)^i v(a) \quad \text{for any } v \in V_i \tag{10}$$

from (4) and (7) it follows that

$$K_i^n(d) = (-1)^i K_i^n(n-d) \quad \text{for any } i, d = 0, 1, \ldots, n$$

and hence

$$K_i^n(x) = (-1)^i K_i^n(n-x) \quad \text{for any } i = 0,1,\ldots,n \text{ and real } x. \tag{11}$$

Taking account of the fact that

$$\mathcal{B}_i'(C) = \frac{1}{|C|^2} \sum_{a,b \in C} K_i^n(d(a,b))$$

and (7) and (10) we have $\mathcal{B}_i'(C) = 0$ for any odd i and any self-complementary code C. This gives the following corollary.

Lemma 1 *For any self-complementary code $C \subseteq F_2^n$ the dual distance is even and*

$$|C| \geq 2\sum_{i=0}^{k-1} \binom{n-1}{i} \quad \text{if } d'(C) = 2k. \tag{12}$$

Below we use some known properties of discrete systems of orthogonal polynomials [Sz]. Let $\mathcal{R}$ be a set of $N+1$ nonnegative integers, let $r(d)$ and $w(d)$ be two functions with positive values on $\mathcal{R}$ and let $t(d)$ be a strictly monotone real function on the interval $\left[\min_{d\in\mathcal{R}} d, \max_{d\in\mathcal{R}} d\right]$. Then polynomials $P_i(y)$ in a real variable y of degree $i, i = 0,1,\ldots,N$ are uniquely determined by the following (orthogonality) conditions :

$$r(i) \sum_{d\in\mathcal{R}} P_i(t(d))P_j(t(d))w(d) = \delta_{i,j} \tag{13}$$

We denote by $P_{N+1}(y)$ the polynomial $\prod_{d\in\mathcal{R}} (y-t(d))$ which is orthogonal with respect to (13) to any polynomial $P_i(y)$, $i = 0,1,\ldots,N$. For this reason if for a polynomial $f(y)$,

$$f(y) \equiv \sum_{i=0}^{N} f_i P_i(y) \quad (mod P_{N+1}(y)),$$

then

$$f_i = r(i) \sum_{d\in\mathcal{R}} f(t(d))P_i(t(d))w(d). \tag{14}$$

Any polynomial $P_i(y), i = 1,\ldots,N$ has i simple zeroes inside interval $\left[\min_{d\in\mathcal{R}} t(d), \max_{d\in\mathcal{R}} t(d)\right]$, and the successive zeroes of $P_i(y)$ lie between ones of $P_{i+1}(y)$. Denote by m_k the rate of polynomials $P_k(y)$ and $P_{k+1}(y)$ leading coefficients. For the system $\{P_i(y), i = 0,1,\ldots,N\}$ the following recurrence relation holds :

$$r_i m_i P_{i+1}(y) = r_i(y+l_i)P_i(y) - r_{i-1}m_{i-1}P_{i-1}(y) \tag{15}$$

where $r_i = r(i), m_{-1} = r_{-1} = 0, P_{-1}(t) \equiv 0$, and the following Christoffel-Darboux formulas are valid :

$$\sum_{i=0}^{k} r_i P_i(x) P_i(y) = r_k m_k \frac{P_{k+1}(x)P_k(y) - P_k(x)P_{k+1}(y)}{x-y}, \quad \text{if } x \neq y, \tag{16}$$

$$\sum_{i=0}^{k} r_i P_i(y) P_i(y) = r_k m_k (P'_{k+1}(y)P_k(y) - P'_k(y)P_{k+1}(y)).$$

The latter formula shows that

$$m_k \frac{P_{k+1}(y)}{P_k(y)} \quad \text{increases with } y \text{ if } P_k(y) \neq 0. \tag{17}$$

As main objects we consider the decreasing substitution function $t_n(x) = n - 2x$ and polynomials $Q_i(z)$ in a real z defined by

$$\binom{n}{i} Q_i^n(t_n(x)) = \binom{n}{i} Q_i^n(n-2x) = K_i^n(x), i = 0, 1, \ldots, n \tag{18}$$

($Q_{n+1}^n(z)$ is assumed to be $(z - t_n(0))(z - t_n(1)) \ldots (z - t_n(n))$). If $z_k(n)$ is the largest root of $Q_k^n(z)(z_0(n) = -\infty)$ then

$$z_k(n) = n - 2d_k(n), k = 0, 1, \ldots, n+1. \tag{19}$$

For the system $\{Q_i^n(z)\}$ there hold the following four properties : the orthogonality conditions

$$\binom{n}{i} \sum_{d=0}^{n} Q_i(t_n(d)) Q_j(t_n(d)) \binom{n}{d} 2^{-n} = \delta_{i,j}; \tag{20}$$

the recurrence relation

$$(n-i) Q_{i+1}^n(z) = z Q_i^n(z) - i Q_{i-1}^n(z), Q_0^n(z) = 1; \tag{21}$$

the properties

$$Q_i^n(z) Q_j^n(z) \equiv \sum_{l=0}^{n} p_{i,j}^l Q_l^n(z) (mod Q_{n+1}^n(z)) \quad \text{with } p_{i,j}^l \geq 0 \tag{22}$$

and

$$Q_i^n(-z) = (-1)^i Q_i^n(z) \quad \text{for any } i = 0, 1, \ldots, n+1. \tag{23}$$

By (23) there exist two systems of polynomials $\left\{R_i^{n,0}(z), i = 0, 1, \ldots, \lfloor \frac{n}{2} \rfloor\right\}$ and $\left\{R_i^{n,1}(z), i = 0, 1, \ldots, \lfloor \frac{n-1}{2} \rfloor\right\}$ such that

$$Q_{2i+\alpha}^n(z) = z^\alpha R_i^{n,\alpha}(z^2), i = 0, 1, \ldots, \lfloor \frac{n-\alpha}{2} \rfloor, \alpha \in \{0, 1\}. \tag{24}$$

In particular,

$$Q_0^n(z)=1;Q_1^n(z)=\frac{z}{n};Q_2^n(z)=\frac{z^2-n}{n(n-1)};Q_3^n(z)=\frac{z^3-z(3n-2)}{n(n-1)(n-2)};$$

$$Q_4^n(z)=\frac{z^4-2z^2(3n-4)+3n(n-2)}{n(n-1)(n-2)(n-3)};Q_5^n(z)=\frac{z^5-10z^3(n-2)+z(15n^2-50n+24)}{n(n-1)(n-2)(n-3)(n-4)};$$

$$R_0^{n,0}(z)=1;R_1^{n,0}(z)=\frac{z-n}{n(n-1)};R_2^{n,0}(z)=\frac{z^2-2z(3n-4)+3n(n-2)}{n(n-1)(n-2)(n-3)};$$

$$R_0^{n,1}(z)=\frac{1}{n};R_1^{n,1}(z)=\frac{z-3n+2}{n(n-1)(n-2)};R_2^{n,1}(z)=\frac{z^2-10z(n-2)+15n^2-50n+24}{n(n-1)(n-2)(n-3)(n-4)}.$$

By (20) and (24) the system $\left\{R_i^{n,\alpha}(z), i=0,1,\ldots,\lfloor\frac{n-\alpha}{2}\rfloor\right\}$ satisfy the following orthogonality conditions :

$$\binom{n}{2i+\alpha}\sum_{d=0}^{\lfloor\frac{n-\alpha}{2}\rfloor} R_i^{n,\alpha}((t_n(d))^2)R_j^{n,\alpha}((t_n(d))^2)w(d)=\delta_{i,j} \tag{25}$$

with weight function $w(d)=(t_n(d))^{2\alpha}u_n(d)$ where

$$u_n(d)=\frac{\binom{n}{d}}{2^{n-1}(1+(1-\alpha)\delta_{2d,n})}.$$

On the other hand, the equalities

$$\binom{n-2}{i}Q_i^{n-2}(n-2x)=K_i^{n-2}(x-1), i=0,1,\ldots,n-2,$$

(23) and (8) imply that

$$\binom{n-2}{2i+\alpha}\sum_{d=1}^{\lfloor\frac{n-\alpha}{2}\rfloor} Q_{2i+\alpha}^{n-2}(t_n(d))Q_{2j+\alpha}^{n-2}(t_n(d))(n^2-(t_n(d))^2)u_n(d)=\delta_{i,j}n(n-1)$$

and hence

$$\binom{n-2}{2i+\alpha}\sum_{d=1}^{\lfloor\frac{n-\alpha}{2}\rfloor} R_i^{n-2,\alpha}((t_n(d))^2)R_j^{n-2,\alpha}((t_n(d))^2)\bar{w}(d)=\delta_{i,j}$$

with the weight function

$$\bar{w}(d)=\frac{n^2-(t(d))^2}{n(n-1)}(t(d))^{2\alpha}u_n(d). \tag{26}$$

Using (25), (16) and the equality $n^\alpha R_i^{n,\alpha}(n^2) = 1$ we obtain that the polynomial

$$\sum_{i=0}^{k} R_i^{n,\alpha}(z)\binom{n}{2i+\alpha} = n^\alpha \sum_{i=0}^{k} R_i^{n,\alpha}(z) R_i^{n,\alpha}(n^2)\binom{n}{2i+\alpha}$$

is equal up to a constant factor to the polynomial

$$\frac{R_{k+1}^{n,\alpha}(z) - R_k^{n,\alpha}(z)}{z - n^2}. \tag{27}$$

which, by (25), is orthogonal with respect to the weight function (26) to any polynomial of degree $k-1$ and in particular to the polynomials $R_i^{n-2,\alpha}(z), i = 0, 1, \ldots, k-1$. Hence

$$R_k^{n-2,\alpha}(z) = c\sum_{i=0}^{k} R_i^{n,\alpha}(z)\binom{n}{2i+\alpha}.$$

Since $n^\alpha R_i^{n,\alpha}(n^2) = 1$ and

$$\binom{n-2}{2k+\alpha} n^\alpha R_k^{n-2,\alpha}(n^2) = \binom{n-2}{2k+\alpha} Q_{2k+\alpha}^{n-2}(n) = K_{2k+\alpha}^{n-2}(-1) = \sum_{j=0}^{k}\binom{n}{2j+\alpha}, \tag{28}$$

for any $\alpha \in \{0, 1\}$ we have

$$\binom{n-2}{2k+\alpha} R_k^{n-2,\alpha}(z) = \sum_{i=0}^{k} R_i^{n,\alpha}(z)\binom{n}{2i+\alpha}, k = 0, 1, \ldots, \lfloor\frac{n-\alpha}{2}\rfloor - 1. \tag{29}$$

Notice also that (21), (24) and (29) imply that

$$\binom{n-2}{2k-1+\alpha} zQ_{2k-1+\alpha}^{n-2}(z) = nz^\alpha\left(\binom{n-2}{2k-2+\alpha} R_{k-1}^{n-2,\alpha}(z) + \binom{n-1}{2k-1+\alpha} R_k^{n,\alpha}(z)\right) \tag{30}$$

At last we take account of the fact that, by (23), the coefficients $p_{i,j}^l$ in (22) are equal to zero if $i+j+l$ is odd, and (24) implies that for any $0 \leq i, j \leq \lfloor\frac{n-\alpha}{2}\rfloor$

$$z^\alpha R_i^{n,\alpha}(z) R_j^{n,\alpha}(z) \equiv \sum_{l=0}^{\lfloor n/2\rfloor} p_{2i+1,2j+1}^{2l} R_l^{n,0}(z)\ (mod R_{\lfloor n/2\rfloor+1}^{n,0}(z)). \tag{31}$$

Theorem 1 *Let*

$$D_n(z) = \begin{cases} \sum_{i=0}^{k-1}\binom{n}{2i} - \binom{n-2}{2k-2}\frac{Q_{2k-2}^{n-2}(z)}{Q_{2k}^{n}(z)} & \textit{if } z_{2k-1}(n-2) < z \leq z_{2k}(n-2), \\ \sum_{i=0}^{k-1}\binom{n}{2i+1} - \binom{n-2}{2k-1}\frac{Q_{2k-1}^{n-2}(z)}{Q_{2k+1}^{n}(z)} & \textit{if } z_{2k}(n-2) < z \leq z_{2k+1}(n-2). \end{cases}$$

The function $D_n(z)$ is continuous, grows in z for $0 \le z \le n-2$ and

$$D_n(z_j(n-2)) = \sum_{i=0}^{j} \binom{n-1}{i} \quad \textit{for any } j = 1,2,\ldots,n-1. \tag{32}$$

For any code $C \subseteq F_2^n$ such that $\rho(C) < n$,

$$|C| \le D_n(\rho(C)). \tag{33}$$

Proof. Since $\rho = \rho(C) < n$ then $0 \le \rho \le n-2$. Taking account of that $z_1(n) = 0$ and $z_{n+1}(n) = n$, for any ρ, $0 \le \rho \le n-2$ there exists a unique integer k and a unique number $\alpha \in \{0,1\}$ such that $1 \le 2k + \alpha \le n-1$ with

$$z_{2k-1+\alpha}(n-2) < \rho \le z_{2k+\alpha}(n-2). \tag{34}$$

For any interval (34) we consider the polynomial

$$f^{\rho}(z) = z^{\alpha} \frac{(R_k^{n-2,\alpha}(z) R_{k-1}^{n-2,\alpha}(\rho^2) - R_{k-1}^{n-2,\alpha}(z) R_k^{n-2,\alpha}(\rho^2))^2}{z - \rho^2} \tag{35}$$

of degree $2k-1+\alpha$. The plan of the proof of the Theorem is as follows. First we consider the expansion

$$f^{\rho}(z) = \sum_{i=0}^{\lfloor n/2 \rfloor} f_i R_i^{n,0}(z) (mod R_{\lfloor n/2 \rfloor+1}^{n,0}(z)) \tag{36}$$

and prove that $f_i \ge 0, i = 0,1,\ldots,\lfloor n/2 \rfloor$. Using (36), (24), (18), (7) and the facts that $R_{\lfloor n/2 \rfloor+1}^{n,0}(n-2d) = 0$ for $d = 0,1,\ldots,\lfloor n/2 \rfloor$, $f^{\rho}(z) \le 0$ for $0 \le z \le \rho^2$ and $|n-2d(a,b)| \le \rho$ for any distinct $a,b \in C$, we have

$$|C| f^{\rho}(n^2) \ge \sum_{a,b \in C} f^{\rho}(|n-2d(a,b)|^2) = \sum_{i=0}^{\lfloor n/2 \rfloor} \frac{f_i}{\binom{n}{2i}} \sum_{v \in V_{2i}} \left| \sum_{a \in C} v(a) \right|^2 \ge f_0 |C|^2$$

This gives rise to the Delsarte inequality [D] :

$$|C| \le \frac{f^{\rho}(n^2)}{f_0}.$$

Then we show that

$$\frac{f^{\rho}(n^2)}{f_0} = \sum_{i=0}^{k-1} \binom{n}{2i+\alpha} - \binom{n-2}{2k-2+\alpha} \frac{R_{k-1}^{n-2,\alpha}(\rho^2)}{R_k^{n,\alpha}(\rho^2)} = D_n(\rho) \tag{37}$$

and investigate the function $D_n(\rho)$.

In order to determine signs of coefficients in (36) we notice that the general properties of zeroes of the orthogonal system $\{Q_i^{n-2}(z)\}$ and (24) yield that for the interval (34) and $k \geq 1$

$$R_i^{n-2,\alpha}(\rho^2) > 0, i = 0, 1, \ldots, k-1 \quad \text{and } R_k^{n-2,\alpha}(\rho^2) \leq 0. \tag{38}$$

By (16) and (29) the polynomials

$$\frac{R_k^{n-2,\alpha}(z)R_{k-1}^{n-2,\alpha}(\rho^2) - R_{k-1}^{n-2,\alpha}(z)R_k^{n-2,\alpha}(\rho^2)}{z-\rho^2}$$

and

$$R_k^{n-2,\alpha}(z)R_{k-1}^{n-2,\alpha}(\rho^2) - R_{k-1}^{n-2,\alpha}(z)R_k^{n-2,\alpha}(\rho^2)$$

are equal up to constant factors to polynomials

$$\sum_{i=0}^{k-1} R_i^{n-2,\alpha}(z)R_i^{n-2,\alpha}(\rho^2)\binom{n-2}{2i+\alpha} \tag{39}$$

and

$$R_k^{n,\alpha}(z)R_{k-1}^{n-2,\alpha}(\rho^2) - R_{k-1}^{n-2,\alpha}(z)R_k^{n,\alpha}(\rho^2) \tag{40}$$

respectively. Now the nonnegativity of coefficients f_i in (36) follows from (35), (38),(39),(29) and (31). In order to prove (37) we can again use (39), (29) and also (28), (40) and the orthogonality condition (25). From (29) it follows that

$$\sum_{i=0}^{k-1}\binom{n}{2i+\alpha} - \binom{n-2}{2k-2+\alpha}\frac{R_{k-1}^{n-2,\alpha}(z)}{R_k^{n,\alpha}(z)} = \sum_{i=0}^{k}\binom{n}{2i+\alpha} - \binom{n-2}{2k+\alpha}\frac{R_k^{n-2,\alpha}(z)}{R_k^{n,\alpha}(z)}$$

and hence (37) equals $\sum_{i=0}^{2k+\alpha}\binom{n-1}{i}$ for $\rho = z_{2k+\alpha}(n-2)$. On the other hand, if ρ in (34) tends to $z_{2k-1+\alpha}(n-2)$, then (37) tends to $\sum\limits_{i=0}^{2k-1+\alpha}\binom{n-1}{i}$ by (30). Thus the function $D_n(z)$ is continuous and has the property (32). Finally the polynomial $R_k^{n-2,\alpha}(z)$ up to a positive factor equals the polynomial (27). Since the relation $R_{k+1}^{n,\alpha}(z)/R_k^{n,\alpha}(z)$ increases on z (see (17)) and $R_{k+1}^{n,\alpha}(n^2) = R_k^{n,\alpha}(n^2)$, the successive zeroes of $R_k^{n-2,\alpha}(z)$ lie between ones of $R_{k+1}^{n,\alpha}(z)$. But then the relation $R_k^{n-2,\alpha}(z)/R_{k+1}^{n,\alpha}(z)$ decreases (see, for example, lemma 3.7 of [L4]). This completes the proof of the Theorem.

The bound (5) for self-complementary codes is an evident corollary to Theorem ($B_n(d) = D_n(n-2d)$). It is attained for some classes of codes, in particular, for the Kerdock codes of length $n = 2^{2l}$ and for the codes of length $n = 2^{2l+1}$ that are dual to BCH codes with designed distances 3 and 5.

For small $\rho = \rho(C)$ the bound (33) of the Theorem takes the following form :

$|C| \le \frac{n^2-\rho^2}{n-\rho^2}$, if $0 \le \rho^2 \le n-2$ (the Grey-Rankin bound [MS]) ;
$|C| \le \frac{n(n^2-\rho^2)}{3n-2-\rho^2}$, if $n-2 \le \rho^2 \le 3n-8$ (the Sidelnikov bound [S1]) ;
$|C| \le \frac{n^2-\rho^2}{2} \cdot \frac{(n-2)(n^2-3n+8)-(n^2-n+2)\rho^2}{3n(n-2)-2(3n-4)\rho^2+\rho^4}$,
if $3n-8 \le \rho^2 \le 3n-10+\sqrt{6n^2-42n+76}$ [L2] ;
$|C| \le \frac{n(n^2-\rho^2)}{6} \cdot \frac{3n^3-23n^2+90n-136-(n^2-3n+8)\rho^2}{15n^2-50n+24-10(n-2)\rho^2+\rho^4}$,
if $3n-10+\sqrt{6n^2-42n+76} \le \rho^2 \le 5(n-4)+\sqrt{10n^2-90n+216}$.
Since $D_n(z)$ increases with z we have

$$\rho(n,M) = \min_{C \subseteq F_2^n, |C|=M} \rho(C) \ge \rho, \quad \text{if } D_n(\rho) = M. \tag{41}$$

The following asymptotic results follow from (33) and (41) (see [L2,L3]) :

$$\rho^2(n,M) \gtrsim c(\gamma)n \quad \text{if } \tfrac{\log M}{\log n} \to \gamma = const > 1, \tag{42}$$

where $c(2) = 2, c(3) = 5, c(4) = \frac{1}{12}(59+\sqrt{1393})$ and $c(\gamma) \sim 4\gamma$ as $\gamma \to \infty$;

$$\rho^2(n,M) \gtrsim \frac{4n \log M}{\log n - \log\log M} \quad \text{if } \tfrac{\log M}{\log n} \to \infty \text{ and } \tfrac{\log M}{n} \to 0; \tag{43}$$

$$\rho^2(n,M) \gtrsim 4\mu(1-\mu)n^2 \quad \text{if } \tfrac{\log M}{n} \to H(\mu) = -\mu\log\mu - (1-\mu)\log(1-\mu). \tag{44}$$

3 Some applications of the bound

1. An upper bound for the minimum ditance of self-dual codes.
Self-dual binary codes of length n exist only for even n and are self-complementary. Let $d(n)$ be (a unique) root of the equation $B_n(d) = 2^{n/2-1}$. Then from (5) it follows that the minimum distance of any self-dual binary code does not exceed $d(n)$. This bound can be easily calculated and is better than the bound that follows from (3). It is also better asymptotically than the Conway-Sloane bound [CS]. However as it was shown by A. Lundqvist from the Linkoping University the improvement happens for too large n.

2. Lower bound for sums of Legendre symbols.
Let

$$T(k,p) = \max \left| \sum_{x=0}^{p-1} \left(\frac{f(x)}{p} \right) \right|,$$

where $p \ge 3$, $\left(\frac{z}{p}\right)$ denotes the Legendre symbol ($\left(\frac{0}{p}\right) = 1$), and the maximum is taken over all polynomials $f(x)$ of degree at most k over F_p that cannot be represented as the product of a constant and the square of a polynomial . The cyclic code $W \subseteq F_2^p$ of

$p^{[k/2]}+1$ vectors $a=(a_1,\ldots,a_p)$ with coordinates $a_i=\left(\frac{g(i)}{p}\right)$, $i=1,\ldots,p$, where $g(x)$ is a square-free monic polynomial over F_p of degree $h, 0\leq h\leq [k/2]$ was considered in [L2,L3]. The definition allows us to prove that

$$T(k,p)\geq \rho(p,p^{[k/2]}+1)-k. \tag{45}$$

Using (41)-(45) gives rise to the best known number theoretic lower bounds on $T(k,p)$.

3. Lower bounds for the cross-correlation of codes.
The cross-correlation of a code $C\subseteq F_2^n$ is defined by

$$\Theta(C)=\max_{a,b\in C}\left|\sum_{i=1}^{n} a_i b_{i+\tau(mod n)}\right|,$$

where the maximum is taken over all $a=(a_1,\ldots,a_n)\in C$, $b=(b_1,\ldots,b_n)\in C$ and $\tau\in\{0,\ldots,n-1\}$ such that $\tau\neq 0$ if $a=b$. For a code $C\subseteq F_2^n$, the following inequality holds : $\Theta(C)\geq\rho(n,|C|n)$. Using the above given lower bounds on $\rho(n,M)$ with $M=|C|n$ we arrive at the best known lower bounds on $\Theta(C)$ for codes $C\subseteq F_2^n$. In particular, for codes C of cardinality at least n^γ we have $\Theta(C)\gtrsim\sqrt{c(\gamma+1)n}$ as $n\to\infty$ (compare with [S1]).

4. Applications to the randomization problem.
First notice that for $k=d'(C)-1$, the lower bounds in (2), (3), (12) are the best known ones on the size of a k-wise independent set (sample space) $C\subseteq F_2^n$. Another randomization problem [AGHP] is to find, for given k and ϵ, the minimum length $n=n(k,\epsilon)$ such that there exists a linear $[n,k]$ code $C\subseteq F_2^n$ with

$$\left(\frac{1}{2}-\epsilon\right)n\leq d(a,b)\leq\left(\frac{1}{2}+\epsilon\right)n \quad \text{for any } a,b\in C, a\neq b.$$

Since for such a code C $\rho(C)\leq 2\epsilon n$ and $|C|=2^k$, the bounds (33) and (42)-(44) lead to the best known lower bounds on $n(k,\epsilon)$.

References

[AGHP] Alon N., Goldreich O., Hastad J., Peralta R., *Simple constructions of almost k-wise independent random variables*, Proc. of the 31st Annual Symposium on the Foundations of Computer Science, 1991.

[BI] Bannai E., Ito T., *Algebraic Combinatorics I*, Association Schemes, Benjamin/Cummings, London, 1984.

[BCN] Brouwer A.E., Cohen A.M., Neumaier A., *Distance-regular graphs*, Springer-Verlag, Berlin, 1989.

[CS] Conway J.H., Sloane N.J.A., *A new upper bound on the minimal distance of self-dual codes*, IEEE Trans. Inform. Theory, IT-36 (1990), 1319-1333.

[D] Delsarte Ph., *An algebraic approach to the association schemes of coding theory*, Philips Res. Reports Suppl. 10 (1973).

[L1] Levenshtein V.I., *On choosing polynomials to obtain bounds in packing problems (in Russian)*, in Proc. Seventh All-Union Conf. on Coding Theory and Information Transm., Part II, Moscow-Vilnius, 1978, 103-108.

[L2] Levenshtein V.I., *Bounds to the maximum size of code with limited scalar product modulus*, Soviet Math. Doklady, vol. 25 (1982), N.2, 525-531.

[L3] Levenshtein V.I., *Bounds for packings of metric spaces and some their applications (in Russian)*, Problemy Kiberneticki, Issue 40, Moscow, "Nauka" 1983, 43-110.

[L4] Levenshtein V.I., *Designs as maximal codes in polynomial metric spaces*, Acta Applicandae Mathematicae, 29, (1992), 1-82.

[MRRW] McEliece R.J., Rodemich E.R., Rumsey H., jr., and Welch L.R., *New upper bounds on the rate of a code via the Delsarte-MacWilliams inequalities*, IEEE Trans. Inform. Theory, IT-23 (1977), 157-166.

[MS] MacWilliams F.J., Sloane N.J.A., *The theory of error-correcting codes*, North Holland Publ. Co., Amsterdam, 1977.

[S1] Sidelnikov V.M., *On mutual correlation of sequences (in Russian)*, Problemy Kiberneticki, Issue 24, Moscow, "Nauka" 1971, 15-42 (a short description in English in Soviet Math. Doklady, 12, N1 (1971), 197-201).

[S2] Sidelnikov V.M., *On extremal polynomials used to estimate the size of codes*, Problems of Information Transmission, 16, N3 (1980), 174-186.

[SZ] Szego G. *Orthogonal polynomials*, Vol. XXII, AMS Col. Pub., Providence, Rhode Island, 1939.

3

CRYPTOGRAPHY

ASYMPTOTIC ANALYSIS OF PROBABILISTIC ALGORITHMS FOR FINDING SHORT CODEWORDS

F. Chabaud
Ecole Normale Supérieure, Paris, France

Abstract

We present the asymptotic analysis of two algorithms [3, 7] for finding short codewords in linear binary codes. For the first of these algorithms, the results are confirmed by implementation, and this allows extrapolation for larger codes.

Résumé

Il a été démontré [1] que le problème de trouver dans un code linéaire binaire un mot de poids donné (en particulier de poids faible) est NP-complet. Ceci permet d'envisager des procédés cryptographiques fondés sur ce fait [5]. Il est donc nécessaire pour assurer la sécurité de tels systèmes d'évaluer les algorithmes existants dont le but est justement de trouver de tels mots. Par exemple, dans [5], trouver l'erreur aléatoire introduite pour chiffrer un mot revient à trouver le mot le plus court d'un code linéaire.

Nous présentons ici l'analyse théorique de deux de ces algorithmes [3, 7]. Ces résultats sont confirmés, tout au moins pour l'un d'eux, par l'implantation, et autorisent des extrapolations pour les grandes dimensions.

Introduction

It has been shown that the problem of finding a codeword of given weight in a linear binary code is NP-complete [1]. This property can be used in cryptographic coding [5]. Nonetheless, to improve the security of such a coding, it is necessary to experiment the algorithms which find such words in random codes. For instance, to find the random error introduced in order to code the message in the McEliece's cryptosystem [5], is equivalent to find the shortest word of a linear code.

*On leave from Ecole Polytechnique, 91128 Palaiseau Cedex, France

1 Theory

1.1 Notations

The general notations that we use are those of [4]. We always consider truly random binary codes. For an (n, k) code, we pose $\beta = n/k$. As introduced in [3], we define

$$\mathrm{minwt}(C) = \min_{m \in C}\{w(m) > 0\},$$

and $w|_S$ will be the restriction of the word w to the set of columns S.

We also use the entropy function [4, page 308] :

$$H_2(x) = -x \log_2 x - (1 - x) \log_2(1 - x).$$

1.2 J.S. Leon's algorithm

The algorithm of J.S. Leon [3] is a probabilistic algorithm. It introduces two parameters s and ℓ.

1.2.1 Principle

This algorithm uses a generator matrix. It searches the zero-bits of minimum weight words. More precisely, at each iteration, one has to :

1. Choose a random selection S of s columns of the matrix, which are put by permutations to the right end of the matrix.

2. Apply Gaussian elimination so that the resulting matrix has the form

$$G_2 = \left(\begin{array}{c|c|c} B & Z & I_e \\ D & 0 & 0 \end{array}\right),$$

where B is a $(n - s, e)$ matrix, Z is a $(s - e, e)$ matrix and D is a $(n - s, k - e)$ one.

3. Look for the linear combinations that lead to codewords m such that $w(m|_S) \leq \ell$. This can be achieved by considering the single matrix Z. In case $w(m|_S) \leq \ell$, compute the corresponding n-bit word and verify if its weight is less than the weight of the previous shortest word obtained.

1.2.2 Asymptotic analysis

If a codeword m of weight w exists, the probability of not finding it with the selection S is :

$$\pi_\ell(w) = \sum_{t=0}^{\ell} \pi(w, t),$$

where

$$\pi(w, t) = \frac{\binom{n-s}{w-t}\binom{s}{t}}{\binom{n}{w}}.$$

Hence, the number of iterations necessary to be certain, with probability $1-\epsilon$, that such a codeword doesn't exist is :

$$r \geq \frac{\ln(\epsilon)}{\ln(1-\pi_\ell(w))}.$$

1.2.3 Complexity

The parameter s must grow with k. First, we suppose that this growth has the asymptotic form $s \sim_\infty \gamma k$, where γ is a positive constant. To evaluate the complexity in the worst case, we suppose that the algorithm fails in finding a codeword of weight $d = \alpha n$. Then

$$\pi(d,t) = \frac{\binom{(\beta-\gamma)k}{\alpha\beta k - t}\binom{\gamma k}{t}}{\binom{\beta k}{\alpha\beta k}}.$$

Clearly, we have

$$\binom{\gamma k}{t} =_\infty O(k^t),$$

and it is easy to prove that

$$\binom{(\beta-\gamma)k}{\alpha\beta k - t} \sim_\infty \binom{(\beta-\gamma)k}{\alpha\beta k}.$$

By using the asymptotic evaluation

$$\binom{k}{ak} \sim_\infty \left[\frac{1}{a^a(1-a)^{1-a}}\right]^k,$$

we have

$$\binom{(\beta-\gamma)k}{\alpha\beta k - t} =_\infty O\left[\left(\frac{(\beta-\gamma)^{\beta-\gamma}}{(\alpha\beta)^{\alpha\beta}[(1-\alpha)\beta-\gamma]^{(1-\alpha)\beta-\gamma}}\right)^k\right],$$

and

$$\binom{\beta k}{\alpha\beta k} \sim_\infty \left(\left[\frac{1}{\alpha^\alpha(1-\alpha)^{1-\alpha}}\right]^\beta\right)^k.$$

Therefore

$$\pi(d,t) =_\infty O\left(k^t \Omega_\alpha^k(\gamma)\right),$$

where, by definition

$$\Omega_\alpha(\gamma) = \frac{(1-\alpha)^{(1-\alpha)\beta}(\beta-\gamma)^{\beta-\gamma}}{\beta^{\alpha\beta}[(1-\alpha)\beta-\gamma]^{(1-\alpha)\beta-\gamma}}.$$

As $\pi_\ell(d) = \sum_{t=0}^{\ell} \pi(d,t)$, we obtain

$$\pi_\ell(d) =_\infty O\left(k^\ell \Omega_\alpha^k(\gamma)\right).$$

The maximization of $\Omega_\alpha(\gamma)$ leads to $\gamma = 1$, and therefore we have $s =_\infty k + o(k)$. The experiments seem to show that the best results are obtained for $s = k + \sigma$ with σ a constant depending on ℓ. We obtain that

$$\pi_\ell(d = \alpha n) =_\infty O\left(k^\ell \Omega_\alpha^k\right),$$

with

$$\boxed{\Omega_\alpha = \frac{(1-\alpha)^{(1-\alpha)\beta}(\beta-1)^{\beta-1}}{\beta^{\alpha\beta}\left[(1-\alpha)\beta-1\right]^{(1-\alpha)\beta-1}}.}$$

As a consequence, the number of iterations is $r = O\left(\frac{(1/\Omega_\alpha)^k}{k^\ell}\right)$. Besides, each iteration requires a certain number of operations :

- $O(k)$ permutations,
- A Gaussian elimination in $O(k^3)$ operations,
- Computing the $\binom{k}{1} + \binom{k}{2} + \ldots + \binom{k}{\ell} =_\infty O(k^\ell)$ linear combinations of Z-codewords in $O(k^\ell)$ operations,
- Computing the $\sum_{i=1}^{\ell}\left(\binom{k}{i}\frac{\sum_{j=0}^{\ell-i}\binom{\sigma}{j}}{2^\sigma}\right) = O(k^\ell)$ vectors m satisfying $w(m|_S) \leq \ell$. The length of those vectors is n, therefore this requires $O(k^{\ell+1})$ operations.

Hence, because of the Gaussian elimination, the optimal parameter ℓ is greater than 2, and the number of operations of the algorithm is

$$\boxed{\omega =_\infty O(k(1/\Omega_\alpha)^k).}$$

The experiments seem to show that the best parameter is in fact $\ell = 2$, and that in this case, we must have $s = k + 2$ to obtain the best results.

These results are similar to those of P.J. Lee and E.F. Brickell [2] on their attack algorithm of McEliece's cryptosystem. In fact the principles of the two algorithms are similar : they try to find a selection of columns where all the bits of the codeword are zeros. If they fail, then they expect the number of ones to be small (less than ℓ), which implies that the number of iterations with expensive Gaussian elimination can be reduced.

β	4	2	4/3
$1/\Omega_\alpha$	1.33	1.18	1.08
$\frac{\ln(1/\Omega_\alpha)}{\ln 2}$	0.40	0.24	0.11

Figure 1: Variations of Ω_α

1.2.4 Performances of the algorithm

Using the asymptotic evaluation of Gilbert-Varshamov for the expected minimum weight of a random linear binary code ($H_2(\alpha) \sim_{+\infty} 1 - 1/\beta$) [6], we obtain the limit values of $1/\Omega_\alpha$ (see figure 1).

1.3 J. Stern's algorithm

This second algorithm [7], obtained independently from the previous one, attacks the code by using a parity check matrix H. It introduces two parameters p and l.

1.3.1 Principle

1. The first step of the algorithm is similar to the algorithm of J.S. Leon : choose a random selection S of $n-k$ columns of the matrix, which are put by permutations to the right end of the matrix. The degenerative cases are eliminated in order to obtain a matrix of the form :

$$\Gamma_1 = (\, Q \mid I_{n-k} \,).$$

2. Randomly split the columns of matrix Q in two subsets in order to obtain, after permutations, a matrix of the form :

$$\Gamma_2 = (\, X \mid Y \mid I_{n-k} \,).$$

3. Randomly choose l lines of the matrix and perform permutations on lines in order to obtain a matrix of the form :

$$\Gamma_3 = \left(\begin{array}{c|c|c} X_l & Y_l & \multirow{2}{*}{J} \\ \hline X_{n-k-l} & Y_{n-k-l} & \end{array} \right).$$

4. For each group P_X of p columns of X_l, compute their sum $\sigma_l(P_X)$, and do the same for each P_Y. If $\sigma_l(P_X) = \sigma_l(P_Y)$, select the $2p$ columns $P_X \cup P_Y$ and compute the sum V of the $n-k-l$ other lines of this columns. If the weight of V is $w-2p$, then it is possible to build a codeword of weight w.

1.3.2 Complexity

The probability of success of this algorithm (see [7]) is

$$\pi(w) = \pi_1(w)\pi_2\pi_3(w),$$

where

$$\pi_1(w) = \frac{\binom{k}{2p}\binom{n-k}{w-2p}}{\binom{n}{w}},$$

$$\pi_2 = \frac{\binom{2p}{p}}{4^p},$$

and

$$\pi_3(w) = \frac{\binom{n-k-l}{w-2p}}{\binom{n-k}{w-2p}}.$$

Using the result obtained for the analysis of J.S. Leon's algorithm, we have

$$\pi_1(w) =_\infty O(k^{2p}\Omega_\alpha^k),$$

and

$$\pi_2 =_\infty O(1).$$

As in the case of the first algorithm, we suppose $l \sim_\infty \lambda k$, and we get

$$\pi_3(w) \sim_\infty \frac{\binom{(\beta-\lambda-1)k}{\alpha\beta k}}{\binom{(\beta-1)k}{\alpha\beta k}}.$$

For reasons similar to those used in the analysis of the previous algorithm, to optimize $\pi_3(w)$, λ has to grow to 0 as a function of k. Therefore, $l = o(k)$, and

$$\pi_3(w) =_\infty O(1).$$

Finally we obtain

$$\pi(w) =_\infty O(k^{2p}\Omega_\alpha^k),$$

with the same constant Ω_α as for the algorithm of J.S. Leon. The number of iterations is $r =_\infty O\left(\frac{(1/\Omega)^k}{k^{2p}}\right)$, and each iteration requires :

- $O(k)$ permutations,
- A Gaussian elimination in $O(k^3)$ operations,
- To compute the sums of $\{P_X\}\{P_Y\}$ which needs $2pl\binom{k/2}{p}$ operations,
- To compute the vector V which needs $2p\frac{(n-k-l)}{2^l}\binom{k/2}{p}^2$ operations.

To minimize the number of operations of an iteration, we have to minimize

$$2pl\binom{k/2}{p} + 2p\frac{(n-k-l)}{2^l}\binom{k/2}{p}^2.$$

Which reduces to the equation

$$\boxed{\binom{k/2}{p}\frac{1+\ln 2(n-k-l)}{2^l} = 1.}$$

Since $l = o(k)$ when $k \to +\infty$, when the parameter p is a constant, we have

$$\boxed{2^l \sim_\infty k^{p+1},}$$

which implies

$$\boxed{l =_\infty O(\ln k).}$$

Therefore the second stage of the algorithm needs $O(\ln k k^p)$ operations, and the third one $\frac{k^{2p+1}}{k^{p+1}} = O(k^p)$. If we suppose that p is greater than 3, the number of operations of the algorithm is

$$\boxed{\omega = O\left(\frac{\ln k(1/\Omega_\alpha)^k}{k^p}\right).}$$

1.3.3 Interest of the algorithm

This result shows that the algorithm of J. Stern is probably even more efficient than J.S. Leon's one. Also we have supposed that p is a constant, while the above form suggests that better results could be obtained if we increase p with k. A more accurate asymptotic study and an implementation would be necessary to make this precise, because the memory required for the sums of $\{P_X\}\{P_Y\}$ grows with $2\binom{k}{p}$. Hence, a linear growth of p with k is impossible. The optimal growth of p might be $O(\ln k)$ or perhaps $O(\ln\ln k)$.

2 Experiments using J.S. Leon's algorithm

2.1 Control of the theory

In order to verify the theory, we made several experiments based on the hypothesis that the CPU time C is proportional to the number of operations. Then $C = O(k(1/\Omega_\alpha)^k)$, and thus $\ln C - \ln k$ must be asymptotically of the form $a + k\ln(1/\Omega_\alpha)$. All the results[1] look like those shown in figure 2.

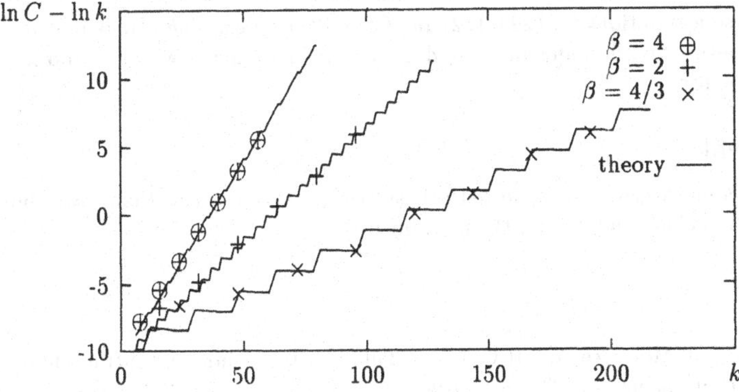

Figure 2: Comparison between theory and experiments

The theoretical curve equation is $a' + k\ln(1/\Omega_\alpha(k))$ where a' is the constant which minimizes the distance to experimental points[2]. To obtain the theoretical curves, we compute $\alpha = d/\beta k$ for each value of k, where $d = \min\{w \,/\, \frac{2^k\binom{n}{w}}{2^n} \geq 1\}$ is the probable minimum weight of the $(\beta k, k)$ code [6]. Then we compute $\Omega_\alpha(k)$ and the constant a'.

[1]These results were obtained on nine DECstations 5000/200 at the LIX *(Laboratoire d'Informatique de l'Ecole Polytechnique, Ecole Polytechnique, 91128 Palaiseau Cedex, France).*

[2]except the first ones on too small dimensions.

2.1.1 Conclusion

Since these results confirm the validity of the theory, they allow us to make previsions for larger codes.

n	224	256	288	384	512
k	112	128	144	192	256
w	26	30	33	44	58
time	1	24.3	253	$1.4.10^6$	$7.3.10^{10}$
w'	20	22	25	33	44
time	8.10^{-3}	4.10^{-2}	0.38	187	9.10^5

Figure 3: Extrapolation of computational times for finding short codewords of larger codes

In figure 3, the computation times are indicated in an arbitrary scale. The unity is given by the time required to find a codeword of weight 26 (the most probable minimum weight)[3] in a truly random $(224,112)$ code. Experiments which have already been performed confirm these estimations for $(224,112)$ and $(256,128)$ codes. The attack of codes such as $(512,256)$ seems for the moment beyond hope, but introducing words of too low weight may be dangerous.

Acknowledgement

I'm grateful to Jacques Stern who suggested this subject. I also want to thank Antoine Joux for his precious help in optimizing the programs.

References

[1] E.R. Berlekamp, R.J. McEliece, and H.C.A. Van Tilborg. On the inherent intractability of certain coding problems. *IEEE Trans. Inform. Theory*, pages 384–386, 1978.

[2] P.J. Lee and E.F. Brickell. An observation on the security of McEliece's public-key cryptosystem. In *Advances in Cryptology - EUROCRYPT '88*, pages 275–280. Lecture Notes in Computer Science, Springer 1989. #330.

[3] J.S. Leon. A probabilistic algorithm for computing minimum weights of large error-correcting codes. *IEEE Trans. Inform. Theory*, IT-34(5):1354–1359, September 1988.

[4] F.J. MacWilliams and N.J.A. Sloane. *The Theory of Error-correcting Codes*. North-Holland, 1983.

[3] It takes nowadays approximatively 4.10^5 s CPU on a DECstation 5000/200, that is to say about 5 days.

[5] R.J. McEliece. A public-key cryptosystem based on algebraic coding theory. *DSN progress report*, 42-4, 1978.

[6] J.N. Pierce. Limit distribution of the minimum distance of random linear codes. *IEEE Trans. Inform. Theory*, IT-13(17):595–599, 1967.

[7] J. Stern. A method for finding codewords of small weight. In *Coding Theory and Applications*, pages 106–113. Lecture Notes in Computer Science, Springer 1989. #388.

RATIONAL INTERVAL MAPS AND CRYPTOGRAPHY

S. Harari
University of Toulon and Var, La Garde, France
P. Liardet
University of Provence, Marseille, France

1 Introduction

In Eurocrypt 1991 Habutsu, Nishio, Sasase, Mori [4] introduced a cryptosystem using tent maps (denoted HNSM in the sequel). Though the functions that are used have good chaotic properties this system seems to has many weaknesses [7]. A general framework for such cryptosystems can be described as follow. Let $\mathcal{E} = \{E_i\,;\ i = 0,\ldots,M\}$ be a family of so-called *plaintext-spaces* and for each $i = 1,\ldots,M$, let L_i be an integer ≥ 2 and let $\mathcal{T}_i = \{T_{i,\ell}\,;\ \ell = 1,\ldots,L_i\}$ be a family of cipher-maps $T_{i,\ell} : E_{i-1} \to E_i$. We assume that there exist maps $S_i : E_i \to E_{i-1}$ such that for all $\ell \in \{1,\ldots,L_i\}$ one has $S_i \circ T_{i,\ell} = Id_{E_{i-1}}$. The key is given by $\{S_1,\ldots,S_M\}$ and the cryptogram of a given paintext $y \in E_0$ is any element in

$$\tau(y) = \bigcup_{(\ell_1,\ldots,\ell_M)} T_{M,\ell_M} \circ \ldots \circ T_{1,\ell_1}(y)$$

where the union is taken over all finite sequences $(\ell_1,\ldots,\ell_M)$ in $\prod_{i=1}^{M}\{1,\ldots,L_i\}$. Usually all the plaintext-spaces are identical to a same space E and we choose the same family of enciphering maps. Therefore, the system is given by a deciphering map $\Sigma : E \to E$ and a family of right inverse maps of $\Sigma : T_1,\ldots,T_L$, $L \geq 2$. The integer M corresponds to the number of iterations and the key is (Σ, M). To avoid attacks on the enciphering algorithm, the choice of the cryptogram depends on a random process which produces a uniform-like distribution of the ciphertext in E_M. It seem that the firt use of such a scheme was considered in terms of cellular automata by S. Wolfram in Crypto'85 [8] and recently H. Gutowitz in [3] proposed a scheme, according to this model, of an enciphering/deciphering system at high rate. The HNSM system uses tent maps Φ_t for keys, namely t is a parameter in $]0,1[$ and Φ_t is the piecewise linear map defined by $\Phi_t(x) = x/t$ for $0 \leq x \leq t$ and $\Phi_t(x) = \frac{1-x}{1-t}$ for $t < x \leq 1$.

In Section 2 of this paper we present, according to the above general framework, cryptosystems based on interval maps called markovian. In order to obtain a good cryptosystem based on this principle, a priori knowledge is needed on the transformation of the Lebesgue measure by the enciphering and deciphering transformations. This is done through two theorems from the theory of dynamical systems, which are recalled and are applied. A classification using the topological conjugation is examined. The HNSM system is then seen to be a particular case and in section 3 we show first how, when using the HNSM system in the recommended manher, one can break the system. Then we strengthen the system so as to minimize the probability of success of this attack and show why the Biham attack [7] in fact does not work well.

[0]Author partially supported under contrat DRET 901636/A000 DRET/DS/SR

The Section 4 and 5 are devoted to the studies of a particular markovian cryptosystem denoted by HARALIA obtain from interval maps which are not piecewise linear but have good known distribution properties of the transform of the Lebesgue measure. These maps are piecewise fractional linear for which an ergodic measure, absolutely continuous with respect to the Lebesgue measure, can be computed exactly, using a previous work of Schweiger [8]. Distribution of cryptograms are determined. This cryptosystem can have very efficient implementations in software or hardware. The computational noise is studied in means and solutions are proposed to avoid errors in the deciphering part of the cryptosystem. The proposed recommandations in oder to use this system will also show that probability for the classical Biham attack to be successfull is very small. The attack on fixed points is also examined.

2 Interval Maps and cryptosystems

2.1 Interval Maps

Let I be the unit interval $[0,1]$ and $\mathcal{J} = \{I_1, \ldots, I_q\}$ (with $q \geq 2$) be a finite partition of I into q subintervals of length > 0. A map $f : I \mapsto I$ is said to be a ***markovian interval map*** for $\mathcal{J}$ if the following properties are satisfied:

- The restriction $f|_{I_i}$ $(i = 1, \ldots, q)$ of f to I_i can be extended to the closure $\overline{I_i}$ of I_i into a strictly monotone continuous map $f_i : \overline{I_i} \to I$;
- For all $j \in \{1, \ldots, q\}$ there exists $\Lambda_j \subset \{1, \ldots, q\}$ such that $f_j(\overline{I_j}) = \bigcup_{k \in \Lambda_j} \overline{I_k}$;
- There exists an integer $p > 0$ such that $f^p(\overline{I_i}) = I$ for all i.

We refer to [1] for the proof of the following.

Theorem 1. *(The folklore theorem) Let $f : I \mapsto I$ be a markovian interval map for $\mathcal{J}$ and assume that the maps f_i $(i = 1, \ldots, q)$ are C^2. Moreover assume that f is eventually expansive i.e. there exists a natural number m and a real number $\theta > 1$ such that*

$$\left|\frac{df^m}{dx}\right| \geq \theta$$

for all point x where f^m (the m-th iterate of f) has a derivative. Then there exists a piecewise continuous map $\rho : [0,1] \mapsto [0, +\infty[$ such that the measure $\nu(dx) = \rho(x)dx$ on $[0,1]$ is a probability measure which is invariant and ergodic under f. Moreover, there exists $D > 0$ such that

$$\frac{1}{D} \leq \rho \leq D.$$

□

The next result is due to R. Adler (see [1]):

Theorem 2. *Let f be as in the folklore theorem and assume that there exists a constant $M > 0$ such that*

$$\frac{f''(x)}{f'(x)^2} \leq M\,, \quad x \in I_j\,, \quad 1 \leq j \leq q.$$

Then the constant D satisfies $D \leq \exp\left(\frac{M\theta}{\theta - 1}\right)$ where θ is a constant such that $|f'(x)| \geq \theta$ for all x in the unit interval (for extremities of subintervals in $\mathcal{J}$ we consider right and left derivatives). □

The interest of this theorem is to give a practicable bound on the constant D.

2.2 Symbolic Representation

The partition $\mathcal{I}$ associated to the interval map f leads to a useful symbolic expansion of points in I and we recall some basic facts refering to [2] for usual definitions and properties on dynamical systems. Let $\Omega = \mathcal{A}^{\mathbb{N}}$ where $\mathcal{A} = \{1, \ldots, q\}$. We consider Ω as a topological space endowed with the compact product topology which arises from the discrete topology on $\mathcal{A}$. Let σ denote the shift transformation on Ω defined by $\sigma(\omega_0, \omega_1, \omega_2, \ldots) = (\omega_1, \omega_2, \omega_3, \ldots)$. A symbolic dynamical system is a pair $(K, \sigma|_K)$ where K is a compact σ invariant ($\sigma(K) = K$) subset of Ω. For short we still denote by σ its restriction $\sigma|_K$ to K. Let f be a given markovian interval map. Using property (ii) of f we define the following oriented graph on $\mathcal{A} : \Gamma_f = \{(i,j) \in \mathcal{A}^2 \,;\, f_i(\mathring{I}_i) \subset \mathring{I}_j\}$ and we consider the topological Markov chain (K_f, σ) where

$$K_f = \{\omega \in \Omega \,;\, \forall i \geq 0, (\omega_i, \omega_{i+1}) \in \Gamma_f\}.$$

Due to property (iii), (K_f, σ) is aperiodic and irreducible.

Let $a = a_0 \ldots a_n$ be a f-allowable string (or word) i.e., such that $(a_k, a_{k+1}) \in \Gamma_f \; \forall k = 0, \ldots, n$ Then the closure

$$I_{a_0 \ldots a_n} = \overline{\mathring{I}_{a_0} \cap f^{-1}(\mathring{I}_{a_1}) \cap \ldots \cap f^{-n}(\mathring{I}_{a_n})}$$

is a non empty closed interval and from (i), (ii) and (iii) we get $f(\mathring{I}_{a_0 \ldots a_n}) = \mathring{I}_{a_1 \ldots a_n}$.

Let $\mathcal{W}_n$ denote the set of all f-allowable strings in $\mathcal{A}^n$ then the family of intervals $\mathcal{J}_n = \{I_a \,;\, a \in \mathcal{W}_n\}$ is a covering of I such that two different intervals of this family are disjoint or have one extremity in common. Intervals in $\mathcal{J}_n$ will be said fundamental subintervals of order n. Notice that f^n is also a markovian interval map. Now we assume that f satisfies assumption of Theorem 1, in particular f is eventually expansive so that if we put $\delta_k = \max_{a \in \mathcal{W}_k} |I_a|$ where $|J|$ denotes the length of an interval J, then $\lim_{k \to \infty} \delta_k = 0$. Therefore we can define a map $\phi : K_f \to I$ by

$$(1) \qquad \{\phi(\omega)\} = \bigcap_n I_{\omega_0 \ldots \omega_n}.$$

Notice that ϕ is continuous and $\phi(\omega) = \lim_{n \to \infty} f_{\omega_0}^{-1} \circ \ldots \circ f_{\omega_n}^{-1}(y)$ uniformly in $y \in I$. Let Ω_f be the set of symbolic sequences $\omega = (\omega_n)_{n \geq 0}$ in K_f such that $\bigcap_n \mathring{I}_{\omega_0 \ldots \omega_n} \neq \emptyset$. A straightforward computation gives

$$\phi \circ \sigma(\omega) = f \circ \phi(\omega)$$

for all $\omega \in \Omega_f$ and $\sigma(\Omega_f) = \Omega_f$. Let D_n be the set of extremities of subintervals in $\mathcal{J}_n$ and let $D_\infty = \bigcup_{n=1}^{\infty} D_n$. By construction and the fact that $f(D_1) \subset D_1$ we get $f^{-1}(D_\infty) = D_\infty$. Let $X_f = I \backslash D_\infty$ then $f(X_f) = X_f$ and the map ϕ realizes a bijection between Ω_f and X_f. For each $x \in I$ define its *symbolic orbit* $s(x)$ in Ω by

$$s(x)_n = i \iff f^n(x) \in I_i.$$

Clearly $s \circ f = \sigma \circ s$ and for all $x \in X_f$ one has $s(x) \in \Omega_f$. Moreover $s(\cdot)$ is continuous at each point x of X_f and

$$\phi(s(x)) = x.$$

Finally, we also have $s(\phi(\omega)) = \omega$ far all $\omega \in \Omega_f$ whenever f is eventually expansive. In other words, the dynamical systems (K_f, σ) and (I, f) are quasi-conjugate.

Remark 1. *Each point $x \in D$ has a symbolic orbit $s(x)$ in Ω but not necessarily in K_f. Indeed, if t tends to x on the left (resp. on the right) then the orbit $s(t)$ tends to an ultimately periodic orbit of K_f. Therefore it is natural to consider two different orbits for each $x \in D_\infty$ (except for $x = 0, 1$).*

2.3 Markovian Cryptosystems

We first explain how to use a markovian interval map f as basis for a cryptosystem. Then we study, in general, ergodic properties of such systems. This leads to a natural classification which shows how to construct, from a computational point of view, very different cryptosystems.

The plaintext space is the unit interval $[0,1]$ and the deciphering map is a given markovian interval maps that will be specified later. Let $M > 0$ be a given integer corresponding to the number of iterations of f. For all y in the plaintext space, we consider the set of preimages $f^{-M}(y)$ which corresponds to all different encipherings of y. Any element x in this set is defined by a symbolic sequence $(i_1, \ldots, i_M)$ such that

$$x = f_{i_M}^{-1} \circ \ldots \circ f_{i_1}^{-1}(y).$$

Here we denote by f_i^{-1} the inverse map of $f_i : I_i \to I$, defined on the interval $f(I_i) = f_i(I_i)$. By construction the symbolic orbit of x is

$$s(x) = (i_M, \ldots, i_1, \ldots)$$

and $f^M(x) = y$. For $y \in X_f$ the number of elements in $f^{-M}(y)$ corresponds to the number of M-uples of symbols $(i_M, \ldots, i_1)$ such that $(i_{k+1}, i_k) \in \Gamma_f$, for $k = 1, \ldots, M-1$. Therefore the complexity of this system is mesured in some sense by the topological entropy $h(f)$ of (K_f, σ) given by (see [2]), $h(f) = \lim_{n\to\infty} \frac{1}{n} \log \operatorname{card}(\mathcal{W}_n) = \log \beta$, where β is the spectral radius of the incidence matrix of Γ_f. Following W. Parry [6], there exists a unique Borel measure μ for the topological Markov chain (K_f, σ) which is equivalent to a (unique) shift invariant Borel probability μ^* such that the dynamical system (K_f, σ, μ^*) is ergodic of maximal entropy (which is the topological entropy) and μ is linear i.e., for all f-allowed word $a_0 \ldots a_k$ we have

$$\mu(\sigma(\{x \in K_f\,;\, x_0 \ldots x_k = a_0 \ldots a_k\})) =$$

$$\beta\mu(\{x \in K_f\,;\, x_0 \ldots x_k = a_0 \ldots a_k\}).$$

Moreover (f, I) is topologically conjugate to a piecewise linear map L with slope $\pm\beta$ and $\Gamma_L = \Gamma_f$. Such a linear system will be said *canonical* and therefore, piecewise linear maps play a fundamental role in markovian cryptosystems.

Now the cryptographic algorithm run as follow:
- Take f (in a given family),
- Fix the number M of iterations,
- Let y be an element in the plaintext space.

Then to encipher y operate as follows:
- Generate a sequence of random variables x_n in $\mathcal{A}$ (the random generator will be specified later),
- Encipher y by

$$X_n(y) = f_{x_n}^{-1} \circ \ldots \circ f_{x_1}^{-1}(y). \tag{2}$$

with $n = M$. In order to have good cryptographic properties, the distribution law of the random sequence X_n must be uniform on I, eventually up to a finite set of points y. That is to say

$$\lim_{N\to\infty} \frac{\operatorname{card}\{n\,;\, 1 \le n \le N \;\&\; X_n(y) \le x\}}{N} = x \quad \text{a.e.} \tag{3}$$

In practical applications we shall assume that for all $I_i \in \mathcal{J}$ we have $f_i(\overline{I_i}) = I$. Such a system will be called q-fold markovian. To simplify we also assume that $\mathcal{J}$ has only two intervals, namely $I_1 = [0, t[$ and $I_2 = [t, 1]$. We denote this partition by $\mathcal{J}(t)$ and put $t(f) = t$. For our purpose we restrict the family of interval maps f (which are trivially markovian) by assuming assumptions of Theorem 1 and, for example, that f is decreasing on both intervals of the partition. Let $\mathcal{F}$ denote this particular family. For any element $f \in \mathcal{F}$ we have $\mathcal{A} = \{1,2\}$, $\Omega = K_f$ and if we put $\hat{1} = (1,2,1,2,\ldots)$, $\hat{2} = \sigma(\hat{1}) = (2,1,2,1,\ldots)$ then $\Omega_f = \{\omega \in \Omega;\ \exists k \geq 0,\ \sigma^k(\omega) \notin \{\hat{1}, \hat{2}\}\}$ For $t \in]0,1[$ we consider the following interval map in $\mathcal{F}$:

$$\Psi_t(x) = \begin{cases} 1 - \frac{x}{t} & \text{if } 0 \leq x < t, \\ \frac{1-x}{1-t} & \text{if } t \leq x \leq 1. \end{cases}$$

All maps in $\mathcal{F}$ are conjugate to any Ψ_t, more precisely

Proposition 1. *Let $f \in \mathcal{F}$ and let $t \in]0,1[$, then there exists a unique homeomorphism $h_t : I \to I$ such that $h_t(0) = 0$ and $f = h_t^{-1} \circ \Psi_t \circ h_t$.*

Proof. Part of this result follows from the general theory explained above. For sake of completeness we give a direct proof. Since all Ψ_t are in $\mathcal{F}$, it is enough to prove the result only for $t = \frac{1}{2}$ and we write H in place of $h_{\frac{1}{2}}$. Let $t_0 = 0 < t_1 < \ldots < t_{2^n} = 1$ be the extremities of intervals in $\mathcal{J}_n$. If H is well defined, then for all integers $n \geq 1$ we have $f^n = H^{-1} \circ \Psi_{\frac{1}{2}}^n \circ H$. Therefore discontinuity points of f^n correspond by H to discontinuity points of $\Psi_{\frac{1}{2}}$ and then $H(t_k) = k/2^n$ for $k = 0, 1, \ldots, 2^n$. This proves the uniqueness of H. Let H_n the continuous map such that H_n is linear on each interval $[t_k, t_{k+1}]$ and $H_n([t_k, t_{k+1}]) = [k2^{-n}, (k+1)2^{-n}]$. It is easy to see that the map H_n is strictly increasing and the sequence $(H_n)_n$ converges uniformly to a strictly increasing continuous map which has the required property. In fact the construction in [6] gives $H(x) = \mu(s([0,x])$ where μ is the Parry measure for the full shift (Ω, σ) (which is also the Bernoulli measure given by the infinite product of the equidistribution on $\{1,2\}$). □

Since $f \in \mathcal{F}$ has a piecewise continuous $f-$invariant ergodic measure $\nu(dx) = \rho(x)dx$, a natural question is to ask if this measure corresponds to the Lebesgue measure through the conjugate map h_t i.e., $f = h_t^{-1} \circ \Psi_t \circ h_t$ and $\nu = \lambda \circ h_t$ or, equivalently,

$$h_t(x) = \int_0^x \rho(u)du. \tag{4}$$

Proposition 2. *Let $f \in \mathcal{F}$ and assume there exists $x \in I$ such that $h_t(x) \neq t$ with $t = t(f)$ and $f'(x) \neq \Psi_t'(h_t(x)) \frac{\rho(x)}{\rho(f(x))}$. Then the measure $\nu^* = \lambda \circ h_t$ is singular with respect to λ. In particular h_t is a monotonic singular map.*

Proof. Take the derivative of $h_t \circ f = \Psi_t \circ h_t$ at the point x. If (4) is true, then $f'(x) = \Psi_t'(h_t(x)) \frac{\rho(x)}{\rho(f(x))}$ in contradiction with the assumption. Hence $\nu^* \neq \nu$. Moreover ν^* is $f-$invariant and ergodic. Therefore ν and ν^* are mutually singular. □

3 The HNSM system revisited and the Biham attack.

Distribution of cryptograms

We first use the HNSM as it is recommended by the authors and show that the distribution function of the cryptograms is singular. In fact, an analogous Proposition 1 can be stated for the family of tent map Φ_t of the HNSM system. Here, the canonical map is $\Phi_{\frac{1}{2}}$ and all maps Φ_t are topologically conjugate to $\Phi_{\frac{1}{2}}$ as in Proposition 1. We fixe $t \in]0,1[$ and simply denote Φ_t by f. Let $\phi_t : \Omega \to I$ be the corresponding map associated to f and defined by (1). On $\Omega = \mathcal{A}^{\mathbb{N}}$ (with $\mathcal{A} = \{1,2\}$) we consider the Bernoulli probability given by the infinite power $\mu_t = \bigotimes_{i=0}^{\infty} P_t$ where $P_t(\{1\}) = t$ and $P_t(\{2\}) = 1-t$. It is easy to see that the map ϕ_t gives rise to an automorphism of the measure preserving system (m.p.s.) (Ω, σ, μ_t) to the m.p.s. (I, f, λ), where λ denotes the Lebesgue measure on I. This is due to the fact that for any $a_1 \ldots a_n \in \mathcal{A}^n$ one has $|I_a| = t^u(1-t)^v$ where u (resp. v) is the number of indices i such that $a_i = 1$ (resp. $a_i = 2$). Let $(i_n)_{n \geq 1}$ be a sequence of random variables independently identically distributed (i.i.d.). As it is specified by the authors we choose the equidistribution law on $\mathcal{A} = \{1,2\}$ and put $Z_n(y) = f_{i_n}^{-1} \circ \ldots \circ f_{i_1}^{-1}(y)$. Then

Proposition 3. *Let i_n be a sequence of random variables i.i.d. of uniform distribution law in $\mathcal{A}$. Then*

$$\lim_{N\to\infty} \sup_{y \in I} \left| \frac{\operatorname{card}\{n \leq N;\ Z_n(y) \leq x\}}{N} - F_t(x) \right| = 0 \quad \text{a.e.} \tag{5}$$

where $F_t(x)$ is the distribution function of the probability $m_t = \mu_{\frac{1}{2}} \circ \phi_t^{-1}$ which is continuous and singular if $t \neq \frac{1}{2}$. Moreover $m_{\frac{1}{2}} = \lambda$ and for $0 < t < t' < 1$ the measures m_t and $m_{t'}$ are mutually singular.

Proof. Suppose $1 \leq k \leq n$. From the definition of $Z_n(\cdot)$ we have $Z_n(y) \in J_{i_n \ldots i_{n-k+1}}$ where

$$J_{i_n \ldots i_{n-k+1}} = I_{i_n} \cap f^{-1}(I_{i_{n-1}}) \cap \ldots \cap f^{-(k-1)}(I_{i_{n-k+1}}).$$

Consider the partition $\mathcal{P} = \mathcal{J}(t)$ and define the partition $\mathcal{P}_k = \bigvee_{s=0}^{k} f^{-s}\mathcal{P}$. Elements of $\mathcal{P}_k$ are intervals of the form J_a, $a \in \mathcal{A}^k$, and one see that

$$\forall n \geq k,\ X_n(y) \in J_a \iff i_n i_{n-1} \ldots i_{n-k+1} = a. \tag{6}$$

For any string a in $\mathcal{A}^k$ define the cylinder set $C_a = \{\omega \in \Omega;\ \omega_0 \ldots \omega_{k-1} = a\}$. From $C_a \cap \Omega_f = \phi_t^{-1}(J_a \cap X_f)$ and $\mu_{\frac{1}{2}}(C_a \cup \Omega_f) = \mu_{\frac{1}{2}}(C_a) = 2^{-k}$ we get

$$m_t(J_a) = \mu_{\frac{1}{2}}(\phi_t^{-1}(J_a)) = 2^{-k}. \tag{7}$$

For any $x \in I$, let $b(x)$ be the string b in $\mathcal{A}^k$ such that J_b contains x and let $\mathcal{A}^k(x) = \{a \in \mathcal{A}^k;\ J_a \subset [0,x]\}$. Hence

$$2^{-k}\operatorname{card}(\mathcal{A}^k(x)) \leq m_t([0,x]) \leq 2^{-k}(\operatorname{card}(\mathcal{A}^k(x)) + 1). \tag{8}$$

Put

$$\zeta_{N,y}(x) = \operatorname{card}\{n;\ 1 \leq n \leq N \,\&\, Z_n(y) \leq x\}$$

and

$$\eta_{k,N}(a) = \operatorname{card}\{n;\ k \leq n \leq N \,\&\, i_n \ldots i_{n-k+1} = a\}.$$

From (6) we derive

$$\sum_{a \in \mathcal{A}^k(x)} \eta_{k,N}(a) \leq \zeta_{N,y}(x) \leq \eta_{k,N}(b(x)) + \sum_{a \in \mathcal{A}^k(x)} \eta_{k,N}(a)$$

and by (8) we get the estimate

$$\sup_{y\in I}\left|\frac{1}{N}\zeta_{N,y}(x) - m_t([0,x])\right| \leq$$
$$\left|\frac{1}{N}\sum_{a\in\mathcal{A}^k(x)} \eta_{k,N}(a) - \mathrm{card}(\mathcal{A}^k(x))2^{-k}\right| +$$
$$\frac{1}{N}\eta_{k,N}(b(x)) + 2^{-k} + \frac{k}{N}.$$

Using the assumption on the random variables i_n we obtain

$$\limsup_{N\to\infty}\ \sup_{y\in I}\left|\frac{1}{N}\zeta_{N,y}(x) - F_t(x)\right| \leq 2^{1-k} \quad \text{a.e.} \tag{9}$$

for all integers $k \geq 1$. This proves the first part of the proposition. Now we remark that the m.p.s. (I, f, m_t) is ergodic (in fact it is isomorphic to the Bernoulli shift $(\Omega, \sigma, \mu_{\frac{1}{2}})$). Since $m_t([0,t]) = \frac{1}{2}$ by construction, we have $m_t \neq m_{t'}$ if $t \neq t'$ and we can conclude, by a classical theorem, that m_t and $m_{t'}$ are mutually singular. The particular case $t = \frac{1}{2}$ follows from $m_{\frac{1}{2}}(J_a) = |J_a|$ for all $a \in \mathcal{A}^k$ and all $k \geq 1$. □

Since the location of $Z_n(y)$ mainly depends on the last values $i_n, i_{n-1}, \ldots, i_{n-k+1}$, Proposition 3 can be viewed as a theoretical estimate when the number of iterate M tends to infinity. For a fixed number of iterates we have the following:

Proposition 4. *Let $v_n = (i_M^{(n)}, \ldots, i_1^{(n)})$ be a sequence of random variables i.i.d. of uniform distribution law in $\mathcal{A}^M$. Put $f = \Phi_t$ and*

$$T_n(y) = f^{-1}_{i_M^{(n)}} \circ \ldots \circ f^{-1}_{i_1^{(n)}}(y).$$

Then

$$\lim_{N\to\infty}\sup_{y\in I}\left|\frac{\mathrm{card}\{n;\ 1 \leq n \leq N \ \&\ T_n(y) \leq x\}}{N} - F_t(x)\right| \leq 2^{1-M}. \tag{10}$$

Proof. Using (6) and the uniform distibution of the sequence of random variables v_n in $\mathcal{A}^M$, we obtain (10) by the same arguing as in Proposition 3. □

3.1 Improvement of the HNSM system

Let us point out a weakness of the HNSM system. In Proposition 3 and 4 we have $F_t(t) = \frac{1}{2}$. Therefore, if the random generator used to produce the sequence $i_1, i_2, \ldots$ has good random properties then the key t is given by the value θ such that the half part of the set of observed cryptograms (whatever is the plaintext) is contained in $[0, \theta]$. To avoid this attack we change the random part of the enciphering algorithm in order to get the uniform distribution (3) in place of a singular one.

Proposition 5. *Let x_n be a sequence of random variables i.i.d. of law P_t and let $f = \Phi_t$, $(t \in]0,1[)$ as above and let $X_n(y)$ be defined by (2) . Then (3) holds. More precisely*

$$\lim_{N\to\infty}\sup_{y\in I}\left|\frac{\mathrm{card}\{n;\ 1 \leq n \leq N \ \&\ X_n(y) \leq x\}}{N} - x\right| = 0 \quad \text{a.e.} \tag{11}$$

Proof We have already noticed that

$$\lambda = \mu_t \circ \phi_t^{-1}.$$

Then, analogous arguing leads to the inequality

$$\sup_{y\in I}\left|\frac{1}{N}\zeta_{N,y}(x) - x\right| \leq$$

$$\left|\frac{1}{N}\sum_{a\in\mathcal{A}^k(x)} \eta_{k,N}(a) - \sum_{a\in\mathcal{A}^k(x)} |J_a|\right| + \frac{1}{N}\eta_{k,N}(b(x)) + |J_{b(x)}| + \frac{k}{N}$$

(we have used the same notations as in the proof of Proposition 1). Fortunately, from the assumption on the random variables i_n, one has

$$\lim_{N\to\infty}\frac{1}{N}\mathrm{card}(\{n;\ 1\leq n\leq N\ i_n i_{n-1}\dots i_{n-k+1} = a\} =$$

$$\mu_t(C_a) = |J_a| \quad \text{a. e.}$$

for all $a \in \mathcal{A}^k$. Now the result follows from (12). □

Proposition 6. *Let $v_n = (i_M^{(n)}, \dots, i_1^{(n)})$ be a sequence of random variables i.i.d. of distribution law $\bigotimes_{k=1}^M P_t$ in $\mathcal{A}^M$. Put $f = \Phi_t$ ($t \in]0,1[$) and $T_n(y) = f^{-1}_{i_M^{(n)}} \circ \dots \circ f^{-1}_{i_1^{(n)}}(y)$. Then*

(13) $\lim_{N\to\infty}\sup_{y\in I}\left|\frac{\mathrm{card}\{n;\ 1\leq n\leq N\ \&\ T_n(y)\leq x\}}{N} - x\right| \leq 2(\max\{t, 1-t\})^M.$

Proof A simple modification of the above proof shows that (12) turns into

$$\sup_{y\in I}\left|\frac{\mathrm{card}\{n;\ 1\leq n\leq N\ \&\ T_n(y)\leq x\}}{N} - x\right| \leq$$

$$\left|\frac{1}{N}\sum_{a\in\mathcal{A}^M(x)} \mathrm{card}\{n\leq N;\ v_n = a\} - \sum_{a\in\mathcal{A}^M(x)} |J_a|\right| +$$

$$\frac{\mathrm{card}\{n\leq N;\ v_n = a\}}{N} + |J_{b(x)}|.$$

equation 14

Passing to the limit, the left member in (14) tends to $2|J_{b(x)}|$ and (13) follows from $|J_{b(x)}| \leq (\max\{t, 1-t\})^N$.

3.2 Remark 2

Behind the above results is the notion of natural extension $\tilde{f}$ of f in terms of measure preserving systems (see [9]). It is easy to see that $\tilde{f}$ can be defined explicitely on the probability space $(I\times I, \tilde{\lambda})$ (with $\tilde{\lambda} = \lambda\otimes\lambda$) by

$$\tilde{f}(x,y) = (f(x), f^{-1}_{i(x)}(y))$$

where $i(x) = j$ if $x \in I_j$. It is a simple fact that the m. p. s. $(\tilde{f}, I\times I, \tilde{\lambda})$ is ergodic. So that a weak form of Proposition 5 can be obtained directly from the ergodic theorem of Birkhoff as soon as we have noticed that

$$\tilde{f}^n(x,y) = (f^n(x), f^{-1}_{i_{n-1}} \circ \dots \circ f^{-1}_{i_0}(y))$$

where $s(x) = (i_0, i_1, \ldots)$. One attack of the HNSM system uses the linearity of the functions Φ_t. There is a simple way to change this by using a homeomorphism H of I with $H(0) = 0$ and replace each map Φ_t by $\Gamma_t = H^{-1} \circ \Phi_t \circ H$. This map is continuous, strictly increasing on the interval $I_1 = [0, H^{-1}(t)[$ and strictly decreasing on $I_2 = [H^{-1}(t), 1]$. Moreover H corresponds to the distribution function of the measure $\nu = \lambda \circ H$. which is ergodic under Γ_t. It is suitable to choose H of class C^1 and moreover to assume that $\frac{1}{D} \leq H' \leq D$ for a constant $D > 0$. Put $f = \Phi_t$ and $F = \Gamma_t$ for short and let F_i be the inverse of F restricted to I_1, $i = 1, 2$. For a given sequence $(x_n)_n$ of random variables in $\mathcal{A}$, let $y' \in I$ and $X'_n(y') = F_{x_n}^{-1} \circ \ldots \circ F_{x_1}^{-1}(y')$. A straightforward computation gives $X'_n(y') = H^{-1} \circ f_{x_n}^{-1} \circ \ldots \circ f_{x_1}^{-1} \circ H(y')$. Consequently, the distribution of $(X'_n(y'))_n$ derives from the distribution of $X_n(y) = f_{x_n}^{-1} \circ \ldots \circ f_{x_1}^{-1}(y)$ with $y = H(y')$ so that Proposition 6 corresponds to the following:

Proposition 7. With the above notations we have

$$(11') \qquad \lim_{N \to \infty} \sup_{y' \in I} \left| \frac{\operatorname{card}\{n;\ 1 \leq n \leq N \ \&\ X'_n(y') \leq x\}}{N} - H(x) \right| = 0 \quad \text{a.e.}$$

We leave the reader to write the proposition which corresponds to Proposition 4 after the introduction of H.

3.3 The Biham attack

Another attack using the linearity of the HNSM system was proposed by Biham [7], using the fixed point 0. We first consider the attack from the enciphering side.

Assume that an opponent know the plaintext y and tries to compute the parameter t which defines Φ_t. The chosen plaintext attack using this weakness is the following: choose a small plaintext y (say $10^{-3}t$) and observe its cryptogram by repeating enciphering until we get a cryptogram $X_m(y) = f_{x_n}^{-1} \circ \ldots \circ f_{x_1}^{-1}(y)$ which is very small. Then we can expect that $X_m(y) = t^m y$ and then we can compute t. But this event occurs with the theoretical probability $P(x_1 = 1, \ldots, x_m = 1) = t^m$. With $0 < t < 1/2$ and $m = 124$ this probability is less than 10^{-37}. This means that the average number of enciphering requiered in order to obtained this event with probability 1 is 10^{37}. This is too big by today's computer standards.

Now we consider the attack from the deciphering side. This one uses a small cryptogram x and observes the value $\bar{y}$ given by the deciphering modulus. Since $\bar{y}$ is only an estimate of $f^m(x)$, the success depends on the data representation. Assume that the plaintext uses n bits and the ciphertext k bits. As we shall see in the next section to avoid computational noise effect, we have to choose k larger than the number m of iteration (but not too large). To perform the attack we may assume that $\alpha < t < 1/2$ and choose $x = \alpha^{-m}$. Then we observe the deciphering value $\bar{y}$ with the estimation $|xt^{-m} - \bar{y}| \leq 2^{-n}$. Thus $t = \alpha \bar{y}^{-\frac{1}{m}} + \mathcal{O}(2^{-n})$. This does not give t with enough precision to describe completely the secret key t.

4 HARALIA Cryptosystem

The family of maps The interval maps used as a basis for defining a new markovian cryptosystem, according to the preceeding arguments, will satisfy some extra conditions dictated by cryptographic and computational requirements. These maps will be non linear, even restricted to any subinterval of the unit interval, and should be rational since the plaintext and the cryptogram must be rational numbers. The computation noise can also be kept at a low enough level only for such maps. We have chosen the family $\mathcal{H} = \{F_a\ ;\ 0 < a \leq \infty\}$ where the interval map $F_a : I \to I$ is defined by:

$$(15)\qquad F_a(x) = \begin{cases} \frac{a-(2a+1)x}{x+a} & \text{if } 0 \leq x < \frac{a}{2a+1} \\ \frac{2a-2ax}{x+a} & \text{if } \frac{a}{2a+1} \leq x \leq 1, \end{cases}$$

and $F_\infty = \Psi_{\frac{1}{2}}$ is the canonical map of this family. An elementary computation gives the two inverses of $f = F_a$, namely

$$f_1^{-1}(y) = \frac{2a - ay}{2a + y} \quad \text{and} \quad f_2^{-1}(y) = \frac{a - ay}{2a + 1 + y}.$$

The parameter a will be used as the secret key. Notice that $\mathcal{H}$ is a subfamily of $\mathcal{F}$. The following figures illustrate the graph of F_a for $a = 0.2$.

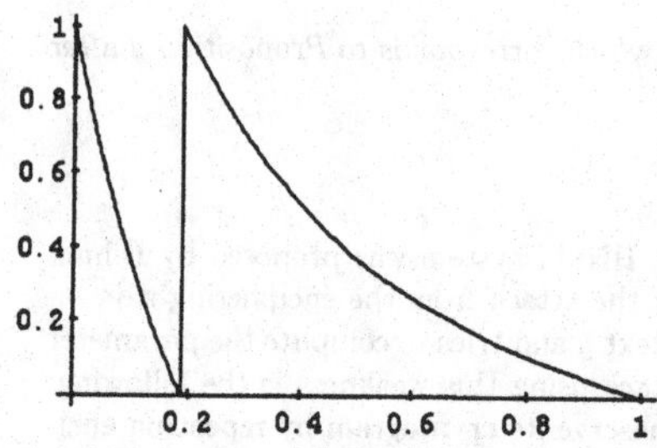

Fig.1. A rational interval map F_a

4.1 Ergodic properties of F_a.

This study is required for understanding both the distribution of cryptograms and the choice of the random process used by the enciphering modulus. In this section, we choose f in $\mathcal{H}$ of parameter $a \in]0, +\infty[$ i.e. $f = F_a$. For short, we omit the reference to the parameter a if there is no ambiguity. In particular $I_1 = [0, \frac{a}{2a+1}[$, $I_2 = [\frac{a}{2a+1}, 1]$. The restriction of f on I_i will be denoted by f_i. Maps f_i and f_i^{-1} are restriction of rational linear maps that we denote with the same notations. From the work of F. Schweiger [8] the ergodic measure $\nu(dx) = \rho(x)dx$ for f given by Theorem 1 can be computed explicitely. In fact we get (thanks to Mathematica!)

$$\rho(x) = \frac{2}{(3a + x)(2 + 3a + x)\log\left(\frac{(1+3a)(2+3a)}{9a(1+a)}\right)}.$$

Proposition 8. *Let $t \in]0, 1[$ and let $h : I \to I$ be the homeomorphism defined by $h(0) = 0$ and $f = h^{-1} \circ \Psi_t \circ h$. Then the measure $\mu = \lambda \circ h$ is singular with respect to the Lebesgue measure. In particular h is singular.*

Proof By assumption μ is ergodic for f so that μ is singular with respect to $\nu(dx) = \rho(x)dx$ (which is equivalent to λ) or $\mu = \nu$. In this latter case, h is given by (4). Then, $\rho(1)f'(0) = \frac{-1}{t}\rho(0)$ and $\rho(0)f'(1) = \frac{-1}{1-t}h'(1)$. This implies $f'(0)f'(1) = \frac{1}{t(1-t)}$. But $f'(0)f'(1) = 4$ so that $t = \frac{1}{2}$. Now we can write $\frac{\rho(1)}{\rho(0)} = \frac{-2}{f'(0)}$ and using the expression of ρ we get $\frac{a(3a+2)}{(3a+1)(a+1)} = \frac{a}{a+1}$, an equality which is never satisfied. Therefore $\mu \neq \nu$ and the result follows as in Proposition 2. □

4.2 Remark 3

The above proposition shows that the family $\mathcal{H}$ is not computationally equivalent to the family $\{\Psi_t \; ; \; 0 < t < 1\}$. These maps can be used in the same way as the tent maps of the HNSM system.

4.3 Distribution properties of cryptograms

This part gives more properties of maps $f \in \mathcal{H}$ and constitute the heart of the HARALIA system.

Let x_n be a sequence of random variable in $\{1,2\}$. Our aim is to find distribution properties of x_n such that for all $y \in I$ the sequence $X_n(y)$ given by (2) satisfies (3). In fact, we shall see how to get (Corollary 4.13)

$$\lim_{N\to\infty} \sup_{y\in I} \left| \frac{\mathrm{card}\{n \, ; \, 1 \leq n \leq N \;\&\; X_n(y) \leq x\}}{N} - \nu([0,x]) \right| = 0. \tag{16}$$

Following [8] we introduce the dual map of any fractional linear map $g : x \mapsto \frac{\alpha+\beta x}{\gamma+\delta x}$ as being the linear map $g^* : x \mapsto \frac{\delta+\beta x}{\gamma+\alpha x}$. Here, rational linear maps are defined on $\mathbf{R}\cup\{\infty\}$ as usual and we have from the definition $(g^*)^{-1} = (g^{-1})^*$. The reason of this definition comes from the following dual relation

$$K(g(x),y)|g'(x)| = K(x, g^*(y))|(g^*)'(y)| \tag{17}$$

where K is given by $K(x,y) = \frac{1}{(1+xy)^2}$. For $f \in \mathcal{H}$ we consider the dual map f_i^* of each rational map f_i corresponding to each restriction $f|_{I_i}$ and we define the dual map f^* on the interval $L = [\frac{1}{3a+2}, \frac{1}{3a}]$ by

$$f^* = \begin{cases} \frac{1-(2a+1)x}{a(1+x)} \; (= f_1^*(x)) & \text{if } \frac{1}{3a+2} \leq x < \frac{1}{3a+1} \\ \frac{1-2ax}{a(1+2x)} \; (= f_1^*(x)) & \text{if } \frac{1}{3a+1} \leq x \leq \frac{1}{3a} \,. \end{cases}$$

We set $L_1 = [\frac{1}{3a+2}, \frac{1}{3a+1}[$ and $L_2 = [\frac{1}{3a+1}, \frac{1}{3a}]$. Now we introduce the transformation $T^* : I \times L \to I \times L$ defined by

$$T^*(x,y) = (f(x), (f^*_{i(x)})^{-1}(y))$$

where $i(x) = j$ iff $x \in I_j$. From (17) we derive (as in [8]) that the probability

$$\gamma^*(dx,dy) = C_a \frac{1}{(1+xy)^2} dxdy$$

on $I \times L$ is invariant and ergodic under the action of T^* (The constant C_a = - $1/\log\left(\frac{(3a+1)(3a+2)}{9a(a+1)}\right)$ is a constant of normalisation). Moreover the m.p.s. $(T^*, I \times L, \gamma^*)$ is also the natural extension of f and the marginal measure $\nu^* = \rho^*(y)dy = C_a \int_0^1 \frac{1}{(1+xy)^2} dx = \frac{C_a}{1+y}$ is invariant and ergodic under the map $f^* : L \to L$.

Proposition 7. *Let $h : I \to L$ be defined by $h(x) = \frac{a+x}{a(3a+2+x)}$. Then $f^* \circ h = h \circ f$ and $\nu^* \circ h = \nu$.*

Proof Using the method of Proposition 1, it is easy to see that f and f^* are topologically conjugate, say by $h : I \to L$. If $\nu^* \circ h = \nu$ then we see that h is differentiable and $\rho^* \circ h \cdot h' = \rho$. This differential equation can be solved and the solution leads to the expression of h given in the proposition. □

Theorem 3. *Let $T : I \times I \to I \times I$ be the map defined by*

$$T(x,y) = (f(x), f_{i(x)}^{-1}(y))$$

and let

$$\gamma(dx, dy) = \frac{2a(a+1)C_a}{(2a(a+1)+(x+a)(y+a))^2} dx\, dy\,.$$

Then the m.p.s. $(T, I \times I, \gamma)$ is ergodic.

Proof From the above proposition we see that the map $(x,y) \to (x, h(y))$ gives rise to an isomorphism between $(T, I \times I, \gamma)$ and $(T^*, I \times L, \gamma^*)$. In fact the expression of γ in the theorem follows easily from the fact that if we replace $\frac{1}{(1+xy)^2}$ by

$$W(x,y) = \frac{|h'(y)|}{(1+xh(y))^2} \quad (= \frac{|h'(x)|}{(1+h(x)y)^2}),$$

then

$$W(f(x), y)|f'(x)| = W(x, f(y))|f'(y)|.$$

This ends the proof. □

Since $(T, I \times I, \gamma)$ is isomorphic to the natural extension of (f, I, ν) we get from the ergodic : Corollary *For $\lambda-$almost all x in I formula (16) holds.* □

By a simple application of Proposition 1, we have

Corollary 4. *Any two different elements in $\mathcal{H}$ are topologically conjugate by a unique map which is singular.* □

5 The HARALIA Cryptosystem: computational point of view

Data Representation

In order to use this algorithm one has to associate to an n tuple of bits $\lambda_0, \ldots, \lambda_{n-1}$ a rational number y in the unit interval in a bijective and computationally efficient way. The rational numbers that are used in the computation are writen on k bits with $k > n$. The precise value of k wil be discussed in the next section. The cryptogram wil be a rational number written on k bits. Thus this algorithm is data expansive.

Choose a pair of relatively prime integers r and s such that $r \leq s$. Given an n-tuple of bits $\lambda_0, \ldots, \lambda_{n-1}$, one can associate a rational number p/q in the unit interval by the following operation:

$$\frac{p}{q} = \sum_{i=1}^{n} \lambda_i . 2^{-i} \frac{r}{s}$$

The floating point representation of the plaintext is obtained by making a long division of p by q on n bits. The tailing $k-n$ bits can be modified in any way. They are initially set to zero.

The inverse transform, which allows the extraction of n bits from a rational number in the unit interval, is the following. Given such a rational number p/q, one must first compute $x = (p/q).(s/r)$. The set of n bits is obtained by writing x, which is a number written on k bits, as $x = \sum_{i=1}^{n} \lambda_i 2^{-i} + \rho$ where $\rho < 2^{-n}$ by using a substract and discard algorithm, which is the analogous to the one used in obtaining the solution of a superincreasing knapsack.

Therefore the $k-n$ least significant bits of x are discarded.

Computation data

The enciphering algorithm consists in the iterated computation of rational operations. The deciphering algorithm is analogous. Before implementing the algorithm one has to determine the number k of bits on which the rational numbers will be represented as a function of the number of bits of the plaintext n and the number m of iterations so that the enciphering and deciphering operations will not be affected by computation noise. This issue is fundamental. If k is too small, the deciphering of a cryptogram will not yield the original plaintext. If k is too large, computation time will be increased. Furthermore the algorithm will be less efficient and the probability of success of certain attacks will be increased as it will be shown later on.

Two different implementations are possible. The first is to implement a floating point representation of positive rational numbers which are less than 1, with $log_2 k$ decimal values. The second one is to consider such a rational as fraction, and represent it as as a couple of integers (numerator and denominator), and program the corresponding usual operations for such numbers. The first representation was used for implementing the algorithm.

The computation noise in the final result originate in the fact that one of the intermediate points of the unit interval that is handled in the computation is close to a point where the enciphering or deciphering function is not continuous. If the computation noise is to high then one of the n most significant bits of the rational number will eventually be altered and will give rise to a deciphering error.

This problem is very hard to handle from a theoretical point of view. However it has a practical solution.

Experimental results show that, in average, $k = m + n$ is sufficient to obtain a noise free enciphering and deciphering operation on the n most significant bits of the plaintext, where m is the number of iterations of the system.

In the case where for a given plaintext, and a given key, the enciphering deciphering cycle did not yield the original plaintext, repeated experiments show that a single random modification of the $k-n$ least significant bits of the rational number associated to the plaintext gives the desired result.

The secret key

The key to the system will be the rational number a, written on k bits, used as a parameter for the interval map in $\mathcal{H}$. The choice of different keys yield independant cryptograms. The number m of iterations must also be kept secret.

Enciphering transformation

Let f be the map in $\mathcal{H}$ of rational parameter a. For all y in $]0,1[$, $f^{-m}(y)$ is a set having 2^m rational elements all of which are in the unit interval. If y is rational $f^{-m}(y)$ is a set of rational numbers. Let y be an element of the unit interval associated to a plaintext. The cryptogram of y using the secret key a will be computed as follows. Let $\tilde{f}^{-1}(y)$ be an element chosen in the set $f^{-1}(y)$ by using a probability of choice induced by the Lebesgue measure on the unit interval and above results. Iterating this method m times it yields an element denoted by $\tilde{f}_a^{-m}(y)$ which will be the cryptogram of the plaintext y. More precisely, choose at random a point ξ in I according to the uniform law (or – better– the law of density ρ_a) and take

$$\tilde{f}^{-m}(y,\xi) = f^{-1}_{i(f^{m-1}(\xi))} \circ \ldots \circ f^{-1}_{i(\xi)}(y)$$

Therefore a plaintext $y \in]0,1[$ enciphered with the key a can have 2^m possible distinct cryptograms. With the help of the Corollary 4.11 one can assert that the cryptograms will have a distribution which, though not uniform on the unit interval, will not have any accumulation point within the interval. The computation of x such that $x = \tilde{f}^{-1}(y)$ is done by computing: $x = \frac{2a.(a+1)}{y+2a} - a$ or by computing $x = \frac{2a.(a+1)}{y+2a+1} - a$. In practice, the choice between the two values is done with the probability distribution $\{\frac{a}{2a+1}, \frac{a+1}{2a+1}\}$.

The Deciphering Transformation

The relation:

$$f^m \circ \tilde{f}^{-m}(y) = y$$

is always true independantely of the choices done at each of the m steps of the computation of the cryptogram. Given a cryptogram c and the secret key a, m the number of iterations, the rational number y is computed by the equation $y = f^m(c)$. The n bits of the cryptogram are then extracted from y. All these computations will be done using k bits of precision on the rational numbers.

Fixed points and number of iterations

The mappings under consideration have fixed points. If an opponent to the system manages to obtain the coordinates of a fixed point, he will have necessary information for obtaining the equation of one of the two mappings involved, and from there deduce the secret key a if he knows the number of iterations m. On the other hand, the functions used being chaotic, the approximate knowledge of the value of one of those fixed points does not yield information on the exact value of this fixed point. Let m be the number of iterations used in the cryptosystem. There are 2^m distinct fixed points for the enciphering function, which are also the fixed points for the deciphering map. It is legitimate to suppose that they are randomly distributed on the unit interval, that is, in the binary expansion of any one of those points in negative powers of two, the coefficients that appear are random variables with equiprobability. The set of those points being of cardinality 2^m the above coordinates are binary sequences of m' bits with $m' \geq m$. Suppose that the number of iterations m satisfies $m > n$ where n is the number of bits of the plaintext and let P be one of those fixed points. Let

$$D(P) = (p_1, p_2, \ldots, p_n, \ldots, p_{m'})$$

be the coefficients of the binary expansion of P. In order for a fixed point P' to be known without error by a system yielding n bit numbers it is necessary and sufficient that this point P' must have the least significant bits of $D(P')$ equal to zero, namely the coefficients $p_{n+1}, \ldots, p_m$ must all be equal to 0. The probability for this to happen is, following the above considerations, equal to:

$$P_c = 2^{-(m-n)}.$$

If $m - n$ is equal to 60, one can consider this event to be of negigible probability. This result gives a lower bound on the number of iterations that are necessary to have a safe system. If $n = 64$ then m must be at least 124.

The detailed algorithm The above theoretical study leads to the following enciphering algorithm:

- Choose the key $a \in]0, +\infty[$ writen on k bits and the number m of iterations.
- Determine $f = F_a$, $g_1 = f_1^{-1}$, and $g_2 = f_2^{-1}$.
- Associate a rational number y on k bits to the plaintext.
- Encipher y by applying the enciphering function and obtain x.
- Check that x is deciphered into y. If this is not the case modify the $k - n$ least significant bits of y and goto 4
- x is the cryptogram of the plaintext.

The Biham attack

The attack consists of finding a pair of rational elements of the unit interval, one being a plaintext the other a cryptogram in relation on one branch of the iterated enciphering map. By obtaining a formal relation between the two points one may deduce the secret key a of the system. It is less likely to succeed than in the case of the HNSM system for the the following reasons.

- The formal relation between the two points, when the rational map is iterated m times, is a rational fraction whose numerator and denominator are polynomials in the variable a of degree $m/2$.
- The nature of the distribution of the cryptograms (uniform like with no accumulation points) makes it very difficult, in terms of probability, to find such a pair.
- The precision on the plaintext is n bits, the precision on the cryptogram is on k bits. Therefore solving an equation, to find the secret key a, involving these quantities yields a result which will be precise on at most n bits. The precision on the key a will be at most n bits, which is not precise enough since k is much larger than n.

6 Conclusion

The *HARALIA* system that is presented is a particular implementation of the class of markovian cryptosystems that have been introduced and been shown to be safe by theoretical arguments. The resulting practical requirements that are needed are compatible with an efficient implementation.

REFERENCES

[1] ADLER R. and FLATTO L. : *Geodesic flows,universal maps,and symbolic dynamics*, Bulletin of the AMS, **25**, No 2 (1991), 229–334.

[2] DENKER M., GRILLENBERGER C. : SIGMUND K.: *Ergodic theory on compact spaces*, Lecture Notes in Math., **525** (1976), Springer–Verlag.

[3] GUTOWITZ H. : A cellular automaton cryptosystem; specification and call for attack, preprint 1992.

[4] HABUTSU T., NISHIO Y., SASASE I., MORI S. : A secret key cryptosystem by iterating chaotic maps. Eurocrypt' 91.

[5] KNUTH D. : *Semi numerical algorithms*, Adisson Wesley (1980).

[6] PARRY W. : Symbolic Dynamics and transformations of the unit intervals. Trans. Amer. Math. Soc. **122** (1966), 368–378.

[7] RUMP SESSION, Eurocrypt 91: Biham's attack.

[8] SCHWEIGER F. : Ergodic properties of piecewice fractional linear maps, Arbeitsbericht, Math. Institut Universität Salzburg (1980), 24–32.

[9] SCHWEIGER F. : *Ergodic properties of fibered systems*, Institut für Mathematik der Univ. Salzburg, Draft version, April 1991.

[10] WOLFRAM S. : Cryptography with cellular automata, Proceeding of Crypto'85 (1985), 429–432.

Université de Toulon et du Var
Groupe d'Étude du Codage de Toulon
B. P. 131
Avenue de l'Université
83957 La Garde, France

Université de Provence
URA CNRS No225
Equipe DSA, case 96
3, place V. Hugo
13331 Marseille cedex 3

AN ASYMPTOTIC THEOREM
FOR SUBSTITUTION-RESISTANT AUTHENTICATION CODES

S. Maset and A. Sgarro
University of Trieste, Trieste, Italy

ABSTRACT

We prove a Shannon-theoretic theorem for authentication codes resistant against impersonation and substitution. The code construction is based upon block-designs. We conjecture that this construction is asymptotically the best possible.

1. INTRODUCTION

A Shannon-theoretic frame for authentication theory has been put forward by G. Simmons (cg e.g.[1,2,3]). A *multicode* is a finite random triple XYZ (*message, codeword, key*). Under each key (encoding rule), encoding and decoding are assumed to be deterministic. Key and message are independent random variables. Below we give an example of an encoding matrix and of the corresponding authentication matrix χ; in the latter one has $\chi(z,y)=1$ iff key z authenticates codeword y, that is iff there exists a message x which is encoded to y under key z.

	x1	x2	x3
z1	y1	y2	y3
z2	y3	y4	y1

	y1	y2	y3	y4
z1	1	1	1	0
z2	1	0	1	1

Examples of multicodes are source codes (Z has one value), ciphers, and *authentication codes*. If a zero-error decoding scheme is prescribed, as in Simmons' original formulation, each codeword can appear at most once in each row of the encoding matrix. Authentication theory begins for good when the attacks are described against which an authentication code must be resistant. *Impersonation attacks* are the simplest possible attacks against an authentication code. In this case the opposer chooses a codeword y hoping it to be authenticated by the current key Z, which he ignores. The probability of fraud (successful attack) for the enemy's optimal strategy is:

$$P_I = \max_y \ \mathrm{Prob}\{\chi(Z,y)=1\}$$

More sophisticated attacks are *substitution* and *deception*, whose fraud probabilities are, respectively:

$$P_S = \Sigma_c \Pr\{Y=c\} \max_{y\neq c} \Pr\{\chi(Z,y)=1|Y=c\}, \qquad P_D = \max(P_I,P_S)$$

In the case of substitution, the opposer grabs the legal codeword c and replaces it by y; in the case of deception one has to beware both of impersonation and substitution. (We are using Massey's definition of deception, which is possibly more palatable to information theorists; in Simmons' more involved game-theoretic setting the right side is just a lower bound to P_D).

At Eurocode 90 [4] coding theorems for impersonation codes have been put forward, which are Shannon-theoretic in a sense which is familiar to information theorists; in particular, a "small" probability of erroneous decoding is allowed. The formalization given here (cf also [5]) has been modified so as to explicitely include key material into the available resources, as we did and we do for codeword material. In the negligible-error case we insist on deterministic decoding, *but we do not insist that it should be deterministically successful!* In other words (as is usual in the Shannon-theoretic approach) we allow for a "small" decoding error-probability P_e. (As a general reference to Shannon theory we suggest e.g. [6]). We formally define our model in the case of a generic fraud attack (impersonation, substitution, deception). $\mathcal{X}^n$, $\mathcal{Z}$ and $\mathcal{Y}$ are the message-alphabet, the key-alphabet and the codeword-alphabet, respectively. The random key Z is uniform and independent of the random message X^n of length n. Let f: $\mathcal{Z}\times\mathcal{X}^n\rightarrow\mathcal{Y}$, g: $\mathcal{Z}\times\mathcal{Y}\rightarrow\mathcal{X}^n\cup\{?\}$ be a multicode (an encoder-decoder pair), and let P_e, P_F be the corresponding probabilities of erroneous decoding and of successful fraud; in particular $P_e = \mathrm{Prob}\{g(f(X^n))\neq X^n\}$. We are no longer requiring that codewords appear at most once in each row of the encoding matrix; functions f and g are arbitrary, save that g(z,y)=? iff $\chi(z,y)=0$. We denote by $R_Z=n^{-1}\log|\mathcal{Z}|$, $R_Y=n^{-1}\log|\mathcal{Y}|$ the key-rate and the codeword-rate, respectively; these parameters measure the "cost" of the code, while P_e, P_F measure its "reliability" (logs are to the base 2). We say that a rate-pair (a,b) in the non-negative real quadrant is ε-*achievable* against the specified fraud attack when for any τ (τ>0) and for all n large enough there are codes with $P_e\leq\varepsilon$, $P_F\leq\varepsilon$, $R_Z\leq a+\tau$, $R_Y\leq b+\tau$.

From now on we assume that the message source is stationary and memoryless, and is ruled by the probability vector P, or rather by its product extension P^n. Formerly, in [4], we have proved the following theorem (H(P)=H(X) denotes Shannon entropy):

Theorem. The region of ε-achievable rate-pairs for codes against impersonation is $\{R_z,R_y : R_z \geq 0, R_y \geq H(P)\}$.

Actually, and disappointingly, this is "almost" Shannon's theorem for source-codes (without authentication), as it implies that the best (most economical) achievable rate-pair is (0,H(P))! (Note, however, that, even if $R_z=0$ is achievable, we do not have finitely many keys: the achievability of 0 simply means that exponentially many keys will do, however slow the exponential growth.). It is no surprise, then, that a very naive code construction was enough to achieve the direct part of the theorem (cf [4]).

In this paper we take into account the case of substitution, either by itself or in conjunction with impersonation. The situation appears to be much more involved, and, possibly, much more interesting.

2. THE THEOREM

The best code construction we could devise shows the achievability of $R_z=H(P)$, $R_y=H(P)$ and is based on well-known facts relative to block-designs (we will use blocks as keys, points as codewords, and the incidence relation will stand for the authentication relation χ; block-designs have been largely used to construct good finite-length authentication codes).

Theorem. The pair (H(P),H(P)) is ε-achievable against deception (against the mixed attack impersonation-substitution).

At this point we are only able to conjecture that, asymptotically, this construction is the best possible, even if impersonation is dropped:

Conjecture. The region of achievable rate-pairs for codes against substitution is $\{R_z,R_y : R_z \geq H(P), R_y \geq H(P)\}$.

This conjecture is rather worrying, as it implies that protection against substitution is extremely costly (the key-rate is extremely high), the situation being similar to the case of the one-time-pad in cryptography. At the moment we are able to prove a converse theorem only for specific codes, e.g. Cartesian codes as defined in [1,2].

3. THE CODE CONSTRUCTION

This chapter assumes in the reader some familiarity with block-designs as can be derived, say, from [7]. Unfortunately, the proof of the theorem involves many computations, which are quite uninteresting and quite long to be written down in full. Below, we shall be very careful when describing our constructions, but very light-handed when it comes to computations, which will be only hinted at. However, we refer to [8], where all the computations have been explicitely carried out. By a *canonical* code we mean one with decoding error probability equal to zero, and with uniform random message. The main code construction we are going to exhibit concerns canonical codes; however, it can be soon recycled to meet our needs, using the following lemma. Technically, what we need in the lemma is the standard machinery of typical sequences (for us, typical messages) as defined in [6]; here we just recall that in a typical sequence the relative frequencies of letters are "approximately" equal to their probabilities; typical sequences are "approximately" equiprobable. Below, $\mathcal{T}_n$ is the set of typical sequences included in $\mathcal{X}^n$.

Lemma. If $\{C^*(n)\}$ is a sequence of canonical codes with message set equal to $\mathcal{T}_n$, one can construct a sequence $\{C(n)\}$ of codes for the given random message X^n such that: the key set is the same; the new codeword set is obtained by adding an extra codeword for each key; the decoding-error probability goes to zero; the impersonation probability is the same; the new substitution probability, $P_S(n)$, is upper-bounded by $\alpha(n) + 2^{n\beta(n)} P^*_S(n)$, where $\alpha(n)$ and $\beta(n)$ go to zero with n, and where $P^*_S(n)$ is the substitution probability of the canonical code.

Proof. The new encoding matrix coincides with the old one as for typical columns (for columns corresponding to typical messages); as for untypical columns use the extra codewords, one for each row (for each key). Decode as before when possible, decode arbitrarily in correspondence to the added codewords (recall however that g(z,y)=? iff z does not authenticate y). Easy computations complete the proof.

We recall (cf [7]) that in a 2-design with parameters (v,k,λ) each pair of points is incident to exactly λ blocks; v is the number of points and each block is incident to k points; a further parameter is b, the number of blocks. We need a very technical result, which asserts the existence of 2-designs with parameters

$$v=q^d,\ k=q^{d-1},\ \lambda=\frac{q^{d-1}-1}{q-1},\ b=\frac{q(q^d-1)}{q-1}$$

where q is an arbitrary power of 2 and d is any positive integer ($d\geq 2$) (cf [7]; these block-designs are of the type $\mathcal{A}_{d,d-1}$ obtained by generalizing affine geometries). In our main construction b is the number of keys, v is the number of codewords and k is the number of messages. The annoying fact is that the number of typical messages, $|T_n|$, is not necessarily a power of 2, as k should be, which implies that we have to "adjust" a little our block-designs before coming to the canonical codes C*(n) as required by the lemma. Let L be a fixed integer to be specified later, and let r=r(n) be the least integer such that 2^{rL} is larger than or equal to $|T_n|$. We take $q=2^r$, d=L+1: now we can use our block-design to construct a code C'(n) in the obvious way, as soon as one interprets the incidence relation as authentication (notice that, given the key z, which of the k codewords authenticated by z -or incident to z- should encode which of the k messages is totally irrelevant). The highly symmetric structure of 2-designs allows to easily compute both P'_I and P'_S:

$$P'_I = \frac{k}{v}, \quad P'_S = \frac{k-1}{v-1}$$

To construct C*(n) as required by the lemma, suppress any $2^{rl} - |T_n|$ columns from the encoding matrix of C'(n). A computation shows that $P^*_I = P'_I$, $P^*_S \le 2^L P'_S$, so that both $P^*_I(n)$ and $P^*_S(n)$ go exponentially to zero with n. The number of keys for C*(n) is b as for C'(n), the number of codewords is *at most* v, as some codewords may have got lost after suppressing the $2^{rl} - |T_n|$ columns. Now we use the lemma to construct the codes C(n) which are needed to prove the theorem; for C(n) the number of codewords is upperbounded by twice the number of keys (b being greater than v). One uses the well-known fact that $|T_n| = 2^{n(H(P)+\rho(n))}$, with $\rho(n) \to 0$, to show that the key-rate is upper bounded by $H(P)+H(P)/L+\gamma(n)$, with $\gamma(n) \to 0$; then the same bounds holds also for the codeword-rate, only with a different infinitesimal. Given τ, it will be enough to choose L large enough to prove the theorem, as $P_e(n)$, $P_I(n)$ and $P_S(n)$ all go to zero.

REFERENCES

1. G.J. Simmons, *A survey of information authentication*, Proceedings of the IEEE, May 1988, 603-620
2. J. Massey, *An introduction to contemporary cryptology*, Proceedings of the IEEE, May 1988, 533-549
3. A. Sgarro, *Information-theoretic bounds for authentication frauds*, in "Advances in Cryptology - Eurocrypt '92", ed. by R.A. Rueppel, Springer Verlag, Lecture Notes in Computer Science 658 (1993) 467-471; full version submitted to Journal of Computer Security
4. A. Sgarro, *A Shannon-theoretic coding theorem in authentication theory*, in "Eurocode '90", ed. by G. Cohen, P. Charpin, Springer Verlag, Lecture Notes in Computer Science 514 (1991) 282-291

5. A. Sgarro, *An asymptotic coding theorem for authentication and secrecy*, Workshop on Sequences 91, Positano, 17-21 June 1991
6. I. Csiszár, J. Körner, *Information theory*, Academic Press, New York, 1982
7. D. R. Hughes, F. C. Piper, *Design theory*, Cambridge University Press, 1985
8. S. Maset, *Verso una teoria asintotica dell'autenticazione con impersonazione e sostituzione*, Tesi di laurea, Università di Udine, AA 1990-91

PSEUDOPRIMES: A SURVEY OF RECENT RESULTS

F. Morain
Ecole Polytechnique, Palaiseau, France

1 Introduction

Public key cryptosystems require the use of large prime numbers, numbers with at least 256 bits (80 decimal digits), see for example [12]. One needs to generate these numbers as fast as possible. One way of dealing with this problem is the use of special primes built up using the converse of Fermat's theorem [35, 14, 17, 29]. Another is to use sophisticated primality proving algorithms, that are fast but need a careful implementation [13, 9].

In another direction, one can be happy with a number declared prime by a compositeness test, such as Miller-Rabin's. Numbers which pass this test, but are nevertheless composite, are called pseudoprimes. There are different species of pseudoprimes. This kind of compositeness algorithm requires a very good ratio between the programming work needed and the results achieved.

In this paper, we aim at presenting the most recent results achieved in the theory of pseudoprime numbers. First of all, we make a list of all pseudoprime varieties existing so far. This includes Lucas-pseudoprimes and the generalization to sequences generated by integer polynomials modulo N, elliptic pseudoprimes. We discuss the making of tables and the consequences on the design of very fast primality algorithms for small numbers. Then, we describe the recent work of Alford, Granville and Pomerance, in which they prove that there

*On leave from the French Department of Defense, Délégation Générale pour l'Armement.
†Research partially supported by the Programme de Recherches Coordonnées (PRC) Maths-Info.

exists an infinite number of Carmichael numbers. We also discuss the potential applications of their work to other classes of numbers.

2 All kinds of pseudoprimes

2.1 The ancestor

Let us start with the simplest kind of pseudoprimes. Fermat's little theorem tells us that if p is a prime number and a an integer prime to p, then

$$a^{p-1} \equiv 1 \bmod p. \tag{1}$$

A composite number p for which (1) holds is called a *pseudoprime to base a* (or psp-a for short). The smallest psp-2 is $N = 341 = 11 \times 31$.

It is known that for each value of a, there is an infinite number of psp-a (see [37]). A composite number p for which (1) holds for all values of a prime to p is called a *Carmichael number*. The smallest one is $561 = 3 \times 11 \times 17$. Until recently, it was not known that these numbers formed an infinite set (see section **xx**). Many properties of pseudoprimes are to be found in [37].

2.2 Refinements

A refined test is one due to Miller and Rabin [38]. Write $N - 1 = 2^t N_0$ with $2 \nmid N_0$. Then:

$$a^N - 1 = (a^{N_0} - 1)(a^{N_0} + 1) \times \cdots \times (a^{2^{t-1}N_0} + 1).$$

If N is prime, it divides the left hand side and thus must divide the right hand side, so one of the factors on the right. A composite N which satisfies $a^{N_0} \equiv 1 \bmod N$ or $a^{2^j N_0} \equiv -1 \bmod N$ for some $0 \leq j < t$ is called a *strong pseudo-prime to base a* (in short spsp-a). It is known that a composite N can be a spsp-a for at most 1/4 of the bases [32]. Some improvements to the scheme have been given by Damgård and Landrock [15] and also Davenport [16], who countered an attack of Arnault on the pseudoprimality routine of AXIOM (see [5] and also [6] for the same attack on the routine of MAPLE).

Recently, Atkin [8] has generalized the concept of strong pseudoprimes to that of q strong pseudoprimes, q being a small prime. Precisely, a composite integer N such that $N-1 = q^t N_0$, $q \nmid N_0$, is called a q-strong pseudoprime to base a ($\mathrm{spsp}_q(a)$), if

$$N \mid 1 + B + B^2 + \cdots + B^{q-1}$$

where $b \equiv a^{N_0} \bmod N \not\equiv 1$ and $B = b^{q^{i-1}}$ with i the least integer for which $b^{q^i} \equiv 1 \bmod N$.

2.3 Making tables

Carmichael numbers were tabulated by many authors [25, 24, 33]. The function $C(x)$ which counts the number of Carmichael numbers up to x is of interest and was also tabulated. In particular, $C(10^{12}) = 8241$, $C(10^{15}) = 105212$, $C(10^{16}) = 246683$. Pinch's tables are available via anonymous ftp.

Tables of pseudoprimes also exist. The most recent one contains all psp-2 up to 10^{12}: There are 101629 of them [34]. Using these tables, Schroeppel [40] checked that there are only 37 composite numbers less than 10^{11} which are spsp-a for a in $\{2, 3, 5\}$. Only one remains if we add $a = 7$ (namely 3215031751), and none if $a = 11$ is added next.

2.4 Generalizations

2.4.1 Linear recurrences

The concept of pseudoprimes was generalized with other relations like (1). For instance, one can look at sequences of numbers defined modulo N for a given integer N. Let $f(X)$ be a polynomial with integer coefficients:

$$f(X) = X^m + a_{m-1}X^{m-1} + \cdots + a_0.$$

Let p be a prime number and let β_i be the m roots of f in a suitable extension of GF(p). Define the sequence V_n as

$$V_n = \beta_1^n + \beta_2^n + \cdots + \beta_m^n \bmod p.$$

Denote by L the splitting field of $f(X)$ mod p and G its Galois group. For each σ in G, define

$$V_{\sigma,r} = \sum_{i=1}^{m} \sigma(\beta_i)\beta_i^r$$

for all $0 \leq r \leq m-1$. Such a collection $V(\sigma)$ is called an *admissible signature* for p. The value of a signature depends on the splitting of $f(X)$ modulo p. A pseudoprime for f is now a composite integer N which has an admissible signature $V(\sigma)$ for some σ (in a suitable context).

For example, let f be a polynomial of degree 2: $f(X) = X^2 + a_1X + a_0$. Let D be its discriminant and p a prime number not dividing D. Select $\sigma = 1$. Then, it is easy to see that

$$(V_p, V_{p+1}) = \begin{cases} (a_1, a_1^2 - 2a_0) & \text{if } (D/p) = 1 \\ (a_1, 2a_0) & \text{if } (D/p) = -1. \end{cases}$$

A less trivial example is the following [3]. Let $f(X) = X^3 - rX^2 + sX - 1$ where s and r are integers. Let α, β, γ be the roots of $f(X)$ and p be a prime number. Then

$$(V_{p-1}, V_p, V_{p+1}) = \begin{cases} (3, r, r^2 - 2s) & \text{if } f(X) \text{ has three roots } \bmod p \\ (B, r, C) & \text{if } f(X) \text{ has one root } \bmod p \\ (D, r, s) & \text{otherwise} \end{cases}$$

In case 2, let α be the root of $f(X)$ mod p; then $B \equiv -r\alpha^2 + (r^2 - s)\alpha \bmod p$ and $C \equiv \alpha^2 + 2\alpha^{-1} \bmod p$. In case 3, let $\delta = (\alpha - \beta)(\beta - \gamma)(\gamma - \alpha)$ and $D \equiv (rs - 3 - \delta)/2 \bmod p$ (note that all the quantities B, C, D are integers modulo p). The conditions on f correspond to a particular action of the Galois group of f mod p. In case 1, the splitting field of f is GF(p) and the Galois group acts on the roots as $\alpha^p = \alpha$, $\beta^p = \beta$, $\gamma^p = \gamma$. In case 2, the splitting field is $\mathrm{GF}(p^2)$ and we have $\alpha^p = \alpha$, $\beta^p = \gamma$, $\gamma^p = \beta$. The third case corresponds to the splitting field of f being $\mathrm{GF}(p^3)$ and the Galois action is (say): $\alpha^p = \beta$, $\beta^p = \gamma$, $\gamma^p = \alpha$.

This work has been done by Gurak [21] and generalizes the concept of Lucas pseudoprimes [10] (this corresponds to second-order recurrences) and the work of [3, 1, 26, 2, 7] for third-order recurrences.

2.4.2 Other generalizations

For completeness, let us add that some authors have studied the properties of elliptic pseudoprimes [18, 31, 11, 19].

Let p be a prime number greater than 3. An elliptic curve E over $\mathbf{Z}/p\mathbf{Z}$ is given by two integers a and b such that the quantity $-16(4a^3+27b^2)$ is non-zero modulo p. The set of points of E, denoted by $E(\mathbf{Z}/p\mathbf{Z})$ is the set of pairs (x, y) in $\mathbf{Z}/p\mathbf{Z}$ such that $y^2 \equiv x^3 + ax + b \bmod p$. An abelian law is usually defined on $E(\mathbf{Z}/p\mathbf{Z})$, which is known as the *tangent-and-chord* method. This law is ordinarily noted additively. For details on the law, we refer for example to [23]. Denote by m the number of points on E. Lagrange's theorem tells us that if P is a point on E, then $mP = O_E$. In some particular cases, it is easy to compute m. For instance, let $-D$ be the discriminant of an imaginary quadratic field with class number 1. (We know that $-D \in \{-4, -3, -7, -8, -11, -19, -43, -67, -163\}$.) Then if $(-D/p) = -1$, the associated curve E has cardinality $p + 1$.

We can define elliptic curves over a ring $\mathbf{Z}/N\mathbf{Z}$ for a composite N. Let $-D$ be as above and E an associated curve together with a point P on E. We say N is an elliptic pseudoprime if and only if $(-D/N) = -1$ and $(N+1)P = O_E$. One way of building such a number N is to write it as $N = p_1 \dots p_r$ and impose that $(-D/p_i) = -1$ and $p_i + 1 \mid N + 1$ for all i.

3 There exists an infinite number of Carmichael numbers

3.1 Background

Recall that $\varphi(N)$ is the cardinality of $(\mathbf{Z}/N\mathbf{Z})^\times$ for any integer N and that $\lambda(N)$ is the maximal order of an element of $(\mathbf{Z}/N\mathbf{Z})^\times$. If $N = \prod_{i=1}^r p_i^{\alpha_i}$ is the decomposition of N as a product of disctint primes, one has $\varphi(N) = \prod_i \varphi(p_i^{\alpha_i}) = \prod_i p_i^{\alpha_i - 1}(p_i - 1)$ and $\lambda(N) = \mathrm{lcm}(\lambda(p_i^{\alpha_i}))$; $\lambda(p_i^{\alpha_i}) = \varphi(p_i^{\alpha_i})$ for odd p_i or $\alpha_i \leq 2$ and $\lambda(2^e) = 2^{e-2}$ for $e \geq 3$.

One can show that a squarefree composite number N is a Carmichael number if and only if for all $p_i \mid N$, one has $p_i - 1 \mid N - 1$. Equivalently, N is a Carmichael number if and only if $\lambda(N) \mid N - 1$. For all this, we refer for instance to [39].

We define the number of divisors of N to be $\tau(N)$ and the number of *prime* divisors of N to be $\omega(N)$.

3.2 First ingredient

The basic idea is simple. First choose an integer Λ with a large number of divisors and let k be an integer prime to Λ. Build the set

$$S(k, \Lambda) = \{p \text{ prime}, p \nmid \Lambda, k \mid p - 1 \mid k\Lambda\}.$$

Suppose now that N is a squarefree product of elements of $S(k, \Lambda)$ such that $N \equiv 1 \bmod \Lambda$. Then N is a Carmichael number, since $N \equiv 1 \bmod k$ by construction, and for all p dividing N, one has:

$$p - 1 \mid k\Lambda \mid N - 1.$$

3.3 Second ingredient

Let $(G, \times)$ be a finite Abelian group. Denote by $|G|$ the cardinality of G and by m the maximal order of an element of G. One can prove the following [41, 42, 30].

Theorem 1. Let $g_1, g_2, \ldots, g_n$ be elements of G. If $n > m(1 + \log(|G|/m))$, there exists indices $i_1, i_2, \ldots, i_r$ such that $g_{i_1} \times g_{i_2} \times \cdots \times g_{i_r} = 1$. □

One now remarks that the set $S(k, \Lambda)$ is naturally isomorphic to a subgroup of $G = (\mathbf{Z}/\Lambda\mathbf{Z})^\times$ (the set of elements of $(\mathbf{Z}/k\Lambda\mathbf{Z})^\times$ which are congruent to 1 mod Λ is isomorphic to $(\mathbf{Z}/\Lambda\mathbf{Z})^\times$. The important point is that G does not depend on k, but on Λ.

3.4 Building large sets $S(k, \Lambda)$

The idea of the proof is now simple. One must build a set $S(k, \Lambda)$ so large that

$$|S(k, \Lambda)| > \lambda(\Lambda)(1 + \log(\varphi(\Lambda)/\lambda(\Lambda))).$$

If this is the case, we can use the preceding Theorem to show that necessarily, there exists a product of elements of $S(k, \Lambda)$ which is congruent to 1 modulo Λ and thus a Carmichael number.

Let us choose

$$\Lambda = \mathrm{lcm}(2, 3, \ldots, m)$$

for some integer m. By a standard result in number theory [22], one has $\Lambda \sim \exp(m)$ when m goes to infinity. Let

$$N = \prod_{\substack{p < m^2 \\ p-1 \mid \Lambda}} p$$

where p denote a prime number. It can be shown that the number of prime factors of N is greater than $c_1 m^2 / \log m$ for a fixed constant $c_1 > 0$ and m large enough. It follows that the number of divisors of N satisfies

$$\tau(N) > 2^{c_1 m^2 / \log m}.$$

We now remark that $\tau(N) \gg \lambda(N)$ since $\lambda(N) \mid \Lambda$. There is some hope to build $S(k, \Lambda)$ such that its cardinality is not too far from $\tau(N)$ and thus greater than $\lambda(N)$.

In [4], the authors are able to prove that there exists an integer k, prime to N such that

$$|S(k, \Lambda)| \geq \tau(N)^{c_2}$$

for some positive constant $c_2 > 0$. With this, we have

$$\varphi(N) < N < \prod_{p < m^2} p \sim \exp(m^2)$$

leading to

$$\lambda(N)\log\varphi(N) < \exp(2m) < 2^{c_1c_2m^2/\log m} < \tau(N)^{c_2}$$

and thus the inequality of Theorem 1 applies and there is at least one Carmichael number built up with the elements of $S(k,\Lambda)$.

The final result of [4] is then

Theorem. The function $C(x)$ is larger than x^c for all sufficiently large x and for any positive constant $c < 5/12(1-1/2\sqrt{e}) = 0.290\ldots$ □

3.5 Remarks

As a consequence of this result, one has a better bound for $C(x)$. It is conjectured by Erdős that one should have

$$C(x) = x^{1-(1+o(1))\log\log\log x/\log\log x}$$

for all sufficiently large x. The preceding result is still far from that.

The work of [4] can be extended to show that for all fixed a, there exist infinitely many squarefree composite n such that all prime factors p of n satisfy $p-a \mid n-1$. The same is true for $p^2-1 \mid n$. However, this does not work for $p-a \mid n-b$ for any b other than 0 and 1, or for $p^2+1 \mid n-1$. These cases are important for other pseudoprimality tests (see above). Also, for any finite set $\mathcal{S}$ of positive integers, there are infinitely many integers n which are spsp-a for all a in $\mathcal{S}$, as well as Carmichael numbers. The number of such numbers up to x is greater than $x^{c(\mathcal{S})}$ for some constant $c(\mathcal{S}) > 0$.

From a practical point of view, it is possible to devise fast algorithms for building Carmichael numbers with a large number of prime factors. For this and generalizations to all kind of elliptic pseudoprimes, we refer to [27, 28, 20, 43].

4 Conclusion

Despite the apparition of two powerful primality proving algorithms, pseudoprimes tests are still interesting. The theory of pseudoprimes has seen a renewed attention due to the result of Adlford, Granville and Pomerance. No doubt that further results will follow, enabling one to get a fast and compact primality testing algorithm by combining different pseudoprimality tests.

Acknowledgment. The author would like to thank Carl Pomerance for his sending the papers [4] as well as [36], and for many interesting discussions on his work.

References

[1] W. ADAMS. Splitting of quartic polynomials. *Math. Comp. 43*, 167 (July 1984), 329–343.

[2] W. ADAMS. Characterizing pseudoprimes for third order linear recurrences. *Math. Comp. 48*, 177 (Jan. 1987), 1–15.

[3] W. ADAMS AND D. SHANKS. Strong primality tests that are not sufficient. *Math. Comp. 39*, 159 (July 1982), 255–300.

[4] W. R. ALFORD, A. GRANVILLE, AND C. POMERANCE. There are infinitely many Carmichael numbers. Preprint, July 13th 1992.

[5] F. ARNAULT. Le test de primalité de Rabin–Miller : un nombre composé qui le "passe". Report 61, Université de Poitiers - Département de Mathématiques, Nov. 1991.

[6] F. ARNAULT. Carmichaels fortement pseudo-premiers. Manuscript, 1992.

[7] S. ARNO. A note on Perrin pseudoprimes. *Math. Comp. 56*, 193 (Jan. 1991), 371–376.

[8] A. O. L. ATKIN. Probabilistic primality testing. In *Analysis of Algorithms Seminar I* (1992), P. Flajolet and P. Zimmermann, Eds., INRIA Research Report XXX. Summary by F. Morain.

[9] A. O. L. ATKIN AND F. MORAIN. Elliptic curves and primality proving. Research Report 1256, INRIA, Juin 1990. Submitted to *Math. Comp.*

[10] R. BAILLIE AND S. S. WAGSTAFF, JR. Lucas pseudoprimes. *Math. Comp. 35*, 152 (Oct. 1980), 1391–1417.

[11] R. BALASUBRAMANIAN AND M. R. MURTY. Elliptic pseudoprimes, II. Submitted for publication.

[12] G. BRASSARD. *Modern Cryptology*, vol. 325 of *Lect. Notes in Computer Science*. Springer-Verlag 1988.

[13] H. COHEN AND A. K. LENSTRA. Implementation of a new primality test. *Math. Comp. 48*, 177 (1987), 103–121.

[14] C. COUVREUR AND J. QUISQUATER. An introduction to fast generation of large prime numbers. *Philips J. Research 37* (1982), 231–264.

[15] I. DAMGÅRD AND P. LANDROCK. Improved bounds for the Rabin primality test. In *Proc. 3rd IMA conference on Coding and Cryptography* (1991), M. Ganley, Ed., Oxford University Press.

[16] J. H. DAVENPORT. Primality testing revisited. In *ISSAC '92* (New York, 1992), P. S. Wang, Ed., ACM Press, pp. 123–129. Proceedings, July 27–29, Berkeley.

[17] D. GORDON. Strong primes are easy to find. In *Advances in Cryptology* (1985), T. Beth, N. Cot, and I. Ingemarsson, Eds., vol. 209 of *Lect. Notes in Computer Science*, Springer-Verlag, pp. 216–223. Proceedings Eurocrypt '84, Paris (France), April 9–11, 1984.

[18] D. M. GORDON. On the number of elliptic pseudoprimes. *Math. Comp. 52*, 185 (Jan. 1989), 231–245.

[19] D. M. GORDON AND C. POMERANCE. The distribution of Lucas and elliptic pseudoprimes. *Math. Comp. 57*, 196 (Oct. 1991), 825–838.

[20] D. GUILLAUME AND F. MORAIN. Building Carmichael numbers with a large number of prime factors and generalization to other numbers. Research Report 1741, INRIA, Aug. 1992.

[21] S. GURAK. Pseudoprimes for higher-order linear recurrence sequences. *Math. Comp. 55*, 192 (Oct. 1990), 783–813.

[22] G. H. HARDY AND E. M. WRIGHT. *An introduction to the theory of numbers*, 5th ed. Clarendon Press, Oxford, 1985.

[23] D. HUSEMÖLLER. *Elliptic curves*, vol. 111 of *Graduate Texts in Mathematics*. Springer, 1987.

[24] G. JAESCHKE. The Carmichael numbers to 10^{12}. *Math. Comp. 55*, 191 (July 1990), 383–389.

[25] W. KELLER. The Carmichael numbers to 10^{13}. *AMS Abstracts 9* (1988), 328–329. Abstract 88T-11-150.

[26] G. C. KURTZ, D. SHANKS, AND H. C. WILLIAMS. Fast primality tests for numbers less than $50 \cdot 10^9$. *Math. Comp. 46*, 174 (Apr. 1986), 691–701.

[27] G. LÖH. Carmichael numbers with a large number of prime factors. *AMS Abstracts 9* (1988), 329. Abstract 88T-11-151.

[28] G. LÖH AND W. NIEBUHR. Carmichael numbers with a large number of prime factors, II. *AMS Abstracts 10* (1989), 305. Abstract 89T-11-131.

[29] U. M. MAURER. Fast generation of secure RSA-products with almost maximal diversity. In *Advances in Cryptology* (1990), J.-J. Quisquater, Ed., vol. 434 of *Lect. Notes in Computer Science*, Springer-Verlag, pp. 636–647. Proc. Eurocrypt '89, Houthalen, April 10–13.

[30] R. MESHULAM. An uncertainty inequality and zero subsums. *Discrete Mathematics 84* (1990), 197–200.

[31] I. MIYAMOTO AND M. R. MURTY. Elliptic pseudoprimes. *Math. Comp. 53*, 187 (July 1989), 415–430.

[32] L. MONIER. Evaluation and comparison of two efficient probabilistic primality testing algorithms. *Theoretical Computer Science 12* (1980), 97–108.

[33] R. PINCH. The Carmichael numbers to 10^{16}. In preparation, 1992.

[34] R. PINCH. The pseudoprimes up to 10^{12}. In preparation, Sept. 1992.

[35] D. A. PLAISTED. Fast verification, testing and generation of large primes. *Theoretical Computer Science 9* (1979), 1–16.

[36] C. POMERANCE. Carmichael numbers. To appear in *Nieuw Arch. Wisk.*, 1992.

[37] C. Pomerance, J. L. Selfridge, and S. S. Wagstaff, Jr. The pseudoprimes to 25.10^9. *Math. Comp. 35*, 151 (1980), 1003–1026.

[38] M. O. Rabin. Probabilistic algorithms in finite fields. *SIAM J. Comput. 9*, 2 (1980), 273–280.

[39] P. Ribenboim. *The book of prime number records*, 2nd ed. Springer, 1989.

[40] R. Schroeppel. Richard Pinch's list of pseudoprimes. E-mail to the NMBRTHRY list, June 1992.

[41] P. van Emde Boas. A combinatorial problem on finite abelian groups, II. Tech. Rep. ZW–007, Math. Centrum Amsterdam Afd. Zuivere Wisk., 1969. 60 pp.

[42] P. van Emde Boas and D. Kruyswijk. A combinatorial problem on finite abelian groups, III. Tech. Rep. ZW–008, Math. Centrum Amsterdam Afd. Zuivere Wisk., 1969.

[43] M. Zhang. Searching for large Carmichael numbers. To appear in *Sichuan Daxue Xuebao*, Dec. 1991.

ONE-TIME IDENTIFICATION WITH LOW MEMORY

S. Vaudenay
Ecole Normale Supérieure, Paris, France

Abstract

In this paper, we describe a practical cryptosystem based on an authentication tree. It is an extension of Ralph Merkle's results [1]. The security of the system depends on assumptions on a *one-way function*. This cryptosystem allows to make interactive proofs of knowledge using keys only once. Thus, it can be used to prove one's identity or to sign some message. A very efficient implementation can be done on a low-cost smart card. This cryptosystem is thus an alternative to the *zero-knowledge* algorithms.

1 Introduction

The notion of *authentication tree* has been defined in [2]. It allows to authenticate some values taken from a directory. If we have a list of values, we can assign these values to the leaves of a tree. Using a *collision-free hash function*, we can propagate the values from the leaves to the root of the tree. Then, if we assume the value of the root is authenticated, we can authenticate any value of the tree by producing the values in the *authentication path* (see figure 1). It is just the values assigned to the brothers of the ancestors of the given node.

In [1], authentication trees are used to achieve a signature scheme. It is based on the Diffie-Lamport scheme [3][4]. This scheme is a one-time signature scheme. It

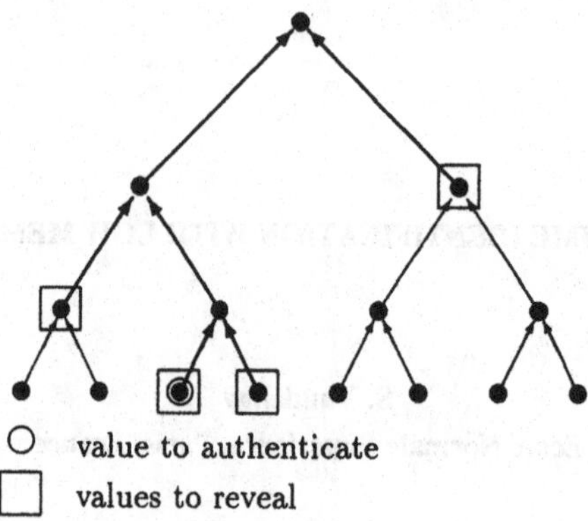

Figure 1: Authentication binary tree

needs a set of keys to sign a message. As it is a one-time signature scheme, it needs as many sets of keys as we have messages to sign. That is the reason why we need an authentication tree to authenticate all the keys.

To achieve such a cryptosystem, we can define an authentication tree that is made with three levels from the top to the bottom :

- in the first-level tree, each leaf corresponds to a user, so we can authenticate every users;
- for a given user, we have a second-level tree in which each leaf corresponds to a set of keys to be used in only one transaction;
- for a set of keys, we have a third-level tree to authenticate each key.

In the third level tree, we can compute all the values quickly with the hash function. In the first level tree, we can store permanently the values of the only authentication path we need. However, in the second level tree, we have to compute successively all the authentication paths.

We wish to sign N messages by using at most M value registers between each transaction. We let T^c be the time needed to compute each authentication path, and C be the number of values of the the authentication path. We can compute without memory, so $M = O(1)$, $T^c = O(N)$ and $C = \log_2 N$. We can keep all the values in memory, so $M = O(N)$, $T^c = O(1)$ and $C = \log_2 N$.

In [1], a path regeneration algorithm was proposed to get :

$$M \sim \frac{1}{2}(\log_2 N)^2,\ T^c_{exp} = O(\log N) \text{ and } C = \log_2 N$$

In this paper, we obtain a family of algorithms whose complexity vary from these figures to :

$$M \sim \frac{1}{2}\left(\frac{\log N}{\log\log N}\right)^2, \; T^c_{exp} = O\left(\frac{\log N}{\log\log N}\right) \text{ and } C_{exp} \sim \frac{(\log N)^2}{2(\log\log N)^3}$$

Thus, we can reduce the memory requirement by a factor $(\log_2 \log N)^2$. However, this is an asymptotic evaluation. In practice, if we need $N \simeq 2000$, we may reduce the memory requirement by a factor of 2.2 which is already quite significant in several applications.

This scheme can be used in other settings than signature. In this paper, an interactive proof of knowledge scheme is proposed. It is not a zero-knowledge scheme : it is a one-time proof of knowledge scheme. We can use this scheme with an arbitrary security parameter with the path regeneration algorithm to obtain a practical cryptosystem.

2 Notations

We assume we have a tree T. Each node is denoted by a word over the alphabet of the non-negative integers. The root is denoted by the empty word Λ. Given a node of degree α denoted by the word a, the words $a0$, $a1$, ..., $a(\alpha-1)$ represent the ordered list of all sons of a.

We assume we have a collision-free hash function $< . >$, and an assignation ϕ which is a map from the set of T leaves to the domain of $< . >$. We can recursively define a map ϕ^* from the set of T nodes by :

$$\phi^*(a) = < \phi^*(a0), \phi^*(a1), \ldots, \phi^*(a(\alpha-1)) >$$

To authenticate $\phi^*(a_0a_1\ldots a_{n-1})$, we have to give the values $\phi^*(a_0a_1\ldots a_{k-1}j)$ where $0 \le k < n$ and $j \ne a_k$.

We assume that the function $< . >$ satisfies Damgård's property :

$$< x_1, x_2, \ldots, x_i > = << x_1, x_2, \ldots, x_{i-1} >, x_i >$$

Therefore, we only have to define the hash of two values to get the function $< . >$. This property will be useful.

A one-way function f will be used in the next section. A pseudo-random generator g will be used too. We can assume that $< . >$, g and f are a set of related cryptographic tools built using the same algorithms such that their implementation can be very efficient. We will assume that hashing two values costs as much time as computing two values of f, and that g and f have the same complexity.

3 Diffie-Lamport scheme

The Diffie-Lamport scheme [4] is a one-time signature scheme. Assume we have a message M to sign. This message is coded by s bits $b_1b_2\ldots b_s$. We assume that we have a set of secret keys $\{K_i; 0 < i \leq s\}$. The values $f(K_i)$ are assumed to be public.

The signature of the message M is the set of keys K_i for each i such that $b_i = 0$ [1]. Then, if keys K_i are the leaves of the authentication tree T, an authenticated signature of the message M is the set of all $f^{b_i}(K_i)$[2].

If a cheater gets an authenticated signature, he can forge a signature of a message M' coded by $b'_1b'_2\ldots b'_l$ if $b'_i \geq b_i$. Thus, we have to use a monotone code (i.e. a code such that for two different codewords $c_1\ldots c_l$ and $d_1\ldots d_l$, there exists i and j such that $c_i < d_i$ and $c_j > d_j$) to be sure that no $b'_1b'_2\ldots b'_s$ is a valid codeword.

A proposed code is the set of $b_1b_2\ldots b_{2l}$ such that $b_i \oplus b_{i+l} = 1$. Another code consists in appending the weight of the message to the codeword.

The first code proposed in [4] is the set of $b_1b_2\ldots b_s$ such that :

$$b_1 + b_2 + \ldots + b_s = \lfloor \frac{s}{2} \rfloor$$

A theorem of Sperner (1928) shows that this code is the shortest monotone binary code of a given cardinality. The main problem was to exhibit an efficient coding scheme, that is the reason why it was abandoned.

In [1], an improvement of Winternitz is described. It simply consists in using a larger alphabet for the code. If the codewords are coded over the alphabet of the set of all the integers from 0 to $\beta - 1$ (thus in base β), and if we assume that the values $f^{\beta-1}(K_i)$ are public, then an authenticated signature is the set of all $f^{b_i}(K_i)$. To verify the signature, we have to apply $\beta - 1 - b_i$ times f to the $f^{b_i}(K_i)$ to get the public keys.

We have to use a monotone code for the same reason. The set of the words $b_1b_2\ldots b_s$ such that :

$$b_1 + b_2 + \ldots + b_s = \lfloor s\frac{\beta - 1}{2} \rfloor$$

is still a good one. However, it is still difficult to exhibit an efficient coding scheme. To get rid of this difficulty, we will not use this algorithm for signatures, but for interactive proofs.

4 A one-time proof of knowledge scheme

We assume we have the three-level authentication tree T described in the introduction. At each level l, we use an assignation ϕ_l.

---

[1] In the original paper, it is the set of keys K_i for each i such that $b_i = 1$, but it is equivalent.

[2] We denote $f^0(K) = K$, $f^1(K) = f(K)$, $f^2(K) = f(f(K))$, ...

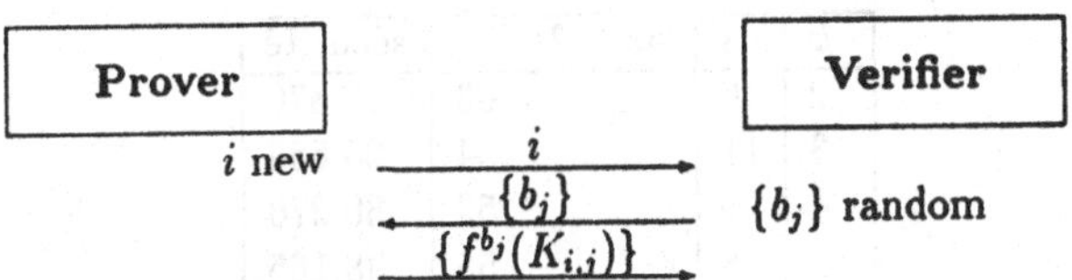

Figure 2: Identification scheme

At the first level, each leaf u corresponds to a user with identity Id and with a second-level tree $T(u)$. We define :

$$\phi_1(u) = < Id, \phi_2^*(\text{root of } T(u)) >$$

to authenticate both the identity of the user and the value assigned to the root of his tree.

At the second level, each leaf a_i corresponds to the ith interactive proof, and a third-level tree T_i. We define :

$$\phi_2(a_i) = \phi_3^*(\text{root of } T_i)$$

At the third level, the tree T_i is a simple one. It consists in a root a_i with s sons $b_{i,j}$ which are the leaves. Each leaf $b_{i,j}$ corresponds to a public key. We define :

$$\phi_3(b_{i,j}) = f^{\beta-1}(K_{i,j})$$

Where β is the one defined in the previous section. Thus, we have :

$$\phi_2(a_i) = < f^{\beta-1}(K_{i,1}), f^{\beta-1}(K_{i,2}), \ldots, f^{\beta-1}(K_{i,s}) >$$

If the user has some secret key K, we may use :

$$K_{i,j} = g(K, i, j)$$

To prove that the user has the identity Id, he has to prove that he knows a set of keys $\{K_{i,j}; 0 < j \leq s\}$ for some i. The prover first chooses i. Then, the verifier chooses s digits b_j in base β such that $b_1 + b_2 + \ldots + b_s = \lfloor s\frac{\beta-1}{2} \rfloor$ Finaly, the prover reveals $f^{b_j}(K_{i,j})$ for all j. The protocol is illustrated in figure 2.

With this protocol, the prover proves his knowledge of the $K_{i,j}$, which is assumed to prove his knowledge of K. We can use this protocol to achieve a signature scheme with the usual method : we simulate the verifier by a pseudorandom generator fed with i and the message to sign. Usually, the security required by the signature is greater than the security required by an interactive protocol. So, in order to sign a message, we will use several i to increase the security of the scheme.

β	s	$\beta s + 2s - 1$	sécurité
2	16	63	12.870
3	11	54	25.653
4	9	53	30.276
5	8	55	38.165
6	7	55	24.017
8	6	59	18.152
12	5	69	12.435
25	4	107	10.425

Figure 3: Possible couples (β, s)

In fact, each i corresponds to a security unit. A user has a fixed number of security units in the beginning, and he can use them as he likes for cryptographic operations with the security required. A proof of identity *costs* one unit, and a signature *costs* several units.

If the prover uses each i once, the knowledge of a cheater is, for each i, a set of $f^{b_j}(K_{i,j})$ where $b_1 + b_2 + \ldots + b_s = \lfloor s\frac{\beta-1}{2}\rfloor$. Then, if a cheater chooses an index i, the probability that he will be able to answer to the challenge of some verifier is the inverse of the number of possible list of b_j which is thus the security parameter of the scheme.

To get a security of $1,000,000$, we can choose $\beta = 2$ and $s = 23$. However, it might not be practical for some applications. A security of $10,000$ is enough for transactions in shops, where the prover is physically there. To sign with a security of 10^{12}, three units are enough. The table of figure 3 shows each combination of (β, s) which has a security greater than $10,000$. The time required to compute the hash value of all $f^{\beta-1}(K_{i,j})$ from K and i is $t = \beta s + 2s - 1$ units with our assumptions. So, t is exactly the complexity of ϕ_2. Thus, we have to minimize t for a given security. Then, we should choose $\beta = 4$ and $s = 9$. We get a security of $30,276 \simeq 2^{15}$.

We also notice that each i, which corresponds to a *cost* of security, can represents a money unit (a *coin*). To give a coin to someone, we can give the $f^{b_j}(K_{i,j})$. It will be a proof that we gave the coin. If a cheater gave the same coin to several people, the proofs would give several combinations of b_j with a probability which depends on the security parameter. These combinations of b_j would prove that the cheater gave more information about the $K_{i,j}$ than he was allowed. It would be a proof that he cheated. Thus, we can imagine to achieve an electronic cash system based on this cryptosystem.

5 The path regeneration algorithm

In this section, T will denote the second-level tree. We assume that it is a well balanced tree of degree α and of depth n such that the number of value to store is almost a constant. ϕ will denote the assignation of T. Thus, a leaf corresponds to a transaction. The algorithm we will describe now shows how to visit the T computing the values of the authentication path.

The authentication of the value of a leaf $a = a_0a_1 \dots a_{n-1}$ consists in revealing the values $\phi^*(a_0a_1 \dots a_{k-1}j)$ for all $0 \leq k < n$ and $j \neq a_k$. Because of Damgård's property, for each k we can remove the set of the values $\phi^*(a_0a_1 \dots a_{k-1}j)$ for $j < a_k$ and keep their hash value instead. Let us define :

$$\phi_l^*(a_0 \dots a_{k-1}a_k) = <\phi^*(a_0 \dots a_{k-1}0), \dots, \phi^*(a_0 \dots a_{k-1}(a_k - 1))>$$

This is the hashed value of the left brothers of $a_0 \dots a_{k-1}a_k$. The authentication now consists in revealing the values $\phi^*(a_0a_1 \dots a_{k-1}j)$ for all $0 \leq k < n$ and $j > a_k$ and the values $\phi_l^*(a_0 \dots a_{k-1}a_k)$ for all $0 \leq k < n$ and $a_k \neq 0$.

Let us define the following :

$$\begin{array}{rcll} M_{k,j} & = & \phi^*(S(a_0a_1 \dots a_{k-1})j) & \text{if } j < a_k \\ M_{k,j} & = & \phi^*(a_0a_1 \dots a_{k-1}j) & \text{if } j \geq a_k \\ A_k & = & \phi_l^*(a_0a_1 \dots a_k) & \text{if } a_k \neq 0 \\ T_{k,i} & = & \phi_l^*(S(a_0 \dots a_{i-1})a_i \dots a_k) & \text{if } a_k \neq 0 \end{array}$$

for $0 \leq k < n$ and $0 < i < k$. The meaning of these formulae will be explained below. $S(x)$ denotes the successor of x in base α. Thus, it denotes the node "immediately on the right" of x. Those values are the ones stored in memory during the visit of the tree.

We define the function $\text{Acc}(x, c, y)$ by :

$$\text{Acc}(x, c, y) = \begin{cases} y & \text{if } c = 0 \\ <x, y> & \text{if } c \neq 0 \end{cases}$$

This function is used to regenerate the stored values A_k and $T_{k,i}$, which are accumulators. An accumulator is used to compute the hashed value of a node by hashing the values from the leftmost son to the rightmost son. In order to accumulate the value of a new son, we use the function Acc. To compute $\phi^*(a_0 \dots a_{k-1}a_k)$, we need $n - 1 - k$ accumulators, one of every tree's floor.

We can show that the algorithm described on figure 4 is a path regeneration algorithm — i.e. it refreshes all the stored values before incrementing a. In the beginning of the life of the authentication tree, we assume that we have the values to authenticate the leaf $00 \dots 0$. So, we can initialize the stored values in step 2. In step 3, we do some operations on the leaf a (the one-time interactive proof scheme for instance) and we

1. $a \longleftarrow 00\ldots0$;
2. initialize $M_{k,j}$ for $0 \leq k < n$ and $0 \leq j < \alpha$;
3. visit a, authenticate $\phi(a)$;
4. if $a_k = \alpha - 1$ for all k then end;
5. $p \longleftarrow \mathrm{Max}\{k; a_k \neq \alpha - 1\}$;
6. $A_p \longleftarrow \mathrm{Acc}(A_p, a_p, M_{p,a_p})$;
7. $M_{n-1,a_{n-1}} \longleftarrow \phi(S(a_0 \ldots a_{n-2})a_{n-1})$;
8. for all $0 < i < n - 1$:

   $T_{n-1,i} \longleftarrow \mathrm{Acc}(T_{n-1,i}, a_{n-1}, \phi(S(a_0 \ldots a_{i-1})a_i \ldots a_{n-1}))$;
9. for $k = n - 2$ downto $k = p$:
   (a) $M_{k,a_k} \longleftarrow T_{k+1,k}$;
   (b) for all $0 < i < k$: $T_{k,i} \longleftarrow \mathrm{Acc}(T_{k,i}, a_k, T_{k+1,i})$;
10. $a \longleftarrow S(a)$;
11. go to 3.

Figure 4: Path regeneration algorithm

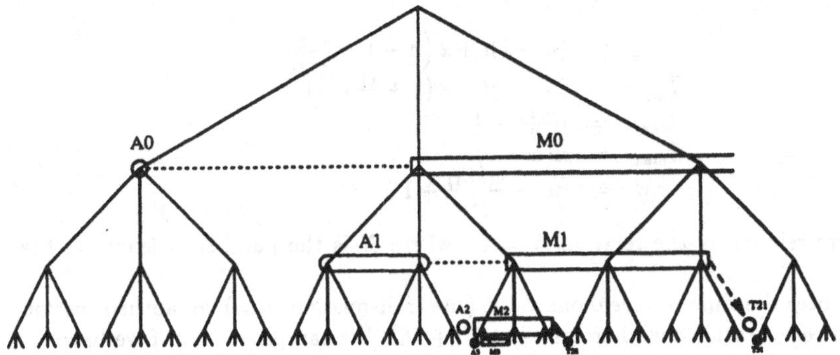

Figure 5: Example of path regeneration

authenticate $\phi(a)$. It simply consists in giving the values $M_{k,i}$ for all $0 \leq k < n$ and $i < a_k$ and the values A_k for all k such that $a_k \neq 0$. The steps 6 to 9 regenerate the stored values for the next a.

From the beginning to the end of the algorithm, $\phi^*(\wedge)$ is computed with the accumulators A_k (one accumulator per floor of the tree). After the visit of the leaf a, the leaves from $0 \ldots 0$ to a are accumulated. These accumulators are used only in the authentication of a.

For each k (at each floor of the tree), the set of $M_{k,.}$ is a window of the α nodes the leftmost of which is $a_0 \ldots a_{k-1}$ (see figure 5). If this node is not useful anymore, the window is shifted to the right (see step 7 if $k = n-1$ and step 9a otherwise). In order to shift the window, we have to compute the value of $a_0 \ldots a_{k-1} + \alpha$, using the accumulators $T_{.,k}$ if $k \neq n-1$.

At the step 5, we compute the index p of the highest floor where the window have to be shifted. Then, the only useful accumulator A_k is modified in step 6, the window of the lowest floor is shifted in step 7, and the other eventual window are shifted in step 9a. At the steps 8 and 9b, the accumulators $T_{k,i}$ are modified. The lowest ones are modified by a computation $\phi(.)$ (step 8). The other ones are modified using lower accumulators (step 9b).

If we denote T^c the computation time, C the number of values to reveal in order to authenticate $\phi(a)$ and M the number of registers required to store all the values, we can be interested either in the average complexity (T^c_{exp} and C_{exp}) or the maximum complexity (T^c_{max} and C_{max}). With our assumptions on $< . >$, we can use the time to compute a value of f as a time unit. We denote by t the time needed to compute a value of ϕ. Then, we have :

$$\begin{array}{rcl} T^c_{exp} & \simeq & (n-1)t + 2\left(n-1+\frac{n-2}{\alpha-1}\right) \\ T^c_{max} & = & (n-1)t + 2\left(1+\frac{n(n-1)}{2}\right) \\ C_{exp} & = & n\frac{\alpha+1}{2} - \frac{n}{\alpha} \\ C_{max} & = & n(\alpha-1) \\ M & = & n\alpha + \frac{n(n-1)}{2} + 1 \end{array}$$

α and n are related by the relation $N = \alpha^n$ where N is the number of leaves in the tree T.

If τ denotes the time to reveal one value (the transmission time), an authentication and a path regeneration together cost $T = T^c + \tau C$. We can prove that if we have :

$$\frac{t}{\tau} < 2\log 2 - 1 \simeq 0.39$$

then the choice $\alpha = 2$ corresponds to the smallest value of T. However, the smallest possible M corresponds to the α given by the implicit equation :

$$\alpha(\log\alpha - 1) - \frac{\log N}{\log \alpha} = -\frac{1}{2}$$

obtained by derivating with respect to α the expression of M, with $N = \alpha^n$. This corresponds to :

$$\alpha \sim \frac{\log N}{(\log\log N)^2} \text{ and } n \sim \frac{\log N}{\log\log N}$$

Thus, we can choose our favorite α between $O(1)$ and $\log N/(\log\log N)^2$ to achieve the space/time tradoff we prefer. We obtain a family of algorithms whose complexity varies from :

$$M \sim \frac{1}{2}(\log_2 N)^2,\ T^c_{exp} = O(\log N) \text{ and } C = \log_2 N$$

to :

$$M \sim \frac{1}{2}\left(\frac{\log N}{\log\log N}\right)^2,\ T^c_{exp} = O\left(\frac{\log N}{\log\log N}\right) \text{ and } C_{exp} \sim \frac{(\log N)^2}{2(\log\log N)^3}$$

Figure 6 shows the several combinations (α, n) where N is larger than 2,000. We assume $t = 53$ (this corresponds to $\beta = 4$ and $s = 9$ in section 4) and $\tau = 16$ which is about the transmission of a 16 byte value at 9,600bps if we assume that f requires about 1ms. These are the actual parameters of a low-cost smart card in which we implemented a software one-way function.

Figure 6 shows that T_{max} is minimal for $\alpha = 5$ and $n = 5$. This corresponds to a low memory : 36 (576 bytes if we assume that a value is a 16 byte value) instead of 78 (1,248 bytes) with binary trees. Moreover, the program which achieves the path

α	n	N	M	C_{max}	C_{exp}	T^c_{exp}	T_{max}	T_{exp}
2	11	2048	78	11	11	568	818	744
3	7	2187	43	14	12	335	586	527
4	6	4096	40	18	14	278	585	502
5	5	3125	36	20	14	222	554	446
7	4	2401	35	24	15	166	557	406
13	3	2197	43	36	21	110	690	446
45	2	2025	92	88	46	55	1465	791
2000	1	2000	2001	1999	1000	0	31986	16000

$t = 53,\ \tau = 16$

Figure 6: Different combinations of (α, n) with $\alpha^n \geq 2000$

regeneration algorithm is very simple, and it does not require much space in ROM. The average time to compute and transmit the authenticating values is 446ms. An implementation of this algorithm with these parameters will be able to have 3,125 security units. If we assume that a session needs 4 security units (one to authenticate, and three to sign a message), and that we have twenty sessions per month, we can use the smart card during more than three years, which is quite more than the expected life of a smart card.

With hardware one-way functions f, g and $\langle , \rangle$, we could get even better results.

Conclusion

Authentication tree is a practical way to authenticate a value in a large directory. It requires a small file to read for the prover, and it requires few computations for the verifier.

The path regeneration algorithm is an efficient algorithm to visit a directory computing the values of the authentication paths. This allows to achieve one-time cryptographic algorithms : a one-time algorithm needs a one-time authenticated key. Several keys can be put in the directory of an authentication path we can visit.

The one-time interactive proof scheme can be used both to authenticate and to sign messages. It consists in revealing an information about the key we use that nobody else will be able to use (up to arbitrary security parameter). As we can prove that someone used the same key several times, because he revealed too much information, this scheme can be used to achieve electronic cash systems.

Finally, it is possible to achieve efficiently such systems with low cost smart cards.

Acknowledgment

I would like to thank GEMPLUS CARD INTERNATIONAL, and particularly DAVID M'RAIHI, for having encouraged this work. I also thank JACQUES STERN without whom this paper would never have been achieved.

References

[1] R. Merkle : A Certified Digital Signature, in : *Advances in Cryptology CRYPTO'89*, Springer Verlag, 1989, 218-238.

[2] R. Merkle : Protocols for Public Key Cryptosystems, in : *Proc. IEEE Symp. on Security and Privacy*, 1980, 122-133.

[3] W. Diffie, M. Hellman : New Directions in Cryptography, in : *IEEE Transaction on Information Theory*, vol. IT-22, 6, 1976, 644-654.

[4] L. Lamport : Constructing digital signatures from a one way function, technical report CSL-98, SRI Intl., 1979.

4

DECODING BLOCK CODES

ON THE EFFICIENT DECODING OF ALGEBRAIC-GEOMETRIC CODES

R. Pellikaan
Eindhoven University of Technology, Eindhoven, The Netherlands

Abstract

This talk is intended to give a survey on the existing literature on the decoding of algebraic-geometric codes. Although the motivation originally was to find an efficient decoding algorithm for algebraic-geometric codes, the latest results give algorithms which can be explained purely in terms of linear algebra. We will treat the following subjects:

1. The decoding problem
2. Decoding by error location
3. Decoding by error location of algebraic-geometric codes
4. Majority coset decoding
5. Decoding algebraic-geometric codes by solving the key equation
6. Improvements of the complexity

1 The decoding problem

Minimum distance decoding (MDD), also called *maximum likelihood decoding* (MLD) under some assumptions on the channel used, is an algorithm which has as input $(C, \mathbf{y})$, where C is a linear code, say given by a generator matrix and $\mathbf{y}$ a word of the same length as C and as output a word $\mathbf{c} \in C$, such that the distance between $\mathbf{y}$ and $\mathbf{c}$ is equal to the distance between $\mathbf{y}$ and C. It is shown by Berlekamp, McEliece and van Tilborg [3] that this problem is NP complete. One may object that this is not a relevant problem in practice, that is usually one knows a great deal in advance of the code C, or one has a lot of time to find the decoding algorithm itself, the essential point in practice is that with input $\mathbf{y}$ one wants to find (one of) the nearest codewords in a short time. So one considers *minimum distance decoding with preprocessing* (MDDP), but also this problem turns out to be hard, see Bruck and Naor [5]. If C is a linear code of minimum distance d and $\mathbf{y}$ a word of distance at most $\lfloor (d-1)/2 \rfloor$ to C, then the nearest codeword is unique. *Decoding up to half the minimum distance with preprocessing*, DHDP for short, has as input a word $\mathbf{y}$ and as output the nearest codeword in case $d(\mathbf{y}, C) \leq (d-1)/2$, and "the received word has more than $\lfloor (d-1)/2 \rfloor$ errors" otherwise. As far as I know it is not known how difficult DHDP is. In the sequel we show that a lot of codes, in particular algebraic-geometric codes, have an efficient algorithm which decodes up to half the minimum distance.

2 Decoding by error location

We have the standard bilinear form defined by $< \mathbf{a}, \mathbf{b} >= \sum_i a_i b_i$. If A is a subset of $\mathbf{F}_q^n$, then we define the dual $A^\perp$ of A in $\mathbf{F}_q^n$ with respect to the standard bilinear form by $A^\perp = \{\mathbf{b} | < \mathbf{a}, \mathbf{b} >= 0 \text{ for all } \mathbf{a} \in A\}$. So in this definition A need not to be linear but $A^\perp$ is always linear.

Let C be a linear code in $\mathbf{F}_q^n$. Suppose we have received a word $\mathbf{y}$ and we have some *erasures*, that is $\mathbf{y} = \mathbf{e} + \mathbf{c}$ for some error $\mathbf{e}$ and codeword $\mathbf{c} \in C$, and we know a set J with at most $d(C) - 1$ elements which contains the set I of *error positions*, that is $I = \{i | e_i \neq 0\}$. Then we can find the error values as follows. Let $\mathbf{v}_1, \ldots, \mathbf{v}_r$ be a basis of $C^\perp$. Consider the following linear equations

$$< \mathbf{x}, \mathbf{v}_i >=< \mathbf{y}, \mathbf{v}_i > \quad \text{for} \quad i = 1, \ldots, r,$$

$$x_j = 0 \quad \text{for } j \notin J.$$

Then $\mathbf{e}$ is the unique solution. It is clear that the error vector is a solution. If $\mathbf{x}$ is another solution, then $\mathbf{x} - \mathbf{e}$ has zero inner product with all the elements of $C^\perp$, so it is an element of C, moreover its weight is at most $|J| \leq d(C) - 1$, so it must be zero, that is to say $\mathbf{x} = \mathbf{e}$. Thus we have shown that we can reduce error decoding to erasure decoding. We still have to find the error positions.

But if we want to decode all words with t errors, then there are $\binom{n}{t}$ possible t sets one has to consider, and this number grows exponentially with n. So we are looking for an efficient way to find the error positions. First we introduce some notation. We define the star multiplication $\mathbf{a} * \mathbf{b}$ of two elements $\mathbf{a}$ and $\mathbf{b}$ of $\mathbf{F}_q^n$ by coordinatewise multiplication, that is $(\mathbf{a} * \mathbf{b})_i = a_i b_i$. For two subsets A and B of $\mathbf{F}_q^n$ we denote the set $\{\mathbf{a} * \mathbf{b} | \mathbf{a} \in A, \mathbf{b} \in B\}$ by $A * B$. The following definition of a t-error correcting pair comes out of the blue for those who have never seen it before. But we shall explain why it works and afterwards show the ubiquity of error correcting pairs and that the examples are abundant.

We will follow our paper [41], an earlier version [39] was never published. The same kind of reasoning was found independently by Kötter [29].

Definition 2.1 Let C be a $\mathbf{F}_q$-linear code of length n. Let A and B be $\mathbf{F}_{q^e}$-linear codes of length n, then (A, B) is called a *t-error correcting pair* for C over $\mathbf{F}_{q^e}$, if

1. $C \subseteq (A * B)^\perp$
2. $k(A) > t$
3. $d(B^\perp) > t$
4. $d(A) + d(C) > n$

Remember that we have defined the dual for any subset of $\mathbf{F}_{q^e}^n$, so $(A * B)^\perp$ is well defined and remark that C is a subspace of $\mathbf{F}_q^n$, so we can write (1') $A * B \subseteq C^\perp$ instead of (1), if we mean by $C^\perp$ the dual of the subset C in $\mathbf{F}_{q^e}^n$ and not in $\mathbf{F}_q^n$.

Suppose (A, B) is a t-error correcting pair for C. We first define the vector space $K_\mathbf{y}$ of a received word $\mathbf{y}$, by

$$K_\mathbf{y} = \{\mathbf{a} \in A | < \mathbf{y}, \mathbf{a} * \mathbf{b} >= 0 \quad \text{for all} \quad \mathbf{b} \in B\}.$$

If $A * B \subseteq C^\perp$ and $\mathbf{y}$ is a word with error $\mathbf{e}$, then $K_\mathbf{y} = K_\mathbf{e}$. Remark that an element which is zero at the error positions is contained in $K_\mathbf{y}$. In more formal terms. Suppose J is a subset of $\{1, \ldots, n\}$. Define

$$A(J) = \{\mathbf{a} \in A | a_j = 0 \text{ for all } j \in J\}.$$

If $I = supp(\mathbf{e})$, then

$$A(I) \subseteq K_\mathbf{y}.$$

Because
$$< \mathbf{y}, \mathbf{a} * \mathbf{b} >=< \mathbf{e}, \mathbf{a} * \mathbf{b} >= \sum_{i \in I} a_i b_i e_i = 0,$$

since $a_i = 0$ for all $i \in I = supp(\mathbf{e})$, if $\mathbf{a} \in A(I)$. So this explains condition (1). Now assume $wt(\mathbf{e}) \leq t$. We need a nonzero element of $K_{\mathbf{y}}$, and the existence of such an element is ensured by assuming that the dimension of A is at least $t+1$, since $K_{\mathbf{y}}$ contains $A(I)$, which is the intersection of t subspaces of codimension 1 and therefore not zero. So this explains condition (2). $K_{\mathbf{y}}$ is the object we know and $A(I)$ is the one we want to know, since it gives an indication where the error positions are located. Condition (3) guarantees that
$$K_{\mathbf{y}} = A(I).$$
We have seen already one inclusion, now suppose $\mathbf{a} \in K_{\mathbf{y}}$, then
$$0 =< \mathbf{y}, \mathbf{a} * \mathbf{b} >=< \mathbf{e}, \mathbf{a} * \mathbf{b} >=< \mathbf{e} * \mathbf{a}, \mathbf{b} >,$$

for all $\mathbf{b} \in B$, so $\mathbf{e} * \mathbf{a} \in B^{\perp}$. But $wt(\mathbf{e} * \mathbf{a}) \leq wt(\mathbf{e}) \leq t < d(B^{\perp})$, hence $\mathbf{e} * \mathbf{a} = 0$, so a_i is zero at all places where e_i is not zero, that is at all $i \in I$, thus $\mathbf{a} \in A(I)$. So we compute $K_{\mathbf{y}}$ and take a nonzero element $\mathbf{a}$ in it; if $K_{\mathbf{y}}$ would be zero, then the received word has more than t errors. We search for the set of zeros of $\mathbf{a}$ and call it J. By the above we know that J contains I and condition (4) implies that J has at most $d(C) - 1$ elements, thus we can find the error values by solving the above mentioned linear equations over the subfield $\mathbf{F}_q$, which has a unique solution. In the algorithm we have to find $K_{\mathbf{y}}$ which is the zero space of a set of linear equations over $\mathbf{F}_{q^e}$ and to find the error values which is done by solving linear equations over $\mathbf{F}_q$, thus the complexity of the algorithm is at most of the order $\mathcal{O}(n^3 e^2)$, since a multiplication in $\mathbf{F}_{q^e}$ is of the order of e^2 multiplications in $\mathbf{F}_q$. Thus we have proved the following theorem.

Theorem 2.2 *If C is an $\mathbf{F}_q$-linear code of length n and (A, B) is a t-error correcting pair for C over $\mathbf{F}_{q^e}$, then one can correct all words with at most t errors with complexity $\mathcal{O}(n^3 e^2)$.*

Example 2.3 Reed-Solomon codes, see [34].
Let $\alpha = (\alpha_1, \ldots, \alpha_n)$ be an n-tuple of n distinct elements of $\mathbf{F}_q$. Let
$$RS_k(\alpha) = \{(f(\alpha_1), \ldots, f(\alpha_n)) | f \in \mathbf{F}_q[x], \deg(f) < k\}.$$

Then $RS_k(\alpha)$ is an $[n, k, n-k+1]$ code, so it is an MDS code and its dual is an $[n, n-k, k+1]$ code. Let $A = RS_{t+1}(\alpha)$ and $B = RS_t(\alpha)$ and $C = RS_{2t}(\alpha)^{\perp}$. Then C has minimum distance $2t+1$. Now A is an $[n, t+1, n-t]$ code, hence $k(A) = t+1$ and $d(A) + d(C) = (n-t) + (2t+1) > n$. Furthermore $d(B^{\perp}) = t+1$. Finally, if $\mathbf{a} \in A$ and $\mathbf{b} \in B$, then there are polynomials f and h of degree at most t and $t-1$, respectively, such that $f(\alpha_i) = a_i$ and $h(\alpha_i) = b_i$ for all i. Now fh has degree at most $2t-1$ and $a_i b_i = fh(\alpha_i)$, so $\mathbf{a} * \mathbf{b} \in C^{\perp}$. Thus $A * B \subseteq C^{\perp}$ and therefore (A, B) is a t-error correcting pair for C.

Example 2.4 BCH codes.
Let α be a primitive element of $\mathbf{F}_{q^e}$ and let $n = q^m - 1$. Let C be the BCH code with the defining set of zeros $\{0, 1, \ldots, \delta-2\}$, that is to say C is the set of all $(c_0, \ldots, c_{n-1}) \in \mathbf{F}_q^n$ such that

$$\sum_{j=0}^{n-1} c_j \alpha^{ij} = 0 \quad \text{for all} \quad 0 \le i \le \delta - 2.$$

Then C is an $\mathbf{F}_q$-linear code of length n and designed minimum distance δ. Let $A = RS_{t+1}(1, \alpha, \ldots, \alpha^{n-1}))$ and $B = RS_t(1, \alpha, \ldots, \alpha^{n-1})$, where $t = \lfloor (\delta-1)/2 \rfloor$. Then A and B are $\mathbf{F}_{q^e}$-linear codes of length n and (A, B) is a t-error correcting pair. This is seen as follows. Let $\mathbf{v}_i = (1, \alpha^i, \ldots, \alpha^{i(n-1)})$ for $0 \le i \le n-1$. Then $\mathbf{v}_0, \ldots, \mathbf{v}_t$ is a basis for A and $\mathbf{v}_0, \ldots, \mathbf{v}_{t-1}$ is a basis for B and $\mathbf{v}_i * \mathbf{v}_j = \mathbf{v}_{i+j}$, so the vector space generated by $A * B$ has basis $\mathbf{v}_0, \ldots, \mathbf{v}_{2t-1}$, and $2t - 1 \le \delta - 2$, so $C \subseteq (A * B)^{\perp}$. The other 3 conditions hold in the same way as in the previous example. With respect to the above bases of A and B the vector space $K_{\mathbf{y}}$ is the nullspace of the matrix

$$\begin{pmatrix} S_0 & S_1 & S_2 & \ldots & S_t \\ S_1 & S_2 & S_3 & \ldots & S_{t+1} \\ S_2 & S_3 & S_4 & \ldots & S_{t+2} \\ \vdots & \vdots & \vdots & \ddots & \vdots \\ S_{t-1} & S_t & S_{t+1} & \ldots & S_{2t-1} \end{pmatrix},$$

where S_i is the syndrome $< \mathbf{y}, \mathbf{v}_i >$.

This line of reasoning for BCH codes was done for the first time by Peterson [43] in 1960 and independently by Arimoto [1] in 1961. Later improvements were found on the minimum distance of cyclic codes, known as the Hartmann-Tzeng, Roos and van Lint-Wilson bounds, see [35]. Feng and Tzeng [18, 19] and Bours et al. [4] gave algorithms decoding up to half the Hartman-Tzeng and in many cases of the Roos designed minimum distance. Duursma and Kötter [12] could explain the just mentioned algorithms by giving error correcting pairs for these cyclic codes, morover they gave error correcting pairs for all but four cases of the table of van Lint and Wilson of all binary cyclic codes of length at most 62.

Example 2.5 Classical Goppa codes.
Let $l = \{\alpha_1, \ldots, \alpha_n\}$ be an enumeration of l, a subset of $\mathbf{F}_{q^e}$ of n elements. Let h be

a polynomial in one variable X with coefficients in $\mathbf{F}_{q^e}$ such that its set of zeros is disjoint from l. Let $\Gamma(l,h)$ be the $\mathbf{F}_q$-linear code of all elements $\mathbf{c} \in \mathbf{F}_q^n$ such that

$$\sum_{i=1}^{n} \frac{c_i}{X-\alpha_i} \text{ is divisible by } h.$$

It is wellknown [37, 12.3] that $\Gamma(l,h)$ has parity check matrix

$$\begin{pmatrix} h(\alpha_1)^{-1} & \dots & h(\alpha_n)^{-1} \\ \alpha_1 h(\alpha_1)^{-1} & \dots & \alpha_n h(\alpha_n)^{-1} \\ \vdots & \ddots & \vdots \\ \alpha_1^{r-1} h(\alpha_1)^{-1} & \dots & \alpha_n^{r-1} h(\alpha_n)^{-1} \end{pmatrix},$$

where r is the degree of the Goppa polynomial h. Then this code has designed minimum distance $r+1$. So if we take $n = q^m - 1$ and $\alpha_i = \alpha^i$ for $i = 0,\dots,n-1$ for some primitive element α of $\mathbf{F}_{q^e}$ and $h = X^{\delta-1}$, then we are back in example 2.4 of the BCH codes. In order to define the pair (A,B) we have to introduce Generalized Reed-Solomon codes. Let $\alpha = (\alpha_1,\dots,\alpha_n)$ be an n-tuple of n distinct elements of $\mathbf{F}_{q^e}$. Let $\mathbf{y}$ be an n-tuple of nonzero elements of $\mathbf{F}_{q^e}$. Define the k-dimensional Generalized Reed-Solomon code over $\mathbf{F}_{q^e}$ by

$$GRS_k(\alpha,\mathbf{y}) = \{(f(\alpha_1)y_1,\dots,f(\alpha_n)y_n) | f \in \mathbf{F}_{q^e}[x], \deg(f) < k\}.$$

Now let $A = RS_{t+1}(\alpha)$ and $B = GRS_t(\alpha,\mathbf{y})$, where $t = \lfloor r/2 \rfloor$ and $y_i = h(\alpha_i)^{-1}$. Then (A,B) is a t-error correcting pair for $\Gamma(l,h)$. Another way to decode these classical Goppa codes is by means of solving the so called key equation with Euclid's algorithm or the Berlekamp-Massey algorithm [2, 36] in the ring of polynomials $\mathbf{F}_{q^e}[X]$. We return to that in section 5.

An important class of codes which can be decoded by error location are algebraic-geometric codes which we will discuss in the next section.

3 Decoding by error location of algebraic-geometric codes

In this section we treat algebraic-geometric codes, also called geometric Goppa codes [20, 21, 22, 23], which can be considered as a generalization of Reed-Solomon codes on the affine line to codes on arbitrary curves. We will not discuss the importance of these codes nor give an exact definition of it and prefer to discuss it by analogy with Reed-Solomon and classical Goppa codes and by an example. The part needed to prove the existence of asymptotically good codes with parameters above the Gilbert-Varshamov bound is very difficult and takes say 5 years to grasp for an outsider. On the other hand for the construction of geometric Goppa codes on explicit curves one needs only the theory of algebraic curves over finite fields, in particular the theorem of Riemann-Roch, and this part a layman can learn in half a year. There are some introductory works on the subject, see [24, 32, 33, 34]. For a selfcontained treatment on algebraic curves over finite fields we recommend the first chapters of the books of Chevalley [8] or Moreno [38]. This suffices to understand the papers concerning the decoding of algebraic-geometric codes. For a more thourough treatment of algebraic curves and their codes we mention the book of Tsfasman and Vlăduţ [59] and the forthcomming book of Stichtenoth [57].

The projective line over $\mathbf{F}_q$ has $q+1$ *points* which are defined over $\mathbf{F}_q$. The field of rational functions on the projective line is denoted by $\mathbf{F}_q(X)$, the elements are quotients of polynomials in $\mathbf{F}_q[X]$. The points of the projective line, not equal to the point at infinity correspond one to one with monic linear polynomials. *Places* are generalizations of points and places which are not the point at infinity correspond one to one with monic irreducible polynomials in $\mathbf{F}_q[X]$, and the degree of a place is the degree of the corresponding irreducible polynomial. Places of degree 1, not equal to the point at infinity, are exactly the points of the affine line. For every place P we introduce a *valuation* v_P at the place on the field of rational functions, which gives the order of zero or pole at this places, that is to say

$$v_P(f) = m \quad \text{if and only if} \quad f = p^m \frac{a}{b},$$

where p is the irreducible polynomial corresponding to the place P and a and b are polynomials not divisible by p. For the point at infinity one can make a coordinate transformation $Y = 1/X$ and consider the valuation defined by the irreducible polynomial Y or equivalently

$$v_\infty(\frac{a}{b}) = \deg(b) - \deg(a),$$

where a and b are polynomials in $\mathbf{F}_q[X]$.

We introduce *divisors* as a way of bookkeeping of the number of zeros and poles of functions and other objects. A divisor is a formal sum of places with integer coefficients, such that at most finitely many coefficients are nonzero. One can add divisors coefficientwise. The *support* of a divisor G is the set of places with a nonzero coefficient in G. One has a partial order on the divisors by comparing them coefficientwise. Thus, if $G = \sum m_P P$ and $H = \sum n_P P$, then $G + H = \sum(m_P + n_P)P$, and $G \le H$ if and only if $m_P \le n_P$ for all P. A divisor is called *effective* if all the coefficients are not negative. The degree of a divisor G of the form $\sum m_P P$ is by definition $\deg(G) = \sum m_P \deg(P)$. A rational function f has a divisor (f), which we call the *principal divisor* of the function, as follows

$$(f) = \sum v_P(f)P.$$

The principal divisor of a rational function f is the difference of two effective divisors, $(f) = (f)_0 - (f)_\infty$, where $(f)_0$ is the *divisor of zeros* and $(f)_\infty$ is the *divisor of poles* of f. Every rational function f has a factorization $f = \prod q_i^{e_i}$, where q_i is an irreducible polynomial and e_i an integer. Let Q_i be the place corresponding to q_i and P_∞ the point at infinity. It is not difficult to see that $(f) = \sum e_iQ_i - e_i\deg(Q_i)P_\infty$. Thus $\deg((f)) = 0$.

For every divisor G one defines the vector space $L(G)$ of rational functions with zeros and poles prescribed by G as follows

$$L(G) = \{f \in \mathbf{F}_q[X] | f = 0 \text{ or } (f) \geq -G\}.$$

So, if $G = mP_\infty$, then $L(G)$ is the vector space of all polynomials of degree at most m, it has basis $1, X, \ldots, X^m$ and dimension $m+1$. Let h be a Goppa polynomial. Let $h = \prod q_i^{e_i}$ be the factorization of h, and Q_i the place corresponding to q_i. If $G = \sum e_iQ_i - P_\infty$, then $L(G)$ is the vector space of rational functions of the form a/h, where a is a polynomial of degree at most $r-1$ and r the degree of h, thus $L(G)$ has a basis of the form $1/h, x/h, \ldots, x^{r-1}/h$ and dimension r. In both examples we have that the dimension of $L(G)$ is equal to $\deg(G)+1$, this is indeed always the case for the projective line.

If $P = (\alpha : 1)$ and G is a divisor which does not have P in its support and $f \in L(G)$, then $v_P(f) \geq 0$, so f has not a pole at P, therefore one can evaluate f at P and get $f(\alpha) \in \mathbf{F}_q$. If $l = \{\alpha_1, \ldots, \alpha_n\}$ are n elements in $\mathbf{F}_q$ and $P_i = (\alpha_i : 1)$, then $P_1, \ldots, P_n$ are n distinct points on the projective line, that is places of degree 1. We denote the divisor $P_1 + \cdots + P_n$ by D. Let G be a divisor with has disjoint support with D. Then we can evaluate f at all $P_i's$ and consider the map

$$\mathrm{ev}_D : L(G) \longrightarrow \mathbf{F}_q^n,$$

defined by $\mathrm{ev}_D(f) = (f(P_1), \ldots, f(P_n))$. This map is $\mathbf{F}_q$-linear and injective if $\deg(G) < n$. We denote the image of this map by $C_L(D,G)$ which is an $\mathbf{F}_q$-linear $[n, m+1, n-m]$ code, where $m = \deg(G)$. In case $G = (k-1)P_\infty$ we get the Reed-Solomon code $RS_k(\alpha)$. Now we consider the classical Goppa code with Goppa polynomial $h \in \mathbf{F}_q[X]$ and locator set $l = \{\alpha_1, \ldots, \alpha_n\}$ in $\mathbf{F}_q$. So we do not consider the subfield subcode case for simplicity, that is $e = 1$ in $\mathbf{F}_{q^e}$. Let $h = \prod q_i^{e_i}$ be the factorization of h and Q_i the place corresponding to q_i. Define the divisor $G = \sum e_iQ_i - P_\infty$. Then $C_L(D,G)$ is the dual of the classical Goppa code $\Gamma(l,g)$, since the above mentioned parity check matrix of $\Gamma(l,g)$ is a generator matrix of $C_L(D,G)$.

For an arbitrary algebraic curve $\mathcal{X}$ over the field $\mathbf{F}_q$ one can consider its fuction field $\mathbf{F}_q(\mathcal{X})$, which is an finite extension of $\mathbf{F}_q(X)$. Similarly to the case of the projective line one can define points, places, valuations, divisors, the principal divisor of a function and the vector space $L(G)$ of a divisor G and the code $C_L(D,G)$. We still have that the degree of a principal divisor is zero, but for the dimension of $L(G)$ we do not have a simple formula. The Theorem of Riemann says that there exists a nonnegative integer l such that for every divisor G

$$\dim(L(G)) \geq \deg(G) + 1 - l,$$

and the smallest nonnegative integer with this property is called the *genus*, and is denoted by $g(\mathcal{X})$ or g. Thus the genus of the projective line is zero. If moreover $\deg(G) > 2g-2$, then $\dim(L(G)) = \deg(G) + 1 - g$. If $2g-2 < \deg(G) < n$, then $C_L(D,G)$ has dimension $\deg(G) + 1 - g$ and minimum distance at least $n - \deg(G)$.

The minimum distance of the dual of a code can be estimated in terms of the code itself, since we have the following property which is easy to verify

$$d(C^{\perp}) > t \quad \text{if and only if } \dim(C(I)) = n - |I| \text{ for all } I \subseteq \{1,\ldots,n\} \text{ and } |I| \leq t,$$

where $C(I)$ is the subcode of words in C which are zero at all $i \in I$, as defined in section 1. If $Q = \sum_{i\in I} P_i$, then $C_L(D,G)(I) = C_L(D,G-Q)$. The dimension of $C_L(D,G-Q)$ is $\deg(G) = |I|$ less than the dimension of $C_L(D,G)$, if $|I| < \deg(G) - (2g-2)$. Thus the dual of $C_L(D,G)$ has at least minimum distance $\deg(G) - 2g + 2$. Another way to get an estimate for the minimum distance of the dual code of $C_L(D,G)$ is with the help of *differentials*. This can be explained by the classical Goppa codes too. Think of differentials as objects of the form fdh, where f and h are rational functions and dh is the differential of h, such that the map which sends h to dh is $\mathbf{F}_q$-linear and the Leibniz rule $d(fh) = fdh + hdf$ holds. One can talk about zeros and poles of differentials. We denote the set of differentials on $\mathcal{X}$ by $\Omega_{\mathcal{X}}$. At every place P there exists a *local parameter* that is a fuction u which has valuation 1 at P, and for every differential ω there exists a function f such that $\omega = fdu$. The valuation $v_P(\omega)$ of ω at P is now by definition $v_P(f)$, so we say ω has a zero of order m if $m = v_P(f) > 0$ and ω has a pole of order m if $m = -v_P(f) > 0$. The divisor (ω) of ω is by definition $(\omega) = \sum v_P(\omega)P$. The divisor of a differential is called *canonical* and has always degree $2g-2$. In the same way as we defined $L(G)$ for functions we now define the vector space $\Omega(G)$ of differentials with zeros and poles prescribed by G as follows

$$\Omega(G) = \{\omega \in \Omega_{\mathcal{X}} | \omega = 0 \text{ or } (\omega) \geq G\}.$$

Now one could have defined the genus as the dimension of the vector space of differentials without poles, that is of $\Omega(O)$, where O is the divisor with coefficient 0 at every place. The Theorem of *Riemann-Roch* states that

$$\dim(L(G)) = \deg(G) + 1 - g + \dim(\Omega(G)).$$

If moreover P is a place of degree 1 and u is a local parameter of P, then f has a formal Laurent series $\sum_{i=m}^{\infty} a_i u^i$, where the coefficients a_i are in $\mathbf{F}_q$ and $m = v_P(\omega)$ and $a_m \neq 0$. The entry a_{-1} is called the *residue* of ω and denoted by $\mathrm{res}_P(\omega)$. If we have n points on the curve, that is n places of degree 1, and a divisor G with none of the P_i's in its support, then we consider the map

$$\mathrm{res}_D : \Omega(G - D) \longrightarrow \mathbf{F}_q^n,$$

defined by $\mathrm{res}_D(\omega) = (\mathrm{res}_{P_1}(\omega), \ldots, \mathrm{res}_{P_n}(\omega))$. Its image we denote by $C_\Omega(D, G)$. If $2g-2 < \deg(G) < n$, then $C_\Omega(D, G)$ is a $\mathbf{F}_q$-linear code of dimension $n - \deg(G) - 1 + g$ and minimum distance at least $\deg(G) - 2g + 2$. We call $\deg(G) - 2g + 2$ the *designed minimum distance* of the code $C_\Omega(D, G)$ and denote it by d^*. Both constructions, that is by evaluating functions or by taking residues of differential forms are called *geometric Goppa* codes or *algebraic-geometric* codes, abreviated by AG codes. The fact that $C_L(D, G)$ and $C_\Omega(D, G)$ are dual codes is seen by the *residue theorem*, which states that the sum of all residues over all places of a fixed differential is zero.

For classical Goppa codes, where we do not consider the subfield subcode case for simplicity, that is $\mathbf{F}_{q^s} = \mathbf{F}_q$, it is as follows. Let $h = \prod q_i^{e_i}$ be the factorization of the Goppa polynomial h and Q_i the place corresponding to q_i. Define the divisor $G = \sum e_i Q_i - P_\infty$, as before. A differential ω is an element of $\Omega(G - D)$ if and only if

$$\omega = \sum_{i=1}^{n} \frac{c_i dX}{X - \alpha_i} \quad \text{and} \quad \sum_{i=1}^{n} \frac{c_i}{X - \alpha_i} \quad \text{is divisible by } h.$$

Moreover $\mathrm{res}_D(\omega) = \mathbf{c}$ whenever ω is of the above form. Thus $C_\Omega(D, G) = \Gamma(l, h)$.

The function field of a curve is a finite extension of $\mathbf{F}_q(X)$, thus it is of the form $\mathbf{F}_q(X)[Y]/(F(X,Y))$, where we may assume that $F(X,Y) \in \mathbf{F}_q[X,Y]$ and is absolutely irreducible, that is irreducible in $k[X,Y]$, where k is the algebraic closure of $\mathbf{F}_q$. We say that $F(X,Y)$ is an *affine equation* of a *plane model* of the curve. If we homogenize F, that is to say we consider $\tilde{F}(X,Y,Z) = Z^m F(X/Z, Y/Z)$, where m is the degree of F, then we say that $\tilde{F}$ is an equation of a *projective plane model*. If $\tilde{F}$ together with all its partial derivatives have no common zero $(a,b,c) \in k^3$, except $(0,0,0)$, we say that the projective plane model has no *singularities*, in this case all the $(a:b:c) \in PG(2,q)$ such that $\tilde{F}(a,b,c) = 0$, are the points of the curve, that is places of degree 1, moreover the genus of the curve is equal to $(m-1)(m-2)/2$.

Example 3.1 Hermitian curves.
The Hermitian curve $H(q)$ is defined by the affine equation

$$U^m + V^m + 1 = 0$$

over $GF(q)$, where $q = (m-1)^2$, for the details we refer to [56, 58]. Let $a, b \in GF(q)$ such that $a^{m-1} + a = b^m = -1$ and $P = (1 : b : 0)$. Define $X = b/(V - bU)$ and $Y = UX - a$. Then we have another equation

$$X^m - Y^{m-1} - Y = 0$$

of the same curve. We prefer the last equation since it has exactly one point at infinity. This plane curve is nonsingular and has $q\sqrt{q}$ points.

Now we return to the decoding of AG codes. The condition $A * B \subseteq C^\perp$ is easy to fulfill for algebraic-geometric codes. Suppose we want to decode the code $C = C_\Omega(D, G)$, we remarked already that the dual of this code is $C_L(D, G)$. Let F be any divisor with disjoint support with D, then $G - F$ has also disjoint support with D. Moreover if $f \in L(F)$ and $h \in L(G - F)$, then $fh \in L(G)$. If $\mathbf{a} = \mathrm{ev}_D(f)$ and $\mathbf{b} = \mathrm{ev}_D(h)$, then $\mathbf{a} * \mathbf{b} = \mathrm{ev}_D(fh)$. Thus

$$C_L(D, F) * C_L(D, G - F) \subseteq C_L(D, G) = C_\Omega(D, G)^\perp.$$

With the above estimates for the dimension and minimum distance of AG codes we have the following proposition and theorem.

Proposition 3.2 *Let F and G be divisors with support disjoint from D.*
Let $A = C_L(D, F)$, $B = C_L(D, G - F)$ and $C = C_\Omega(D, G)$. Then
1. $A * B \subseteq C^\perp$.
2. *If $t + g \leq deg(F) < n$, then $k(A) > t$.*
3. *If $deg(G - F) > 2g - 2$, then $d(A) + d(C) > n$.*
4. *If $deg(G - F) > t + 2g - 2$, then $d(B^\perp) > t$.*

Theorem 3.3 *Every algebraic-geometric code $C_\Omega(D, G)$ of designed minimum distance d^* on a curve of genus g has a $\lfloor (d^* - 1 - g)/2 \rfloor$-error correcting pair whenever $deg(G) > 2g - 2$.*

After the invention of Goppa of AG codes and the work of Tsfasman, Vlăduţ and Zink [60], who proved that there exists curves over finite fields with many points, giving asymptotically good codes, at the beginning of the eighties, it took some time before Justesen, Larsen, Elbrønd Jensen, Havemose and Høholdt [26] found at the end of the eighties a generalization of Peterson's algorithm for AG codes on plane curves. This was generalized to arbitrary curves by Skorobogatov and Vlăduţ [55]. Independently Krachkovskii proved the same result, but his work was only published as a preprint in Russian [30]. These papers are surveyed in this and the previous section. Independently Porter gave another decoding algorithm which we will discuss in section 5. The above algorithm is called the ***basic algorithm*** in [55, 59]. In the ***modified algorithm*** one does not look at one error locating pair only, but at a sequence $(C_L(D,iP), C_L(D,G-iP))$ of pairs, where P is a point not in the support of G nor D, and one looks at the smallest i such that the coresponding vector space $K_{\mathbf{y}}$ is not zero. With the modified algorithm $(d^*-1)/2-s$ errors can be decoded [26, 28, 9], where s is the so called *Clifford defect*, see Duursma [9]. For plane curves s is roughly equal to $g/4$, but only in very special cases s is equal to zero. It can be shown that the modified algorithm does not always decode up to half the minimum distance [28], but the number of error patterns where the algorithm does not decode the error is relatively small [28, 41]. It was my contribution [40] to show that, on so called ***maximal curves***, there exist u divisors $F_1, \ldots, F_u$, where u is of the order $2g$, such that for every received word with at most $\lfloor (d^*-1)/2 \rfloor$ errors, at least one of the pairs $(C_L(D,F_i), C_L(D,G-F_i))$ corrects the errors. This was generalized to all curves by Vlăduţ [61]. For these last results one has to introduce the Zeta function of a curve over a finite field and the Jacobian of the curve. It is a counting argument to show that certain divisors exist, but in order to find these divisors explicitly one has to know quite a lot about the Jacobian and not much is known for this purpose. Rotillon and Thiong Ly [46] considered the problem of finding these divisors for the Klein quartic, and Le Brigand [31] for hyperelliptic curves. Carbonne and Thiong Ly [7] found for some curves the minimal u such that the above procedure works. First Ehrhard [13] considered divisors F_i of the form lP_j. In all these attempts one runs the algorithm for a lot of error locating pairs in parallel. Later Ehrhard [15] gave an explicit solution of the problem of decoding up to half the minimum distance, now the divisor F_i is changed during the algorithm. It is probably not possible to give an explanation of this result purely in terms of linear algebra. Moreover the designed minimum distance should be at least $4g$. Feng, Rao and Duursma gave another solution to the problem of finding an efficient explicit algorithm which decodes up to half the designed minimum distance. Their work on majority coset decoding is treated in the next section.

Any linear code is representable by means of a curve, see [42]. This is a nice mathematical result, but from a pratical coding point of view it puts matters upside down. If we want to construct good codes we have gained nothing by this knowledge. But it is hoped for that by this fact one can prove that any linear code of minimum distance d has a $\lfloor (d-1)/2 \rfloor$-error correcting pair. The property $A * B \subseteq C^{\perp}$ is difficult to fulfill for arbitrary linear codes and comes quite naturally in the geometric situation as the product of functions.

4 Majority coset decoding

In the following we explain the beautiful idea of Feng and Rao [16] which was treated in greater generality by Duursma [10, 11]. This can be explained purely in terms of linear algebra, in the same way as we abstracted from AG codes to linear codes in section 2.

Suppose we have a code C_1 for which we need a decoding algorithm, and a subcode C_2 for which we have a decoding algorithm. *Coset decoding* is an algorithm which has as input a word $\mathbf{y}_1$ such that $\mathbf{y}_1 \in \mathbf{e} + C_1$, and as output $\mathbf{y}_2$ such that $\mathbf{y}_2 \in \mathbf{e} + C_2$.

This was done for instance by Yaghoobian and Blake [62], where they decode Hermitian codes, but the complexity of the algorithm is in general rather great. Feng and Rao apply coset decoding to the code $C_1 = C_\Omega(D, mP)$, which has designed minimum distance d^* and its subcode $C_2 = C_\Omega(D,(m+g)P)$, which has designed minimum distance $d^* + g$, where P is a point not in D and g is the genus of the curve. We have seen in Theorem 4.5 that C_2 has a t-error correcting pair, where $t = \lfloor (d^* - 1)/2 \rfloor$. The coset decoding is done in several steps where the dimension of the subcode drops by one in every step. The whole setup is also used by Feng and Rao in another paper [17] to get an improved lower bound on the minimum distance of AG codes in case the redundancy is between g and $2g$. Furthermore it can be explained in terms of linear algebra only as we will do in the following. Instead of saying in the future that we are talking about three sequences of linear codes which satisfy conditions (5), ..., (8), we prefer to give this notion a name, which is done in the following defnition.

Definition 4.1 An *array* of codes is a triple $(\mathcal{A}, \mathcal{B}, \mathcal{C})$ of sequences of linear codes in $\mathbf{F}_q^n$, enumerated by $\mathcal{A} = (A_i)_{1 \le i \le u}$, $\mathcal{B} = (B_j)_{1 \le j \le v}$ and $\mathcal{C} = (C_r)_{w \le r \le l}$, such that the following holds:

5. $\dim(A_i) = i$, $\dim(B_j) = j$ and $\dim(C_r) = n - r$.

6. $A_i \subseteq A_{i+1}$, $B_j \subseteq B_{j+1}$ and $C_{r+1} \subseteq C_r$.

7. For every i and j there exists an r such that $A_i * B_j \subseteq (C_r)^\perp$. Thus for every i and j we define $r(i,j)$ to be the smallest index r such that $A_i * B_j \subseteq (C_r)^\perp$.

8. If $\mathbf{a} \in A_i \setminus A_{i-1}$ and $\mathbf{b} \in B_j \setminus B_{j-1}$ and $r = r(i,j) \ge w+1$, then $\mathbf{a} * \mathbf{b}$ is an element of $(C_r)^\perp \setminus (C_{r-1})^\perp$.

Remark that $r(i,j)$ is increasing, that is if $i \le i'$ and $j \le j'$, then $r(i,j) \le r(i',j')$. Define the following set

$$N_r = \{(i,j) | 1 \le i \le u, 1 \le j \le v, r(i,j) = r+1\}$$

Let n_r be the number of elements of N_r. Define

$$d_r = \min\{n_{r'} | r \leq r' < l\} \cup \{d(C_l)\}.$$

Theorem 4.2 *For an array of codes we have that $d_r \leq d(C_r)$, for all $w \leq r \leq l$.*

For the proof we introduce bases. Let $\mathbf{a}_1, \ldots, \mathbf{a}_i$ be a basis of A_i and let $\mathbf{b}_1, \ldots, \mathbf{b}_j$ be a basis of B_j. Then $\mathbf{a}_i \in A_i \setminus A_{i-1}$ and $\mathbf{b}_j \in B_j \setminus B_{j-1}$, by conditions (5) and (6). Thus $\mathbf{a}_i * \mathbf{b}_j \in (C_r)^\perp \setminus (C_{r-1})^\perp$, where $r = r(i,j)$, by condition (7). The proof of the theorem goes by decreasing induction on r. If $r = l$, then the conclusion is obvious by the definition of $d_l = d(C_l)$. Suppose we have already proved $d(C_{r+1}) \geq d_{r+1}$. Now we prove that $d(C_r) \geq d_r$. If $\mathbf{y}$ is a nonzero word in C_{r+1}, then $wt(\mathbf{y}) \geq d_{r+1} \geq d_r$, by the induction hypothesis, and the definition of d_r. If $\mathbf{y} \in C_r \setminus C_{r+1}$, then look at the entries $S_{i,j}$ in the $u \times v$ matrix $\mathbf{S}$, where $S_{i,j} = < \mathbf{y}, \mathbf{a}_i * \mathbf{b}_j >$. If $r(i,j) \leq r$, then $S_{i,j} = 0$, by (7). If $r(i,j) = r+1$, then $S_{i,j} \neq 0$. Because $(C_{r+1})^\perp = (C_r)^\perp + < \mathbf{a}_i * \mathbf{b}_j >$, since $\mathbf{a}_i * \mathbf{b}_j \in (C_{r+1})^\perp \setminus (C_r)^\perp$ by the above remark and $\dim(C_r) = \dim(C_{r+1}) + 1$, by (5). Therefore $\mathbf{S}$ has an echelon form with nonzero pivots at all entries $S_{i,j}$ where $r(i,j) = r+1$. Thus the rank of the matrix $\mathbf{S}$ is at least equal to the number of elements of N_r, which we called n_r. On the other hand, we can decompose the same matrix in a product of three matrices $\mathbf{S} = \mathbf{AWB}^t$, where $\mathbf{A}$ is the matrix with $\mathbf{a}_1, \ldots, \mathbf{a}_u$ as rows, $\mathbf{B}$ is the matrix with $\mathbf{b}_1, \ldots, \mathbf{b}_v$ as rows and $\mathbf{W}$ is the $n \times n$ diagonal matrix with $\mathbf{y}$ on the diagonal, and zeros outside the diagonal. Hence $\mathrm{rank}(\mathbf{S}) \leq \mathrm{rank}(\mathbf{W}) = wt(\mathbf{y})$. Combining the two inequalities involving the rank of $\mathbf{S}$ we get $wt(\mathbf{y}) \geq n_r$, which is at least d_r, by definition. Thus $d(C_r) \geq d_r$, and we have proved the theorem by induction.

In order to be able to decode C_w up to $\lfloor (d_w - 1)/2 \rfloor$ errors, we need an extra condition.

Definition 4.3 An array of codes $(\mathcal{A}, \mathcal{B}, \mathcal{C})$ of sequences of linear codes in $\mathbf{F}_q^n$, enumerated by $\mathcal{A} = (A_i)_{1 \leq i \leq u}$, $\mathcal{B} = (B_j)_{1 \leq j \leq v}$ and $\mathcal{C} = (C_r)_{w \leq r \leq l}$, is called a *t-error correcting array* for a code C in $\mathbf{F}_q^n$, if $C = C_w$ and $t \leq (d_w - 1)/2$, and $C_l = 0$ or there exists i and j such that (A_i, B_j) is a t-error correcting pair for C_r, where $r = r(i,j)$.

Remark that we could have taken the codes A_i, B_j and C_r to be $\mathbf{F}_{q^e}$-linear as we have done in the definition of a t-error correcting pair. In that case we assume that C is in $\mathbf{F}_q^n$ and a subfield subcode of C_w.

Theorem 4.4 *If a code has a t-error correcting array, then it has a decoding algorithm of complexity $\mathcal{O}(n^3)$ which corrects t errors.*

Before we prove the theorem we introduce some definitions, see [16, 17]. Let $\mathbf{S}$ be a $(u \times v)$-matrix with entries $S_{i,j}, 1 \leq i \leq u, 1 \leq j \leq v$. Let $\mathbf{S}(i,j)$ be the $(i \times j)$-matrix with entries $S_{i',j'}, 1 \leq i' \leq i, 1 \leq j' \leq j$. Consider the folowing two conditions.

9. $\text{rank}(\mathbf{S}(i-1,j-1)) = \text{rank}(\mathbf{S}(i-1,j)) = \text{rank}(\mathbf{S}(i,j-1))$

10. $\text{rank}(\mathbf{S}(i-1,j-1)) = \text{rank}(\mathbf{S}(i,j))$

Clearly (10) implies (9); conversely if (9) holds, then there exists a unique value of $S_{i,j}$ such that (10) holds. We call (i,j) a *discrepancy* if (9) holds and (10) does not hold. Remark that (9) holds for (i,j) if and only if (10) holds for all $(i',j), 1 \leq i' < i$ and all $(i,j'), 1 \leq j' < j$. Thus in every row there is at most one discrepancy and the same holds for every column. The number of discrepancies of $\mathbf{S}$ is equal to the rank of $\mathbf{S}$.

Now we prove Theorem 4.4. Suppose $\mathbf{y}$ is a received word and has error $\mathbf{e}$ with respect to $C = C_w$ of weight at most $t \leq (d_w - 1)/2$. Let

$$S_{i,j} = <\mathbf{e}, \mathbf{a}_i * \mathbf{b}_j>$$

and let $\mathbf{S}$ be the corresponding matrix with entries $S_{i,j}$. The proof now goes by increasing induction on r. For all (i,j) such that $r(i,j) \leq w$ we know the syndrome $S_{i,j}$, since $\mathbf{y} \in \mathbf{e} + C_w$. Suppose we already know an element $\mathbf{y}_r \in \mathbf{e} + C_r$ and all the syndromes $S_{i,j}$ for (i,j) such that $r(i,j) \leq r < l$. Now we explain how to get an element $\mathbf{y}_{r+1} \in \mathbf{e} + C_{r+1}$ and the syndromes $S_{i,j}$ such that $r(i,j) \leq r+1$. We call a pair (i,j) a *candidate* if $r(i,j) = r+1$ and (9) holds. A candidate is called *true* if (10) holds, and *false* if (10) does not hold. We know the candidates but we do not know which candidates are true or false. We first prove the remarkable property that the number of true candidates, which we will denote by T, is greater than the number of false candidates, which we will denote by F. Afterwards we show how to compute a certain $\lambda \in \mathbf{F}_q$ for every candidate, which is the same for all true candidates and thus gives us $\mathbf{y}_{r+1}$ and new syndromes. We have seen already in the proof of Theorem 4.2 that the rank of $\mathbf{S}$ is at most $wt(\mathbf{e}) \leq t$. Denote the number of *known* discrepancies, that is discrepancies at entries (i,j) such that $r(i,j) \leq r$, by K. The other discrepancies are called *unknown*. Clearly we have that all false candidates are unknown discrepancies. Thus

$$K + F \leq \text{the number of all discrepancies} = \text{rank}(\mathbf{S}) \leq t.$$

Furthermore for every pair (i,j) such that $r(i,j) = r+1$, that is $(i,j) \in N_r$, which is not a candidate, there exists a known discrepancy in the same row or column, possibly in both. For every candidate, true or false, there exist no known discrepancy in the same row nor column. We called n_r the number of elements of N_r, thus

$$n_r \leq T + F + 2K.$$

If we combine the two inequalties above and use that $2t < d_w \leq d_r \leq n_r$, then we get

$$n_r \leq T + F + 2K \leq T + F + (2t - 2F) < T - F + n_r.$$

Therefore

$$F < T,$$

that is the number of true candidates is greater than the number of false candidates. We associate with each candidate an element $\lambda \in \mathbf{F}_q$ in such a way that all true candiates have the same λ, thus giving a way to know the true candidates, by majority as we explained above, and therefore the syndromes $S_{i,j}$ such that $r(i,j) = r+1$. For every candidate, that is for every (i,j) such that (9) holds, there is a unique $S'_{i,j} \in \mathbf{F}_q$ to fill at entry (i,j) in order that (10) holds. This entry $S'_{i,j}$ can be computed using the known syndromes, and is equal to $S_{i,j}$ if and only if (i,j) is a true candidate. The vector space C_r contains the vector space C_{r+1} and the quotient is one dimensional, so there exists a vector $\mathbf{c}_r$ such that $C_r =< \mathbf{c}_r > + C_{r+1}$. Therefore there exists a unique $\lambda_r \in \mathbf{F}_q$ such that $\mathbf{y}_r + \lambda_r \mathbf{c}_r$, which we will call $\mathbf{y}_{r+1}$, is an element of $\mathbf{e} + C_{r+1}$. If $r(i,j) = r+1$, then $C_{r+1}^{\perp} = C_r^{\perp} + < \mathbf{a}_i * \mathbf{b}_j >$. Thus $< \mathbf{c}_r, \mathbf{a}_i * \mathbf{b}_j >$ is not equal to zero. By taking the inner product with $\mathbf{a}_i * \mathbf{b}_j$ we get

$$< \mathbf{y}_r, \mathbf{a}_i * \mathbf{b}_j > + \lambda_r < \mathbf{c}_r, \mathbf{a}_i * \mathbf{b}_j > = S_{i,j}.$$

For every candidate (i,j) we compute $S'_{i,j}$ as explained above and the element $\lambda_{i,j}$ defined by

$$\lambda_{i,j} = \frac{S'_{i,j} - < \mathbf{y}_r, \mathbf{a}_i * \mathbf{b}_j >}{< \mathbf{c}_r, \mathbf{a}_i * \mathbf{b}_j >}.$$

If (i,j) is a true candidate, then $S'_{i,j} = S_{i,j}$ so $\lambda_{i,j} = \lambda_r$. Therefore all $\lambda_{i,j}$ are the same for all true candidates (i,j). Thus the λ which occurs most often among the $\lambda_{i,j}$ of the candidates is equal to λ_r. In this way we get $\mathbf{y}_{r+1} \in \mathbf{e} + C_{r+1}$ and the syndromes $S_{i,j}$ such that $r(i,j) \leq r+1$, that is one step further in the induction. Finaly $C_l = 0$ or there exists i and j such that (A_i, B_j) is a t-error correcting pair for C_r, where $r = r(i,j)$ and $t \leq (d_w - 1)/2$. In the first case we are done, since $\mathbf{y}_l = \mathbf{e}$ when $C_l = 0$, and in the second case we can apply the decoding algorithm of section 2 to the word $\mathbf{y}_r$ with the pair (A_i, B_j) for C_r, which gives $\mathbf{e}$ as output.

Theorem 4.5 *Every algebraic-geometric code $C_\Omega(D,G)$ of designed minimum distance d^* on a curve of genus g has a $\lfloor (d^*-1)/2 \rfloor$-error correcting array, whenever $2g-2 < deg(G) < n$ and n is smaller than the number of points on the curve.*

We give a scetch of the proof. The number of points on the curve is greater than n, so we can take a point P which is not in the support of D. Let a_P be the coefficient of P in G. There are increasing sequences $(m_i)_{i=1}^{u}$ and $(n_j)_{j=1}^{v}$ such that $A_i = C_L(D, m_i P)$ and $B_j = C_L(D, G + n_j P)$ and $C_r = C_\Omega(G + n_r P)$ define a $\lfloor (d^*-1)/2 \rfloor$-error correcting array. Conditions 5 and 6 are immediate by the appropriate choice of the sequences (m_i) and (n_i). Condition 7 is seen as follows. If $\mathbf{a} \in A_i \setminus A_{i-1}$ and $\mathbf{b} \in B_j \setminus B_{j-1}$, then there exist fuctions $f \in L(m_iP) \setminus L((m_i-1)P)$ and $h \in L(G+n_jP) \setminus L(G+(n_j-1)P)$ such that $f(P_l) = a_l$ and $h(P_l) = b_l$ for all l. Thus f has pole order m_i at P and h has pole order $a_P + n_j$ at P, thus fh has pole order $a_P + m_i + n_j$ at P. There exists an r such that $m_i + n_j = n_r$, so $fh \in L(G + n_rP) \setminus L(G + (n_r - 1)P)$. Furthermore $\mathrm{ev}_D(fh) = \mathbf{a} * \mathbf{b}$ and the map ev_D is injective on $L(G + n_rP)$ since $\deg(G) < n$, so $\mathbf{a} * \mathbf{b} \in C_r^{\perp} \setminus C_{r-1}^{\perp}$.

In the papers of Feng and Rao only codes of the form $C_\Omega(D, mP)$ are considered, so there it suffices to take $B_i = A_i$. In Duursma's approach arbitrary diviors G are allowed and one can take B_i different from A_i. The restriction that the number of points on the curve is greater than n remains.

5 Decoding algebraic-geometric codes by solving the key equation

Around the same time as Justesen et al. [26], Porter [44] found another decoding algorithm, which is a generalization of solving the key equation of classical Goppa codes by Euclid's algorithm in the ring of polynomials in one variable. The thesis of Porter contained some mistakes but these were corrected in [45]. Ehrhard [13, 14] also showed that the results of Porter are correct, moreover he proved the equivalence of the modified algorithm and the decoding by solving the key equation.

One can view the ring of polynomials in one variable as the ring of rational functions on the projective line with only poles at the point at infinity. The ring of polynomials in one variable is replaced by the ring $K_\infty(P)$ of rational functions on the curve with only poles at a fixed point P, where P is not equal to one of the points used to construct the geometric Goppa code.

Now we scetch the generalization of the classical Goppa codes. Let l be a subset of $\mathbf{F}_q$ and h a Goppa polynomial. Let $l = \{\alpha_1, \ldots, \alpha_n\}$. Suppose h is not zero at α_i, for all i. We already mentioned that the classical Goppa code $\Gamma(L, h)$ is defined by

$$\Gamma(l, h) = \{\mathbf{y} \mid \sum \frac{y_i}{X - \alpha_i} \equiv 0 \pmod{h}\}.$$

If we let $\varepsilon_i = dX/(X - \alpha_i)$, and take for P the point at infinity on the projective line and for E the divisor of zeros of h, then $\Gamma(l, h) = C_\Omega(D, E - P)$ and

$$\mathbf{y} \in \Gamma(L, h) \text{ if and only if } \sum \frac{y_i}{X - \alpha_i} dX \in \Omega(E - P - D)$$

First it is shown that for arbitrary AG codes one may assume that the divisor G in the definition of the code $C_\Omega(D, G)$ is of the form $E - \mu P$, where E is an effective divisor and μ a positive integer. Secondly one can show that there always exist n independent differentials $\varepsilon_1, \ldots, \varepsilon_n \in \Omega(-D-\mu P)$ such that $\text{res}_{P_i}(\varepsilon_j)$ is 1 if $i = j$ and 0 otherwise, and for every differential $\omega \in \Omega(E - \mu P - D)$ we have

$$\omega = \sum \text{res}_{P_i}(\omega)\varepsilon_i.$$

If we let $\varepsilon(\mathbf{y}) = \sum y_i \varepsilon_i$, then the map ε is the right inverse of res_D and

$$\varepsilon(\mathbf{y}) \in \Omega(E - \mu P - D) \text{ if and only if } \mathbf{y} \in C_\Omega(D, E - \mu P).$$

Now suppose that E is the divisor of zeros of a function $h \in K_\infty(P)$ which is not zero at P_i for all i. We want to define the *syndrome* of a received word. In order to represent the syndrome as a rational function one proves the existence of a particular differential η first. For classical Goppa codes one takes $\eta = dX$. The syndrome $S(\mathbf{y})$ of a received word $\mathbf{y}$ is now defined as follows.

$$S(\mathbf{y})\eta = \sum y_i \frac{h(P_i) - h}{h(P_i)} \varepsilon_i.$$

The syndrome is an element of the ring $K_\infty(P)$, and if E is the divisor of zeros of $h \in K_\infty(P)$, then

$$\mathbf{y} \in C_\Omega(D, E - \mu P) \text{ if and only if } S(\mathbf{y}) \equiv 0 \pmod{h}.$$

For simplicity we assume that η is a differential such that $(\eta) = (2g-2)P$. Now one searches for *solutions* of the *key equation*, that is for pairs (f, r) with $f, r \in K_\infty(P)$ such that there exists an $a \in K_\infty(P)$ with the property

$$fS(\mathbf{y}) = r + ah.$$

A solution is called *valid* if moreover $\deg(r) - \deg(f) \leq 2g - 2 + \mu$. A valid solution (f, r) is called *minimal* if $\deg(f)$ is minimal among all the degrees of f' such that (f', r') is a valid solution. If the received word $\mathbf{y}$ has at most $(d^* - 1)/2 - s$ errors, where s is the Clifford defect [9], then a minimal valid solution (f, r) has the property that $\text{res}_{P_i}(r\eta/f)$ is the error value at place i for all i. Shen [52, 53, 54] computed explicit formulas for the differentials $\varepsilon_1, \ldots, \varepsilon_n$ and the syndromes of codes on Hermitian curves.

Euclid's algorithm gives in the case of classical Goppa codes a sequence of solutions of the key equation, and the first valid solution in this sequence is also a minimal valid solution. The ring $K_\infty(P)$ is not a Euclidean ring whenever the genus is not zero. The sequence of solutions in the Euclidean algorihm is replaced by an algorithm giving the so called *subresultant sequence*, see Porter [44] and Shen [51, 54].

6 Improvements of the complexity

It is not before one has decoding algorithms as fast as Euclid's algorithm or the Berlekamp-Massey algorithm [2, 36] solving the key equation, that AG codes will be implemented for practical purposes.

For polynomials in one variable division with rest gives Euclid's algorithm. Buchberger's algorithm [6] computing Gröbner bases is a generalization of this to polynomials in several variables. Sakata [47, 48] gave a generalization of the Berlekamp-Massey algorithm to several variables. Using Sakata's ideas Justesen, Larsen, Elbrønd Jensen and Høholdt [27] could improve the complexity of their algorithm for codes on plane curves from $\mathcal{O}(n^3)$ to $\mathcal{O}(n^{7/3})$. Dahl Jensen [25] generalized this and obtained complexity $\mathcal{O}(n^{3-\frac{2}{r+1}})$ for curves in projective r-space. In analogy with the work Justesen et al. [27] following Sakata, Shen [53, 54] obtained the same complexity for codes on Hermitian curves using Porter's algorithm.

Up to now nobody succeeded in obtaining the same complexity for decoding AG codes as Euclid's algorithm, that is $\mathcal{O}(n^2)$.

Acknowledgement I want to thank I.M. Duursma and G.-L. Feng for many discussions explaining the mechanism of majority coset decoding.

References

[1] S. Arimoto, Encoding and decoding of p-ary group codes and the correction system, Information Processing in Japan **2** (1961), 320-325. in Japanese.

[2] E.R. Berlekamp, Algebraic coding theory, McGraw-Hill, New York 1968.

[3] E.R. Berlekamp, R.J. McEliece and H.C.A. van Tilborg, On the inherent intractibility of certain coding problems, IEEE Trans. Inform. Theory **24** (1978), 384-386.

[4] P. Bours, J.C.M. Janssen, M. van Asperdt and H.C.A. van Tilborg, Algebraic decoding beyond e_{BCH} of some binary cyclic codes, when $e > e_{BCH}$, IEEE Trans. Inform. Theory **36** (1990), 214-222.

[5] J. Bruck and M. Naor, The hardness of decoding linear codes with preprocessing, IEEE Trans. Inform. Theory **36** (1990), 381-385.

[6] B. Buchberger, Ein Algorithmus zum Auffinden der Basiselemente des Restklassenringes nach einem nulldimensionalen Polynomideal, PhD Thesis, University of Innsbruck, Austria, 1965.

[7] Ph. Carbonne and A. Thiong Ly, Minimal exponent for Pellikaan's decoding algorithm, Proceedings Eurocode 92.

[8] C. Chevalley, Introduction to the theory of algebraic functions in one variable, Math. Surveys VI, Providence, AMS 1951.

[9] I.M. Duursma, Algebraic decoding using special divisors, to appear in IEEE Trans. Inform. Theory.

[10] I.M. Duursma, Majority coset decoding, to appear in IEEE Trans. Inform. Theory.

[11] I.M. Duursma, On the decoding procedure of Feng and Rao, Proceedings ACCT-3, Voneshta Voda, June 1992.

[12] I.M. Duursma and R. Kötter, On error locating pairs for cyclic codes, preprint October 1992.

[13] D. Ehrhard, Über das Dekodieren Algebraisch-Geometrischer Codes, PhD Thesis, University of Düsseldorf, July 1991.

[14] D. Ehrhard, Decoding algebraic-geometric codes by solving a key equation, in the Proceedings AGCT-3, H. Stichtenoth and M.A. Tsfasman (eds.), Luminy 1991, Springer Lect. Notes. **1518** (1992), 18-25.

[15] D. Ehrhard, Achieving the designed error capacity in decoding algebraic-geometric codes, to appear in IEEE Trans. Inform. Theory.

[16] G.-L. Feng and T.R.N. Rao, Decoding of algebraic geometric codes up to the designed minimum distance, to appear in IEEE Trans. Inform. Theory.

[17] G.-L. Feng and T.R.N. Rao, A novel approach for construction of algebraic-geometric codes from affine plane curves, University of Southwestern Louisiana, preprint 1992.

[18] G.-L. Feng and K.K. Tzeng, A generalization of the Berlekamp-Massey algorithm for multisequence shift register synthesis with application to decoding cyclic codes, IEEE Trans. Inform. Theory **37** (1991), 1274-1287.

[19] G.-L. Feng and K.K. Tzeng, Decoding cyclic and BCH codes up to the actual minimum distance using nonrecurrent syndrome dependence relations, IEEE Trans. Inform. Theory **37** (1991), 1716-1723.

[20] V.D. Goppa, Codes associated with divisors, Probl. Peredachi Inform. **13**(1) (1977), 33-39. Translation: Probl. Inform. Transmission **13** (1977), 22-26.

[21] V.D. Goppa, Codes on algebraic curves, Dokl. Akad. Nauk SSSR **259** (1981), 1289-1290, Translation: Soviet Math. Dokl. **24** (1981), 170-172.

[22] V.D. Goppa, Algebraico-geometric codes, Izv. Akad. Nauk SSSR **46** (1982), Translation: Math. USSR Izvestija **21** (1983), 75-91.

[23] V.D. Goppa, Codes and information, Russian Math. Surveys **39** (1984), 87-141.

[24] V.D. Goppa, Geometry and codes, Mathematics and its Applications **24**, Kluwer Acad. Publ., Dordrecht, 1991.

[25] C. Dahl Jensen, Codes and geometry, PhD Thesis, Technical University of Denmark, May 1991,

[26] J. Justesen, K.J. Larsen, H.Elbrønd Jensen, A. Havemose and T. Høholdt, Construction and decoding of a class of algebraic geometric codes, IEEE Trans. Inform. Theory **35** (1989), 811-821.

[27] J. Justesen, K.J. Larsen, H.Elbrønd Jensen and T. Høholdt, Fast decoding of codes from algebraic plane curves, IEEE Trans. Inform. Theory **38** (1992), 111-119.

[28] J. Justesen, H.Elbrønd Jensen and T. Høholdt, On the number of correctable errors for some AG-codes, to appear in IEEE Trans. Inform. Theory.

[29] R. Kötter, A unified description of an error locating procedure for linear codes, Proceedings ACCT-3, Voneshta Voda, June 1992.

[30] V. Yu. Krachkovskii, Decoding of codes on algebraic curves, Odessa, preprint 1988 in Russian.

[31] D. Le Brigand, Decoding of codes on hyperelliptic curves, Proceedings Eurocode 90, G.D. Cohen and P. Charpin (eds.), Lect. Notes in Comp. Sc. **514** (1991), 126-134.

[32] J.H. van Lint, Algebraic geometric codes, in Coding Theory and Design Theory, part I, D. Ray-Chaudhuri (ed.), IMA Volumes Math. Appl. **21**, Springer-Verlag, Berlin etc. (1990),

[33] J.H. van Lint and G. van der Geer, Introduction to coding theory and algebraic geometry, DMV Seminar **12**, Birkhäuser Verlag, Basel Boston Berlin, 1988.

[34] J.H. van Lint and T.A. Springer, Generalized Reed-Solomon codes from algebraic geometry, IEEE Trans. Inform. Theory **33** (1987), 305-309.

[35] J.H. van Lint and R.M. Wilson, On the minimum distance of cyclic codes, IEEE Trans. Inform. Theory **32** (1996), 23-40.

[36] J.L. Massey, Shift-register synthesis and BCH decoding, IEEE Trans. Inform. Theory **15** (1969), 122-127.

[37] F.J. McWilliams and N.J.A. Sloane, The theory of error-correcting codes, North-Holland Math. Library **16**, North-Holland, Amsterdam, 1977.

[38] C. Moreno, Algebraic curves over finite fields, Cambridge Tracts in Math. **97**, Cambridge Un. Press, 1991.

[39] R. Pellikaan, On decoding linear codes by error correcting pairs, preprint Eindhoven University of Technology, 1988.

[40] R. Pellikaan, On a decoding algorithm for codes on maximal curves, IEEE Trans. Inform. Theory **35** (1989), 1228-1232.

[41] R. Pellikaan, On the decoding by error location and the number of dependent error positions, Discrete Math. **106/107** (1992), 369-381.

[42] R. Pellikaan, B.-Z. Shen and G.J.M. van Wee, Which linear codes are algebraic-geometric ?, IEEE Trans. Inform. Theory. IT-**37** (1991), 583-602.

[43] W.W. Peterson, Encoding and error-correction procedures for the Bose-Chauduri codes, IEEE Trans. Inform. Theory **6** (1960), 459-470.

[44] S.C. Porter, Decoding codes arising from Goppa's construction on algebraic curves, Thesis, Yale University, dec. 1988.

[45] S.C. Porter, B.-Z. Shen and R. Pellikaan, On decoding geometric Goppa codes using an extra place, IEEE Trans. Inform. Theory. IT-**38** (1992), 1663-1676.

[46] D. Rotillon and J.A. Thiong Ly, Decoding codes on the Klein quartic, Proceedings Eurocode 90, G.D. Cohen and P. Charpin (eds.), Lect. Notes in Comp. Sc. **514** (1991), 135-150.

[47] S. Sakata, On determining the independent point set for doubly periodic arrays and encoding two-dimensional cyclic codes and their duals, IEEE Trans. Inform. Theory **27** (1981), 556-565.

[48] S. Sakata, Finding a minimal set of linear recurring relations capable of generating a given finite two-dimensional array, Journal of Symbolic Computation, **5** (1988), 321-337.

[49] S. Sakata, Extension of the Berlekamp-Massey algorithm to N dimensions, Information and Computation, **84** (1990), 207-239.

[50] S. Sakata, Decoding binary 2-D cyclic codes by the 2-D Berlekamp-Massey algorithm, IEEE Trans. Inform. Theory **37** (1991), 1200-1203.

[51] B.-Z. Shen, Solving a congruence on a graded algebra by a subresultant sequence and its application, to appear in Journ. of Symbolic Computation.

[52] B.-Z. Shen, On encoding and decoding of the codes from Hermitian curves, to appear in Cryptography and Coding III, the IMA Conference Proceedings Series, M. Ganley (ed.), Oxford University Press.

[53] B.-Z. Shen, Constructing syndromes for the codes from Hemitian curves and a decoding approach, preprint Eindhoven University of Technology, 1992.

[54] B.-Z. Shen, Algebraic-geometric codes and their decoding algorithm, PhD Thesis, Eindhoven University of Technology, September 1992.

[55] A.N. Skorobogatov and S.G. Vlăduţ, On the decoding of algebraic-geometric codes, IEEE Trans. Inform. Theory **36** (1990), 1051-1060.

[56] H. Stichtenoth, A note on Hermitian codes over GF(q^2), IEEE Trans. Inform. Theory **34** (1988), 1345-1348.

[57] H. Stichtenoth, Algebraic function fields and codes, to appear in Universitext, Springer-Verlag, 1993.

[58] H.J. Tiersma, Codes comming from Hermitian curves, IEEE Trans. Inform. Theory **33** (1987), 605-609.

[59] M.A. Tsfasman and S.G. Vlăduţ, Algebraic-geometric codes, Mathematics and its Applications **58**, Kluwer Acad. Publ., Dordrecht, 1991.

[60] M.A. Tsfasman, S.G. Vlăduţ and T. Zink, Modular curves, Shimura curves and Goppa codes, better than Varshamov-Gilbert bound, Math. Nachrichten **109** (1982), 21-28.

[61] S.G. Vlăduţ, On the decoding of algebraic-geometric codes over GF(q) for q$\geq$ 16, IEEE Trans. Inform. Theory **36** (1990), 1461-1463.

[62] T. Yaghoobian and I.F. Blake, Hermitian codes as generalized Reed-Solomon codes, Designs, Codes and Cryptogrphy **2** (1992), 15-18.

GRÖBNER BASES AND ABELIAN CODES

H. Chabanne
INRIA, Rocquencourt, Le Chesnay, France

Abstract

We investigate the links between Gröbner Bases and Abelian Codes. Abelian codes are multivariate generalization of cyclic codes, and our purpose it to show how Gröbner bases can replace the generating polynomial of cyclic codes.

We use these bases to properly define the information symbols of an abelian code in polynomial representation. In particular, we show that Gröbner bases are related with parity check matrices of the code. From this, we are able to generalize permutation decoding to abelian codes. This generalization introduces some choice which allows us to decode more errors than in the cyclic case.

Finally, we examine the connections between Gröbner Basis for an abelian code and those of its reciprocal, annihilator and orthogonal code.

1 Introduction

Cyclic codes can be defined as ideals in the algebra of polynomials modulo $X^n - 1$. More generally, abelian codes are defined as ideals in the algebra of multivariate polynomials modulo $(X_1^{n_1} - 1, \ldots, X_u^{n_u} - 1)$ [1, 3, 11].

An abelian code of length $n = n_1 \ldots n_u$ over F_q, with $\gcd(q, n) = 1$, consists of all multiples of a generating polynomial $p(X_1, \ldots, X_u)$. Thus, for each abelian code, we can associate the set of zeroes of this generating polynomial which are zeroes of any codeword. Finally, Fourier analysis can be performed as in the cyclic case (Section 2).

But, as the underlying algebra is no euclidian ring for abelian codes, the knowledge of a generating polynomial is not as important as for cyclic codes [7]. In particular, the linear structure of the code is hard to conciliate with " abelian shifts = product by monomials ". In order to perform this task properly, we will use Gröbner bases [2, 13, 6, 16].

Gröbner bases possess nice properties. They are related with information symbols [14, 15] and with parity check matrices of the code (Section 3). They can be used to generalize permutation or Kasami's covering polynomials decoding procedure to abelian codes [12, 10, 8, 4]. They allow us to introduce a new extension of permutation decoding methods irrevelant in the cyclic case (Section 4).

In Section 5, we point out geometric relations between Gröbner bases for a code and those of its orthogonal.

2 Some recalls about abelian codes

We recall here some well-known results about abelian codes. Proofs and details can be found in [1, 3, 11].

2.1 Preliminaries

Let G be a finite abelian group. Let F_q be the finite field of order q. We call **G-abelian code** - or **abelian code** when G is well-understood - an ideal of the group algebra $\mathsf{F}_q G = A$.

If G is the product of u cyclic groups $G = G_1 \times \ldots \times G_u$, we have an algebra isomorphism between the group algebra A and the quotient algebra $\mathsf{F}_q[X_1, \ldots, X_u]/(X_1^{n_1} - 1, \ldots, X_u^{n_u} - 1)$ where for every v in $[1, u]$, n_v is the order of the cyclic group G_v. With this isomorphism, we express any element $a \in A$ as a formal sum $a = \sum_{g \in G} a_g T^g, a_g \in \mathsf{F}_q$ or as a polynomial in several indeterminates modulo $(X_1^{n_1} - 1, \ldots, X_u^{n_u} - 1) : a = \sum_{g \simeq (i_1, \ldots, i_u)} a_g X_1^{i_1} \ldots X_u^{i_u}$.

Thus an abelian code $C \subset A$ is a linear code over F_q of length $n = n_1 \ldots n_u$.

Here, we restrict ourselves to the case where the order n of the group G and the characteristic of the field F_q are relatively prime. It is well known [5] that the algebra A is isomorphic to a product of fields $A \simeq \mathsf{F}_{q^{e_1}} \times \ldots \times \mathsf{F}_{q^{e_t}}$. Thus each ideal of A is isomorphic to a sub-product of these fields and is principal.

More precisely, consider the sets $E_g = \{g, qg, \ldots, q^{e_g - 1} g\} \subset G$, where e_g is the smallest integer such that $q^{e_g} g = g$. These sets, called **q-subsets**, form a partition of the group G. They generalize cyclotomic cosets of the cyclic case. There are as many q-subsets as the number of fields in the decomposition of $A : G = E_{g_1} \cup \ldots \cup E_{g_t}$. And for each q-subset E_{g_s}, $s \in [1, t]$, we can associate an unique maximal subfield $\mathsf{F}_{q^{e_s}}$ of A which satisfies

1. $|E_{g_s}| = e_{g_s} = e_s$,
2. any ideal C in A is isomorphic to a product of fields which does not contain $\mathsf{F}_{q^{e_s}}$ if and only if for all $g = (i_1, \ldots, i_u)$ in E_{g_s} and for all c in C, one has $c(\xi_1^{i_1}, \ldots, \xi_u^{i_u}) = 0$ where ξ_v is a primitive n_v-th root of unity for v in $[1, u]$.

The subset of G for which all the elements of the ideal C cancel is called the **set of zeroes of** C and is denoted by Z_C. So we know that this set is a union of q-subsets. Moreover we have $|Z_C| = n - k$ where k is the dimension of C.

2.2 Construction of a parity check matrix

Our purpose here is to construct a parity check matrix of an abelian code. In fact, we need such a matrix in the next sections.

Let L be the smallest extension of F_q which contains ξ_v for $v \in [1, u]$. The degree of L over F_q is the order of q in the multiplicative group of integers modulo $lcm(n_1, \ldots, n_u)$.

If a is in A we define its **Fourier transform** by

$$\mathcal{F}(a) = \sum_{g \simeq (i_1, \ldots, i_u) \in G} a(\xi_1^{i_1}, \ldots, \xi_u^{i_u}) Y_1^{n_1 - i_1} \ldots Y_u^{n_u - i_u}$$

Let L^G be the algebra $L[Y_1, \ldots, Y_u]$ with componentwise multiplication. The Fourier transform is an isomorphism of the algebra A into L^G.

We number the elements of G with the lexicographic order : we put $\eta_L((i_1, \ldots, i_u)) = \eta_L(g)$ for the location of g. Let $\hat{a}$ be the vector obtained from the coefficients of $a \in A$ when the ordering η_L is preserved : $\hat{a} = \left(a_{\eta_L^{-1}}(1), \ldots, a_{\eta_L^{-1}}(n)\right)$.

We number now all elements of G with the reverse order : we put $\eta_{-L}(g) = n - \eta_L(g) + 1$ for the location of g. Let $\tilde{l}$ be the vector obtained from the coefficients of $l \in L^G$ when the ordering η_{-L} is preserved.

Remark 1 The Fourier transform of a in A may be determined by matrix multiplication. For $v \in [1, u]$, let V^v be the $n_v \times n_v$ Vandermonde matrix defined by $V_{ij}^v = \xi_v^{ij}$. Let $V = V^1 \otimes \ldots \otimes V^u$ be the $n \times n$ matrix obtained by Kronecker product $\otimes$ of matrices V^v. Then $V\hat{a} = \widetilde{\mathcal{F}(a)}$.

Proposition 1 *Let C be an abelian code. Let $\mathcal{H}$ be the matrix defined by*

$$\mathcal{H}_{ij} = \begin{cases} 1 & if\ i = j \\ 0 & otherwise \end{cases}$$

where $i = \eta_L(g)$ for g in Z_C and $j = \eta_L(g)$ for g in G (this is a $(n-k) \times n$ matrix).

Then $\mathcal{H}$ is a parity check matrix for $\mathcal{F}(C)$ the image of the code C via the Fourier transform.

Proof : Since the multiplication in L^G is componentwise, the ideal $\mathcal{F}(C)$ admits $\{\delta_{\eta_L(g)}, g \in G \setminus Z_C\}$ as basis of subspace of L^G, where δ_i is the vector with 1 as i-th coordinate and 0 elsewhere . □

Corollary 1 *Let $\mathcal{H}'$ be the matrix obtained from $\mathcal{H}$ by renumbering columns back to front (the j-th column of $\mathcal{H}'$ will be the $(n-j+1)$-th column of $\mathcal{H}$). Let T be the matrix of the linear application of L^{n-k} into F_q^{n-k} defined by $(l_1, \ldots, l_{n-k}) \mapsto (tr(l_1), \ldots, tr(l_{n-k}))$ where $tr(.)$ is the trace function of L over F_q. Let V be - as in the Remark above - the matrix of the Fourier transform of A into L^G.*

Then the matrix $H = T\mathcal{H}'V$ is a parity check matrix for C.

Remark 2 Proposition 1 and Corollary 1 remain true for the cyclic case.

3 Gröbner bases and abelian codes (I)

As in the previous section, we denote by A the group algebra $\mathbb{F}_qG \simeq \mathbb{F}_q[X_1,\ldots,X_u]/(X_1^{n_1}-1,\ldots,X_u^{n_u}-1)$. Let C be an ideal in A. Let H be a parity check matrix for C. For any element a in A, we will call **syndrome** of a the vector Ha using $a \in \mathbb{F}_q^n$. Note that the syndrome of a polynomial is the sum of the syndromes of the monomials which compose it.

3.1 Definition

Let $\trianglelefteq$ be a total ordering defined on $\mathbb{N}^u$ such that if $(i_1,\ldots,i_u)$ and $(i'_1,\ldots,i'_u)$ are u-tuple of non negative integers

- $(i_1,\ldots,i_u) \trianglelefteq (i_1+i'_1,\ldots,i_u+i'_u)$
- $(i_1,\ldots,i_u) \trianglelefteq (i'_1,\ldots,i'_u)$ if and only if $(i_1+j_1,\ldots,i_u+j_u) \trianglelefteq (i'_1+j_1,\ldots,i'_u+j_u)$ for all $(j_1,\ldots,j_u)$ in $\mathbb{N}^u$

such an ordering $\trianglelefteq$ is called **admissible.**

If $a = \sum_{i_1,\ldots,i_u} a_{i_1,\ldots,i_u} X_1^{i_1}\ldots X_u^{i_u}$ is a non zero multivariate polynomial, we define its **headterm** as $\mathrm{ht}(a) = \max\{(i_1,\ldots,i_u) : a_{i_1,\ldots,i_u} \neq 0\}$ where the maximum is taken with respect to the chosen ordering.

For any subset B of $\mathbb{F}_q[X_1,\ldots,X_u]$, let $\mathrm{St}(B) = \{\mathrm{ht}(b)+(i_1,\ldots,i_u) : b \in B,\ b \neq 0,\ (i_1,\ldots,i_u) \in \mathbb{N}^u\}$ be the **staircase** of B. In the same way, for any ideal I in $\mathbb{F}_q[X_1,\ldots,X_u]$, let $\mathrm{St}(I) = \{\mathrm{ht}(b) : b \in I\setminus\{0\}\}$ be the **staircase** of I.

Let C be an abelian code. Let p be a generating polynomial of C in A.

Definition 1 *Let $I_C \subset \mathbb{F}_q[X_1,\ldots,X_u]$ be the ideal generated by $(p, X_1^{n_1}-1,\ldots,X_u^{n_u}-1)$.*
A subset B of I_C is called a **Gröbner basis** *of C (resp. I_C) if $\mathrm{St}(B) = \mathrm{St}(I_C)$.*

An example is given at the end of Section 3.

Remark 3 All the elements of I_C cancel on Z_C and Z_C is the largest set which has this property.

Proposition 2 *Let B be a subset of $\mathbb{F}_q[X_1,\ldots,X_u]$. The following facts are equivalent :*

1. *B is a Gröbner basis of C (resp. I_C).*
2. *The monomials which do not have their headterm in $\mathrm{St}(B)$ have a non zero image in A. And, in this way, the syndromes of these monomials form a $\mathbb{F}_q$-basis of the vector space of the syndromes of C.*

Proof : We need the following result ([13] for a proof).

Lemma 1 *B is a Gröbner basis of I_C if and only if $\{X_1^{i_1}\ldots X_u^{i_u} : (i_1,\ldots,i_u) \notin \mathrm{St}(B)\}$ is a $\mathbb{F}_q$-basis of $\mathbb{F}_q[X_1,\ldots,X_u]/(I_C)$.*

Now, let Γ be the complementary set of $\mathrm{St}(B)$ in $\mathbb{N}^u$.

Since $X_1^{n_1}-1,\ldots,X_u^{n_u}-1$ are elements of I_C, we have $\Gamma \subset [0,n_1-1]\times\ldots\times[0,n_u-1]$. Thus we identify each u-tuple in Γ with an element of G. And Lemma 1 becomes : B is a Gröbner basis of I_C if and only if $\{T^g : g \notin \mathrm{St}(B)\}$ is a basis of $A/(C)$.

On the other hand, let H be a parity check matrix of C. The second part of the Proposition follows from $Ha = Ha'$ if and only if $a-a'$ is in C. □

Definition 2 *The set* $\{g=(i_1,\ldots,i_u)\in G :$ *no element of* C *admits* g *as headterm*$\}$ *is called the* **independent point set** *of* C *for* $\trianglelefteq$.

Remark 4 Proposition 2 implies that

1. the number of independent point set of C is $n-k$ where k is the dimension of C,
2. Gröbner bases of C (resp. I_C) are finite,
3. the elements of a Gröbner basis of C which have their headterm in $[0,n_1-1]\times\ldots\times[0,n_u-1]$ may (will) be taken as elements of A.

Let $g=(i_1,\ldots,i_u)$, $g'=(i'_1,\ldots,i'_u)$ be in G ; we define $g\odot g'$ as $\prod_{v=1}^{u}\xi_v^{i_v i'_v}$.

We number the columns of H the parity check matrix of the G-abelian code C with the elements of G : H_g stands for the column with number g.

Corollary 2 *Let* $Z_C=\{z_1,\ldots,z_{n-k}\}$ *be the set of zeroes of* C. *Then* B *is a Gröbner basis of* C *if and only if the vectors* $(\gamma\odot z_1,\ldots,\gamma\odot z_{n-k})$ *for all* $\gamma\notin\mathrm{St}(B)$, *are linearly independent.*

Proof : Notice first that when the columns H_{γ_s}, $s=1,\ldots,t$ are linearly dependent, there exist $c_1,\ldots,c_t$ in $\mathbb{F}_q$ not all zero such that $\sum_{s=1}^t c_s H_{\gamma_s} == 0$. Thus the word $c=\sum_{s=1}^t c_s T^{\gamma_s}$ is in C because its syndrome is zero. And for all $(i_1,\ldots,i_u)\in Z_C$, we must have $c(\xi_1^{i_1},\ldots,\xi_u^{i_u})=0$. So, we obtain $\left(\sum_{s=1}^t c_s(\gamma_s\odot z_1),\ldots,\sum_{s=1}^t c_s(\gamma_s\odot z_{n-k})\right)=(0,\ldots,0)$ and the vectors $\left(\gamma_s\odot z_1,\ldots,\gamma_s\odot z_{(n-k)}\right)$, $s=1,\ldots,t$ are linearly dependent. The converse is also true.

Let $\Gamma=\{\gamma : \gamma=(i_1,\ldots,i_u)\notin\mathrm{St}(B)\}$. We may consider Γ as a subset of G.

Let k be the dimension of C. Using the equivalence of Proposition 2, we know that B is Gröbner basis of C if and only if the $(n-k)$ syndromes of the monomials which do not have their headterm in $\mathrm{St}(B)$ are linearly independent. Such a syndrome is one of the columns of H. More precisly, the $(n-k)$ syndromes of these monomials will be H_γ with $\gamma\in\Gamma$. And these columns are linearly independent if and only if the vectors $(\gamma\odot z_1,\ldots,\gamma\odot z_{n-k})$ for all $\gamma\in\Gamma$ are linearly independent using the first argument of the proof. □

3.2 Construction of Gröbner bases

We recall now how to compute a Gröbner basis B of an abelian code C for a given admissible ordering $\trianglelefteq$. This algorithm is due to J.C. Faugère, P. Gianni, D. Lazard and T. Mora [6]. As output, we obtain B a Gröbner basis of C for $\trianglelefteq$, and Γ the associated independent point set. Proofs, complexity and details can be found in [13, 6].

Algorithm 1. Computing a Gröbner basis.

Any sentence beginning with a # is a comment

$\Gamma = \emptyset, B = \emptyset, List = \{(0,\ldots,0)\}$

while $List \neq \emptyset$ do

$g =$ the smallest element of $List$ with respect to $\trianglelefteq$

$List = List \setminus \{g\}$

if $g \notin \mathrm{St}(B)$ then

if $\exists\, c = T^g + \sum_{\gamma\in\Gamma} c_\gamma T^\gamma \in C$ then $B = B \cup \{c\}$

in order to check if such word exists, we solve

$$\begin{cases} H_{1,g} + \sum_{\gamma\in\Gamma} c_\gamma H_{1,\gamma} = 0 \\ \ldots \qquad \ldots \\ H_{(n-k),g} + \sum_{\gamma\in\Gamma} c_\gamma H_{(n-k),\gamma} = 0 \end{cases}$$

where $H_{i,g}$ is the i-th row of the column H_g

if d is the minimum distance of the code C and if $|\Gamma| < d-1$ then there can not exist a word $c = T^g + \sum_{\gamma\in\Gamma} c_\gamma T^\gamma$ in C

else

$\Gamma = \Gamma \cup \{g\}$

$List = List \cup \{\delta_v + g :\ v \in [1,u]\}$

as usual δ_i is the vector with 1 as i-th coordinate and 0 elsewhere

□

We can note that the elements of B have only one monomial with exponent in $\mathrm{St}(I_C)$. Moreover, the cardinality of B is minimal.

Now we can give a generalization of ([12],p.191,th.1)

Corollary 3 *Let B be a Gröbner basis of C obtained as output of Algorithm 1. Let $b_1,\ldots,b_t$ be the elements of B which have their headterm in $[0,n_1-1]\times\ldots\times[0,n_u-1]$.*

Let $\Delta_1 = \mathrm{ht}(b_1) + \mathbb{N}^u \subset \mathbb{N}^u$, and for $s = 2,\ldots,t$ let $\Delta_s = (\mathrm{ht}(b_s)+\mathbb{N}^u) \setminus \cup_{s'<s}\Delta_{s'}$, $\Delta_s \subset \mathbb{N}^u$.

There are k polynomials in $C \subset A$, namely $X_1^{i_1}\ldots X_u^{i_u} b_s(X_1,\ldots,X_u)$ whose headterm are in $\Delta_s \cap ([0,n_1-1]\times\ldots\times[0,n_u-1])$ for some s. They form a $\mathbb{F}_q$-basis of C.

Therefore these polynomials expressed in the basis of monomials yield a generator matrix of C.

Remark 5 Polynomials $b_1,\ldots,b_t$ are all elements of B except those of the type $X_v^{n_v}-1$, $v\in[1,u]$.

3.3 Computation of the syndrome

Let B and Γ be defined as above, let $\Gamma_{\mathbf{X}} = \{X_1^{i_1} \dots X_u^{i_u} : (i_1, \dots, i_u) \in \Gamma\}$ and let $\mathrm{Vect}(\Gamma_{\mathbf{X}})$ be the subspace whose basis is $\Gamma_{\mathbf{X}}$. The subspace $\mathrm{Vect}(\Gamma_{\mathbf{X}})$ is a supplementary of C in $A \simeq \mathbf{F}_q^n : C \bigoplus \mathrm{Vect}(\Gamma_{\mathbf{X}}) = \mathbf{F}_q^n$.

Definition 3 *Consider the linear mapping of A into* $\mathrm{Vect}(\Gamma_{\mathbf{X}})$ *whose kernel is C and whose restriction to* $\mathrm{Vect}(\Gamma_{\mathbf{X}})$ *is identity. We call the* **remainder** *of a polynomial a in A by B, and denote by $R(a, B)$, the image of a under this mapping.*

Note that the matrix of the mapping $R(., B)$ written in the canonical basis of monomials is a parity check matrix for C. This fact justifies the use of the word syndrome in the section 's title.

Definition 4 *The* **bordering** *of C is defined by*

$$\partial C = \{X_1^{i_1} \dots X_u^{i_u} \in \mathrm{St}(I_C) : \exists v \in [1, u],\ X_1^{i_1} \dots X_v^{i_v - 1} \dots X_u^{i_u} \in \Gamma_{\mathbf{X}}\}$$

Since the application remainder is linear, in order to compute the remainder of a polynomial a, we may restrict ourselves to compute the remainder of the monomials which compose a.

In order to do this we assume that $R(b, B)$ is known for every b in ∂C. Moreover, for each monomial $X_1^{i_1} \dots X_u^{i_u}$, let $(t_i)_{i=1,\dots,j}$ be a sequence of j monomials, j minimal such that

- $t_1 = X_1^{i_1} \dots X_u^{i_u}$,
- t_j is in ∂C,
- $t_i = X_v t_{i+1}$ for some variable X_v, $v \in [1, u]$.

Then we compute $R(t_i, B)$ using $R(t_i, B) = R(X_v R(t_{i+1}, B), B)$ ($R(t_{i+1}, B) \in \mathrm{Vect}(\Gamma_{\mathbf{X}})$ and $X_v R(t_{i+1}, B)$ is known).

This method [13] allows us to compute the remainder of a polynomial a by B in polynomial time in the weight of a and in the codimension of C in $\mathbf{F}_q^n$.

We conclude this section with an example.

Example Let $u = 2$, $n_1 = 7$, $n_2 = 9$ and let $q = 2$. Let $\xi_1 = \theta^9$, $\xi_2 = \theta^7$ where $\theta^6 + \theta + 1 = 0$ in $\mathbf{F}_2$. Let C be the binary $C_7 \times C_9$ abelian code of dimension 45 such that $Z_C = \{(3,1),(6,2),(5,4),(3,8),(6,7),(5,5),(2,7),(4,5),(1,1),(2,2),(4,4),(1,8),(3,6),(6,3),(5,6),(3,3),(6,6),(5,3)\}$. Its generating polynomial is $1 + X_1^6X_2^6 + X_1X_2^8 + X_1^3X_2 + X_1^4X_2^2 + X_1^5X_2^3 + X_1X_2^6 + X_1^2X_2^7 + X_1^3X_2^8 + X_1^6X_2^2 + X_1^3X_2^6 + X_1^4X_2^7 + X_1^4X_2^4 + X_1X_2^4 + X_1^2X_2^5 + X_1^6X_2 + X_1^2X_2^4 + X_1^6X_2^3 + X_1X_2^5 + X_1^4X_2 + X_1^5X_2^2 + X_1^3X_2^3 + X_1^4X_2^3 + X_1^5X_2^4 + X_1^5X_2^7 + X_1^6X_2^8 + X_1^2X_2^3 + X_1X_2^3 + X_2^2 + X_2^5 + X_1^2X_2^6 + X_1^4X_2^8 + X_1^2X_2^2 + X_1^6X_2^7 + X_1^2X_2 + X_1^4X_2^5 + X_1^5X_2^6 + X_1X_2 + X_2 + X_1^3X_2^5 + X_1^4X_2^6 + X_1^5X_2^5 + X_1X_2^2 + X_1^3X_2^4 + X_1X_2^7 + X_1^2X_2^8 + X_2^7 + X_2^8 + X_2^4$. We choose as admissible ordering the lexicographic ordering. Then we graphically obtain the figure 1 on page 8.

The Gröbner basis here is $\{X_2^8 + X_2^7 + X_2^6 + X_2^5 + X_2^4 + X_2^3 + X_2^2 + X_2 + 1, X_1^2X_2^2 + X_1^2X_2 + X_1^2 + X_1X_2^7 + X_1X_2^2 + X_1X_2^4 + X_1 + X_2^2 + X_1X_2 + X_2^7 + X_2^4 + X_2 + 1, X_1^3 + X_1X_2^7 + X_1X_2^2 + X_2^7 + X_2^5 + X_2^4 + X_2^2 + 1\}$.

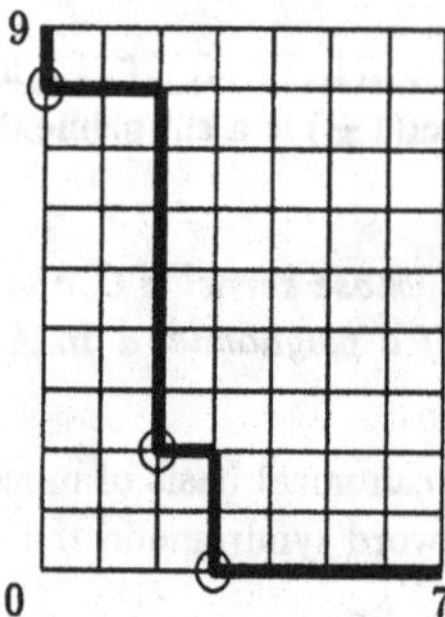

○ : headterm of a Gröbner basis' member,
the thick line stands for ∂C,
Γ is under the thick line,
$\mathrm{St}(I_C)$ is above the thick line.

Figure 1: An example of Gröbner basis

4 Permutation decoding of abelian codes

In this section we restrict ourselves to the case where F_q is a field of characteristic two. Our purpose here is to show how Gröbner bases can be used to decode abelian codes.

As usual, let C be a G-abelian code over F_q, let B be a Gröbner basis of C for an admissible ordering $\trianglelefteq$ and let Γ be the associated independent point set.

By collecting terms, we write a polynomial $a = \sum a_g T^g$, $a_g \neq 0$, in $\mathsf{F}_q G = A$ as $a = a_\Gamma + a_{\mathrm{St}}$ with $a_\Gamma = \sum_{\gamma \in \Gamma} a_\gamma T^\gamma$. And as usual the **support** of such polynomial a will be the set of monomials composing a and the **weight** of a, $wt(a)$, the number of these monomials.

Theorem 1 *Let d be the minimum distance of C. Let $m = c + e$ where c is in C and e in A with $wt(e) \leq [\frac{d-1}{2}]$. Then $R(m, B) = R(e, B) = S$ and $e = S \Leftrightarrow wt(S) \leq [\frac{d-1}{2}]$.*

Proof: Let $t = [\frac{d-1}{2}]$. Assume first that $e_{\mathrm{St}} = 0$. Then $e = e_\Gamma$ and $R(e, B) = e$ by definition. Hence $wt(S) = wt(e) \leq t$ by hypothesis.

Conversely, suppose $e_{\mathrm{St}} \neq 0$ and let $\bar{e} = e_{\mathrm{St}} + R(e_{\mathrm{St}}, B)$. Then $R(\bar{e}, B) = 0$ and $\bar{e}$ is in C. Moreover, note that the support of e_{St} and $R(e_{\mathrm{St}}, B)$ are disjoints.

So $wt(\bar{e}) \geq d = 2t + 1$, $S = e_\Gamma + R(e_{\mathrm{St}}, B)$ and
$wt(S) \geq wt(R(e_{\mathrm{St}}, B)) - wt(e_\Gamma) \geq wt(\bar{e}) - wt(e_{\mathrm{St}}) - wt(e_\Gamma) \geq 2t+1 - wt(e_\Gamma) - wt(e_{\mathrm{St}}) \geq t+1$.
□

Note that this proof is the same that [12, Theorem 19, p. 513].

Let σ be an automorphism of the group G. Let $\hat{\sigma}$ be the linear extension of σ to A. If $\hat{\sigma}(C) = C$ then we say that $\hat{\sigma}$ is an **automorphism** of C. The set of all automorphisms of C is a group, denoted by $Aut(C)$.

We can note that the mappings
$T_v : a = \sum_{g \simeq (i_1, \ldots, i_u)} a_g X_1^{i_1} \ldots X_u^{i_u} \in A \mapsto X_v a$ (modulo $(X_1^{n_1} - 1, \ldots, X_u^{n_u} - 1)$), $v \in [1, u]$ and $\sigma_2 : a \mapsto a^2$ belong to $Aut(C)$.

Let Π be a subset of $Aut(C)$. The previous Theorem yields the following decoding method which generalizes a permutation decoding procedure for cyclic codes to abelian codes [10].

Algorithm 2. Decoding a word $m \in A$.

repeat

take $\pi \in \Pi$ and remove it from Π

$S = R(\pi(m), B)$

until $wt(S) \leq [\frac{d-1}{2}]$ or $\Pi = \emptyset$

decode if you can (if $wt(S) \leq [\frac{d-1}{2}]$ then $\pi^{-1}(\pi(m) + S) = c$ is in C)

□

Using the automorphism group of the code on the received message, we want to move all non zero monomials out of the information places that is, in the independent point set of the code. This method is a powerfull tool to decode abelian codes. Some examples of decoding are given in [4].

Remark 6 1. Kasami 's method [12, 8] using covering polynomials can be generalized as well.

2. If we restrict ourselves to the use the mapping T_v, $v \in [1, u]$ in Algorithm 2 then we choose the automorphisms in Π in a certain order which corresponds to the sequences $(t_i)_{i=1,\dots,j}$ introduced for the computation of the remainder in Section 3.3.

3. The geometry of the independent point set has a great importance here. As its form depends on the non canonical choice of a total ordering $\trianglelefteq$, we can introduce new extension of permutation decoding irrevelant in the cyclic case. If errors are not corrected for a given total ordering, we can try another one.

We conclude this section with an illustrative example.

Example Let C be the abelian code of Example on page 7.
We compute $\mathrm{St}(I_C)$ for different orderings

1. the lexicographic ordering $\trianglelefteq_L$,
2. the inverse lexicographic ordering $\trianglelefteq_{IL}$:
   $(i_1, i_2) \trianglelefteq_{IL} (j_1, j_2)$ if $i_2 < j_2$ or ($i_2 = j_2$ and $i_1 \leq j_1$)
3. the degree ordering $\trianglelefteq_D$:
   $(i_1, i_2) \trianglelefteq_D (j_1, j_2)$ if $i_1 + i_2 < j_1 + j_2$ or ($i_1 + i_2 = j_1 + j_2$ and $i_1 \leq j_1$)
4. and finally with the inverse lexicographic ordering $\trianglelefteq_{ID}$:
   $(i_1, i_2) \trianglelefteq_{ID} (j_1, j_2)$ if $i_1 + i_2 < j_1 + j_2$ or ($i_1 + i_2 = j_1 + j_2$ and $i_2 \leq j_2$) .

Note that many other admissible orderings would be available [16].
We obtain the following headterms for the Gröbner bases' members of C for the ordering

1. $(\trianglelefteq_L)$: {(0,8),(2,2),(3,0)},
2. $(\trianglelefteq_{IL})$: {(0,4),(3,2),(6,0)},
3. $(\trianglelefteq_D)$: {(0,6),(1,4),(3,3),(4,1),(5,0)},
4. $(\trianglelefteq_{ID})$: {(0,6),(1,4),(3,2),(5,1),(6,0)}.

We want to correct the weight 3 errors.

• Using the $C_7 \times C_9$-abelian structure, one of the four previous orderings and the mappings T_1 and T_2, from 16.2% (lexicographic ordering) to 24.4% (inverse lexicographic ordering) of the errors remain uncorrected.

• Using successively the four orderings - if an error is not corrected with an ordering, we try the next one - only 1.1% of the errors remain uncorrected.

Remark 7 Using the chinese theorem, we can express C as a cyclic code. We obtain a BCH [63,45,7] triple error correcting [8]. Then using the cyclic structure and the shift mapping, 13.3% of the errors remain uncorrected.

Thus the choice of the ordering in our method allows us to decode more errors than in the cyclic case.

5 Gröbner bases and abelian codes (II)

5.1 Gröbner basis of the orthogonal code

Definition 5 *Let C be an abelian code, C^0 the* **annihilator code** *of C is defined by*

$$C^0 = \{a \in A : ac = 0, \forall c \in C\}$$

Proposition 3 *The annihilator of the G-abelian code C is a G-abelian code which admits as set of zeroes $Z_{C^0} = G \setminus Z_C$. Thus its dimension as vector space is $n - k$ where k is the dimension of C in $\mathbb{F}_q^n$.*

Proof omitted.

Proposition 4 *Let C be an abelian code and let Γ be the independent point set of C for a given ordering $\trianglelefteq$. Then C^0 the annihilator code of C admits as independent point set for $\trianglelefteq$ the image of $([0, n_1 - 1] \times \ldots \times [0, n_u - 1]) \setminus \Gamma$ under the mapping $(i_1, \ldots, i_u) \mapsto (n_1 - i_1 - 1, \ldots, n_u - i_u - 1)$.*

Proof : Let c be in C and let $\mathrm{ht}(c) = (i_1, \ldots, i_u)$. We claim that there can not exist an element a in C^0 such that $\mathrm{ht}(a) = (n_1 - i_1 - 1, \ldots, n_u - i_u - 1)$. Assume the contrary, then $\mathrm{ht}(ac) = \mathrm{ht}(a) + \mathrm{ht}(c) = (n_1 - 1, \ldots, n_u - 1)$ is stricly greater than other exponents in ac. Thus $ac \neq 0$ and a can not belong to C^0.

For every $(i_1, \ldots, i_u) \in ([0, n_1 - 1] \times \ldots \times [0, n_u - 1]) \setminus \Gamma$, we have $(n_1 - i_1 - 1, \ldots, n_u - i_u - 1)$ in the independent set point of C^0. Thus we have found k points in the independent set point of C^0, using Proposition 4, we conclude. □

Definition 6 *Let C be a abelian code*

1. *the* **reciprocal code** *of C is defined by*

$$C^{-} = \{\sum_{g\in G} a_g T^{-g} : \sum_{g\in G} a_g T^g \in C\} = \{\sum_{g\in G} a_{-g} T^g : \sum_{g\in G} a_g T^g \in C\}$$

2. *the* **orthogonal code** *of C is defined by*

$$C^{\perp} = \{a = \sum_{g\in G} a_g T^g \in A : \sum_{g\in G} a_g c_g = 0,\ \forall c = \sum_{g\in G} c_g T^g \in C\}$$

Proposition 5 *The orthogonal code of an abelian code is equal to the annihilator of its reciprocal.*

Proof omitted.

We conjecture that

An abelian code and its reciprocal have the same independent point set for any ordering.

5.2 Concluding remarks

We have noticed on several, but not for all, examples of computation that Gröbner bases members are generating polynomials in the usual sense. This property could be explainable by the fact that they have a very small total degree among all the elements of their ideal (a polynomial with u indeterminates over F_q with total degree d having at most dq^{u-1} zeroes [9]).

More generally, Gröbner bases methods [2, 13, 6] allow us to manipulate polynomials in several indeterminates and abelian codes. Their use could be helpful in the search of another exceptional elements in abelian codes as words of minimum weight.

References

[1] S. Berman " *Semisimple cyclic and Abelian Codes* " , *Cybernetics, Vol. 3, No. 3, pp. 17-23, 1967.*

[2] B. Buchberger " *Grœbner bases : an algorithm method in polynomial ideal theory* " , *in Recent trends in multidimensional system theory, Bose Ed. , Reidel, 1985.*

[3] P. Camion " *Abelian codes* " *MRC Technical Summary Report No. 1059, University of Wisconsin, Madison, Wisconsin, 1970.*

[4] H. Chabanne " *Permutation Decoding of Abelian Codes* " , *IEEE Transactions on Informations Theory, Vol. IT 38, No. 6, pp. 1826-1829, Nov. 1992.*

[5] C. W. Curtis and I. Reiner " *Representation Theory of Finite Groups and Associative Algebras* ", *New York, Wiley, 1962.*

[6] J.C. Faugère, P. Gianni, D. Lazard, T. Mora *" Efficient computation of zero-dimensional Gröbner bases by change of ordering " , CALSYF 8, Luminy, 1989.*

[7] H. Imai *" A theory of two-dimensional cyclic codes " , Information and control, No. 34, pp. 1-21, 1977.*

[8] T. Kasami *" A decoding procedure for multiple error-correcting cyclic codes " , IEEE Transactions on Informations Theory, 10, pp. 134-138, 1968.*

[9] L. Niederreiter, R. Lidl *" Finite Fields " , Ency. of Math. and its Applic. , Vol. 20, Cambridge Univ. Press*

[10] F.J. Mac Williams *" Permutation decoding of systematic codes ",Bell. Syst. Techn. J. , Vol. 43, pp. 485-505, 1964.*

[11] J. Mac Williams *" Binary codes which are ideals in the group algebra of an Abelian group " , Bell Syst. Tech. J. , Vol. 49, No. 6, pp. 987-1011, July-Aug, 1970.*

[12] F.J. Mac Williams and N. J. Sloane *" the theory of error correcting codes " , North Holland Math. Libr. Vol. 16, 1977.*

[13] M.G. Marinari, H. M. Möller and T. Mora *" Gröbner Bases of Ideals given by Dual Bases " , ISSAC 91, Bonn Germany, 15-17, July 1991, NY ACM 1991.*

[14] S. Sakata *" General theory of doubly periodic arrays over an arbitrary finite field and its applications " , IEEE Transactions on Informations Theory, Vol. IT-24, No. 6, pp. 719-730, Nov. 1978.*

[15] S. Sakata *" On determining the independent point set for doubly periodic arrays and encoding two-dimensional cyclic codes and their duals " , IEEE Transactions on Informations Theory, Vol. IT-27, No. 5, pp. 556-565, Sept. 1981.*

[16] K.P. Schemmel, *" An extension of Buchberger's algorithm to compute all reduced Græbner bases of a polynomial ideal " EUROCAL'87, Leipzig, 1987.*

Acknowledgements. The author thanks Valérie, all the members of project CODES - especially Nicolas Sendrier and Claude Carlet - and the anonymous referees for their remarks.

MULTI-PARAMETER CYCLIC CODES FOR BURST AND RANDOM ERROR CORRECTION

E. Olcayto and J.M. Irvine
University of Strathclyde, Glasgow, Scotland, UK

Abstract

The paper examines some of the properties of some multi-parameter burst error correcting cyclic codes. Some conjectures given earlier are proven, and the codes are shown to be equivalent to single parity check product codes (Gilbert codes) or shortened Gilbert codes. The codes are shown to correct arbitrary burst shapes, and design algorithms are included.

Cet article étudie certaines propriétés des codes cycliques multi-paramètres concernant la correction d'erreurs par lots. Des conjectures avancées précédement sont démontreés ainsi que l'équivalence entre ces codes et les codes de Gilbert (éventuellement raccourcis). On montre que ces codes peuvent corriger des erreurs de répartition arbitraires et des algorithms de résolution sont donnés.

1 Introduction

We consider the mathematical properties of a class of burst error correcting codes. These codes have an H matrix which is constructed from an array of identity matrices.

Two parameter codes were introduced in [4, 5]. The H matrix of these codes has the following format

$$H = \underbrace{\begin{pmatrix} I_{m_1} & \cdots & I_{m_1} \\ I_{m_2} & \cdots & I_{m_2} \end{pmatrix}}_{n=\mathrm{lcm}(m_1,m_2)}$$

where I_{m_i} is an m_i by m_i identity matrix. The parity check matrix has $\mathrm{lcm}(m_1, m_2)$ columns and $(m_1 + m_2)$ rows, of which $m_1 + m_2 - \gcd(m_1, m_2)$ are independent, giving an $(\mathrm{lcm}(m_1, m_2), m_1 + m_2 - \gcd(m_1, m_2))$ code.

The idea of two parameter codes was extended in [6] to N parameters, with parity check matrix

$$H = \underbrace{\begin{pmatrix} I_{m_1} & I_{m_1} & \cdots & I_{m_1} \\ I_{m_2} & I_{m_2} & \cdots & I_{m_2} \\ \vdots & \vdots & \ddots & \vdots \\ I_{m_N} & I_{m_N} & \cdots & I_{m_N} \end{pmatrix}}_{n=\mathrm{lcm}(m_1,m_2,\ldots,m_N)}$$

These codes were shown [4] to be able to correct bursts in multi-dimensional channels. A multi-dimensional channel is a channel where several streams of information are present in parallel. Such N-dimensional channels can be mapped onto a conventional one-dimensional channel by bounding $N-1$ dimensions and 'unfolding' the channel. An example of a three dimensional channel is given in Figure 1. The x and y dimensions are assumed to be bounded to L_x and L_y, whereas $L_z \to \infty$.

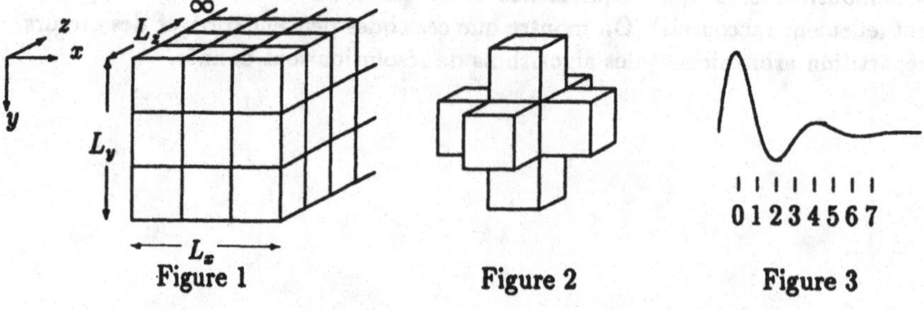

Figure 1 Figure 2 Figure 3

A mapping of this channel with points (x, y, z) onto a one-dimensional channel with points (a) would be $a = x + yL_x + zL_xL_y$. A full description of this process is given in [4].

Contiguous bursts in a multi-dimensional channel will not necessarily give contiguous bursts when mapped onto the one-dimensional case. For example, take the case of a burst in the form of a discrete sphere in a three dimensional channel as shown in Figure 2. Mapping this onto a one-dimensional channel gives the following burst pattern.

			$L_x(L_y-1)-1$				L_x-2						L_x-2				$L_x(L_y-1)-1$			
0	1	0	···	0	1	0	···	0	1	1	1	0	···	0	1	0	···	0	1	0

Non-contiguous bursts may also occur in one-dimensional channels themselves. Take the case of a channel which is subject to ringing of the form shown in Figure 3. If a burst starts in position 0, positions 1, 3, 5 and 7 onwards will have 0 noise and not be corrupted, giving an error shape 1010101.

It is obviously not efficient to use a normal burst correcting code to correct errors in this case, since such a code assumes that the error probability of all points within a burst is equal to some $p_e > 0$, whereas here $p_{e_i} = 0$ for some positions i within the burst. Let us introduce some new concepts for the description of bursts

Definition 1 The *span* of an error burst E is the inclusive length from the first and last points in the burst where $p_{e_i} > 0$.

Definition 2 The *cardinality* of an error burst E, denoted $|E|$, is the number of points in the burst for which $p_{e_i} > 0$.

For a conventional contiguous burst, both the span and cardinality of the burst are equal to its length.

Split-syndrome codes are able to correct such arbitrary burst error shapes, where the cardinality of the burst is upper bounded by $\min(m_1 - d_1, m_2 - d_2)$, where d_1 is the largest factor of m_1 other than itself, and d_2 is the highest factor of m_2 other than itself. The actual cardinality of the burst error shape which can be corrected depends both on the shape of the errors and the code parameters, and an algorithm for this is given below.

2 Generator Polynomial

The parity check matrix for two parameter codes has the form

$$H = \left(\begin{array}{c} \overbrace{\begin{matrix} 1 & 0 & \cdots & 0 \\ 0 & 1 & & 0 \\ \vdots & & \ddots & \vdots \\ 0 & 0 & \cdots & 1 \end{matrix}}^{m_1} \quad \cdots \quad \overbrace{\begin{matrix} 1 & 0 & \cdots & 0 \\ 0 & 1 & & 0 \\ \vdots & & \ddots & \vdots \\ 0 & 0 & \cdots & 1 \end{matrix}}^{m_1} \\ \underbrace{\underbrace{\begin{matrix} 1 & 0 & \cdots & 0 \\ 0 & 1 & & 0 \\ \vdots & & \ddots & \vdots \\ 0 & 0 & \cdots & 1 \end{matrix}}_{m_2} \quad \cdots \quad \underbrace{\begin{matrix} 1 & 0 & \cdots & 0 \\ 0 & 1 & & 0 \\ \vdots & & \ddots & \vdots \\ 0 & 0 & \cdots & 1 \end{matrix}}_{m_2}}_{n} \end{array} \right)$$

The length of the code for which H is the parity check matrix is $\mathrm{lcm}(m_1, m_2) = n$. H can be expressed in polynomial form as follows

$$H = \left(\begin{array}{c} x^{n-1} + x^{n-m_1-1} + x^{n-2m_1-1} + \cdots + x^{m_1-1} \\ x^{n-2} + x^{n-m_1-2} + x^{n-2m_1-2} + \cdots + x^{m_1-2} \\ \vdots \\ x^{n-m_1} + x^{n-2m_1} + x^{n-3m_1} + \cdots + x^0 \\ \\ x^{n-1} + x^{n-m_2-1} + x^{n-2m_2-1} + \cdots + x^{m_2-1} \\ x^{n-2} + x^{n-m_2-2} + x^{n-2m_2-2} + \cdots + x^{m_2-2} \\ \vdots \\ x^{n-m_2} + x^{n-2m_2} + x^{n-3m_2} + \cdots + x^0 \end{array} \right) = \left(\begin{array}{c} \frac{(x^n-1)}{(x^{m_1}-1)} x^{m_1-1} \\ \frac{(x^n-1)}{(x^{m_1}-1)} x^{m_1-2} \\ \vdots \\ \frac{(x^n-1)}{(x^{m_1}-1)} x^0 \\ \frac{(x^n-1)}{(x^{m_2}-1)} x^{m_2-1} \\ \frac{(x^n-1)}{(x^{m_2}-1)} x^{m_2-2} \\ \vdots \\ \frac{(x^n-1)}{(x^{m_2}-1)} x^0 \end{array} \right)$$

In order to find $h(x)$, we must find the shortest non-zero row from H resulting from elementary row operations on H.

The first part of H simply consists of linear combinations (actually cyclic shifts) of $\frac{x^n-1}{x^{m_1}-1}$, while the second part of H consists of linear combinations (cyclic shifts) of $\frac{x^n-1}{x^{m_2}-1}$.

Therefore, the required result is simply the smallest possible linear combination of $\frac{x^n-1}{x^{m_1}-1}$ and $\frac{x^n-1}{x^{m_2}-1}$. A direct consequence of the Euclidean algorithm is that the smallest linear combinations of two polynomials is the greatest common divisor of the two numbers, so that $h(x) = \gcd\left(\frac{x^n-1}{x^{m_1}-1}, \frac{x^n-1}{x^{m_2}-1}\right)$. This gives $g(x) = \mathrm{lcm}(x^{m_1} - 1, x^{m_2} - 1)$.

This argument can be easily extended to N parameters, where

$$g(x) = \text{lcm}(x^{m_1} - 1, \ldots, x^{m_N} - 1)$$

Further work in [7] showed that by ensuring that the code parameters from a set of N primes or powers of primes such that parameter $m_i = \frac{n}{p_i}$, $n = \prod_{j=1}^{N} p_j$, the multi-parameter codes could correct random errors in addition to burst errors, or multiple bursts of errors. The form of the generator matrix given in [7] is equivalent to the least common multiple of $(x^{m_1} - 1, \ldots, x^{m_N} - 1)$, and so agrees with the form given here.

For two parameter codes, the generator polynomial can be simply expressed as

$$g(x) = \frac{(x^{m_1} - 1)(x^{m_2} - 1)}{(x^{\gcd(m_1, m_2)} - 1)}$$

When m_1 and m_2 are relatively prime, this form of the generator polynomial is equivalent to that of Gilbert codes, which have

$$g(x) = \frac{(x^{n_1} - 1)(x^{n_2} - 1)}{(x - 1)}$$

Gilbert codes [3] are formed from the product of two single parity check codes. Such a code is a $n_1 \times n_2$ array with each row of the array forming a codeword of a $(n_1, n_1 - 1$ SPC code, and each column forming a codeword of a $(n_2, n_2 - 1)$ SPC code. This results in a $(n_1 n_2, (n_1 - 1)(n_2 - 1))$ code with minimum distance 4. Gilbert codes are cyclic when n_1 and n_2 are relatively prime.

Where m_1 and m_2 are relatively prime, $\text{lcm}(m_1, m_2) = m_1 m_2$, and split syndrome codes are equivalent to Gilbert codes. For $\gcd(m_1, m_2) > 1$, split-syndrome codes are shortened Gilbert codes since the parity check matrix of the Gilbert code is

$$H = \underbrace{\begin{pmatrix} I_{m_1} & \cdots & I_{m_1} \\ I_{m_2} & \cdots & I_{m_2} \end{pmatrix}}_{n = m_1 m_2}$$

which includes the parity check matrix of the corresponding split-syndrome code as a subset.

The idea of products of single parity check codes can be extended to more than two dimensions in the same way as the multi-parameter codes. In fact, the relatively prime multi-parameter codes are the same as such an extension of Gilbert codes given by Bahl and Chien [1], and the decoder described in [7] can be used to decode these codes.

This correspondence between split-syndrome codes and Gilbert codes means that relatively prime parameter Gilbert codes can also correct arbitrary burst error shapes.

3 Roots of the Code

It can be seen that the roots of $g(x)$ are all the roots of the factors $(x^{m_i} - 1)$, with the multiplicity of each root equal to the maximum of its multiplicity in each $(x^{m_i} - 1)$.

$$g(x) = \prod_j (x - \alpha_j)$$

Therefore the set of roots of $g(x)$ is

$$\bigcup_{i=1}^{N} \{\alpha_j : (x - \alpha_j) \mid (x^{m_i} - 1)\}$$

The roots of each $x^{m_i} - 1$ are $\{\alpha^j : jm_i = p^r - 1\}$, where p^r is the number of elements in the field, so the roots of $g(x)$ are

$$\bigcup_{i=1}^{N} \left\{\alpha_j : \frac{p^r - 1}{m_i} \mid j\right\}$$

Since the set of irreducible polynomials over a field allows prime decomposition of polynomials over the field, we can also write the definition of $g(x)$ as the least common multiple of the minimal polynomials of

$$\alpha^0, \alpha^{\frac{n}{m_1}}, \alpha^{2\frac{n}{m_1}}, \ldots, \alpha^{n-\frac{n}{m_1}}, \alpha^{\frac{n}{m_2}}, \alpha^{2\frac{n}{m_2}}, \ldots, \alpha^{n-\frac{n}{m_2}}, \ldots, \alpha^{\frac{n}{m_N}}, \alpha^{2\frac{n}{m_N}}, \ldots, \alpha^{n-\frac{n}{m_N}}.$$

This definition of the codes show them to be reversible, since if α^i is a root, so too is α^{-i}.

4 Minimum Code Distance

Two parameter codes have minimum distance 4, being either Gilbert codes or shortened Gilbert codes and therefore products of simple parity check codes with distance 2.

N-parameter codes contain two-parameter codes, and so have minimum distance of at least 4. Where, as above, these codes can be expressed in terms of Gilbert codes they have minimum distance 2^N.

5 Method of Error Correction

Product codes can be very simply decoded by using decoders based on their constituent codes. In the case of two-parameter codes, the decoder consists of two

shift registers arranged as round counters, of lengths m_1 and m_2, into which the codeword is clocked. These registers divide by $x^{m_1} - 1$ and $x^{m_2} - 1$, and an error is detected when the syndromes are non-zero and the errors in them coincide. Due to their differing lengths this corresponding errors can only match once within $\text{lcm}(m_1, m_2)$.

Two alterations are required to allow arbitrary burst shapes to be corrected. Firstly, the comparison between the registers must be made on the basis of corresponding error positions. Since the span of the error burst may be greater than the length of the syndromes this is done be comparing bit $a \bmod m_1$ in register one with bit $a \bmod m_2$ of register two for each error position a. Secondly, a programmable counter is added to the circuit which is triggered when the first coincidence, which marks the start of the burst, is detected. This counter has a copy of the burst shape, and it maps a copy of the syndrome into the burst pattern to allow correction to take place.

6 Choice of Code Parameters and Error Correction Capabilities

There has been considerable interest in the burst error correcting capabilities of Gilbert codes and their derivatives (surveyed in [2]), but all such analysis considers the case of simple contiguous bursts.

Error positions are located by the coincidence of the error patterns in the separate syndromes. Since the block length of the split syndrome code is the least common multiple of the syndrome lengths, by the Chinese Remainder Theorem the coincidence of any particular bit in the syndromes is unique within the block length. However, if three or more bits are in error it is possible that errors in one syndrome would coincide with bits in the other syndrome corresponding to different errors, leading to two different error patterns being indistinguishable.

Without loss of generality, let $m_1 < m_2$. Failure will occur if the syndromes coincide after less than $\text{lcm}(m_1, m_2)$ shifts, which would mean that coincidence would not be unique.

The simplest way for coincidence to happen is as follows. Assume a repeating pattern of period q within a syndrome, such that $q = \frac{m_i}{p}$ for some p. A shift of jq steps, where $jq < m_i$ would cause the syndromes to coincide again. For the (10,13) code shown below, the error pattern $E(0,2,4,6,8)$ causes the string to be repeated in the smaller syndrome, producing the same syndrome as error $E(52,54,56,58,60)$. For this reason, no error pattern must be allowed to repeat within a syndrome, which leads to the upper limit on correction length of $m - d$, where m is the syndrome and d is the highest divisor of m other than itself.

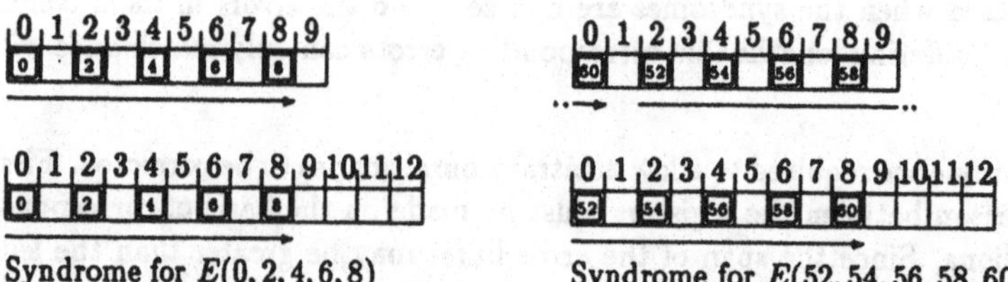

Syndrome for $E(0,2,4,6,8)$ Syndrome for $E(52,54,56,58,60)$

The different lengths of the syndromes can mean that non-repeating syndromes can also be confused. The longer syndrome will contain $m_2 - m_1$ more 0 bits than the shorter one. Denote a block of $m_2 - m_1$ zeros by Δm.

Let the burst pattern to be corrected start in bit x. After x shifts have occured, the burst patterns will then start at position 0 in each syndrome (Figure 4) and the burst is detected. However, if an error pattern has a block of Δm clear bits occuring within it ($A\Delta mB$ for example), the clear bits can match up twice: once correctly as shown in Figure 4 after x shifts with the clear bits in the error pattern, with $x \bmod m_1 = x \bmod m_2 = 0$, and once incorrectly after y shifts as shown in Figure 5 with the Δm clear bits which follows the burst pattern, with $y \bmod m_1 = length(B) - length(A)$, $y \bmod m_2 = length(B) + \Delta m$. Removing the Δm's gives $AB = A_2BA_1 = BA$, which means that $A = P^{(i)}$, $B = P^{(j)}$, for some string P and $i, j \geq 1$. ($P^{(i)}$ is string P repeated i times). Therefore a necessary, but not sufficent, condition for failure is that the correctable burst syndromes repeat when blocks of Δm 0's are removed. If this condition is met, the possible syndromes would have to be checked to see if the combination could give $AB = A_2BA_1 = BA$.

A	Δm	B	
A	Δm	B	Δm

Figure 4

A_2	Δm	B	A_1
B	Δm	A	Δm

Figure 5

Confusion can also occur in a similar way if there is more than one block of Δm clear bits within the burst pattern.

The block of clear bits must occur within the burst, not at either end. Let the pattern which causes failure have a 1 in bit 0. If this condition was not met, a shift of $< m_1$ would always fulfil it, and shifting in the longer register will move the clear bits to the upper Δm part of the syndrome, keeping it clear. The burst pattern cannot therefore start with the clear bits. Since the 1 in bit 0 cyclically follows the Δm block in syndrome m_2, the Δm clear bits in syndrome m_1 must be followed by a 1, so the pattern cannot end with a block of clear bits. This also means that blocks of Δm bits cannot occur together — they must be seperated by at least one error pattern block.

As an example, take the case of a (11,13) code. $\Delta m = 2$ and $11 - 2 = 9$, which can repeat as three strings of length 3 (100 and cyclic shifts and combinations). Depending on the position of the Δm block this gives either 10010000100 $E(0,3,8)$ or 10000100100 $E(0,5,8)$, and similarly for 110, 101, and 111. $11 - (2 \times 2) = 7$, $11-(3\times2) = 5$ and $11-(4\times2) = 3$ are all prime, so their are no other combinations which will cause incorrect decoding. There are therefore eight pairs of syndromes which coincide. As an example, the syndrome of $E(0,3,8)$ is the same as a that for $E(47,52,55)$, and the syndromes produced are as follows. The first error (0 or 47) produces the a bits, the second (3 or 52) the b bits, and the third (8 or 55) the c bits. Although the order of the bits are different between the two sets of syndromes they are indistinguishable, and so the syndromes are confused.

Syndromes for $E(0,3,8)$

0	1	2	3	4	5	6	7	8	9	10
a			b			Δm		c		

0	1	2	3	4	5	6	7	8	9	10	11	12
a			b			Δm		c			Δm	

Syndromes for $E(47,52,55)$

0	1	2	3	4	5	6	7	8	9	10
c			a			Δm		b		

0	1	2	3	4	5	6	7	8	9	10	11	12
b			c			Δm		a			Δm	

We now have a complete set of conditions for the confusion of syndromes and can derive an algorithm to find suitable code parameters for a given error pattern.

Firstly, the mapping between the actual error pattern and the pattern produced in the syndromes must be unique, i.e., the bits in the error pattern must give distinct numbers modulo m_1 and m_2.

Secondly, no possible error pattern may give rise to repetition in either of the syndromes, and, in the case of the shorter syndrome, the syndrome must also not repeat when one or more sets of Δm bits are removed. This can be checked by finding all the divisors q_i of the syndrome length by prime factorisation, and then checking to see whether every q_ith bit in the syndrome belongs to the error pattern. If all do, i.e. $\forall j,\ j = 0$ to $\frac{m}{q_i-1}, jq_i = e \bmod m$ forsome $e \in E$ where E is the error burst and m the syndrome length, there is a distinct possibility that the syndrome could repeat with period q_i. If any bit is not in the error pattern, it will always be 0, and therefore the syndrome will not repeat. For each divisor q_i, a similar check must be carried out on bits $jq_i + k$, for each k in the range 0 to $q_1 - 1$.

The shorter syndrome also has to be checked with Δm block(s) removed. If $m_1 - j\Delta m = xy$ for some j, x, y, then there is a possibility of the repetition of x strings of length y. To see whether such a repetition is actually possible, the $j\Delta m$ blocks are added between the x strings and the syndrome checked for repetition. There are $\binom{x-1}{j}$ such possible ways of inserting the Δm blocks. Both syndromes must repeat in the same way, and any bits in the error pattern in the upper Δm part of the syndrome must be zero. If the repetition is possible, the code is checked to see if it could fail by shifting the syndromes against each other to see if they coincide more than once. These operations need only be carried out only once.

A simple example will illustrate the above concepts.

Example : Take a code which is required to correct 4 errors occurring in positions i, $i+2$, $i+3$ and $i+6$. In this case, $E = (0,2,3,6)$, and the smallest two parameter code to correct such errors has $m_1 = 5$ and $m_2 = 7$. 5 and 7 are both prime, as is $5-2$, so that the parameter pass the checks. The code has $g(x) = \mathrm{lcm}(x^5-1, x^7-1) = x^{11}+x^{10}+x^9+x^8+x^7+x^4+x^3+x^2+x^1+1$ with roots $\alpha^0, \alpha^5, \alpha^7, \alpha^{10}, \alpha^{14}, \alpha^{15}, \alpha^{20}, \alpha^{21}, \alpha^{25}, \alpha^{28}, \alpha^{30}$ (with α a 35th root of unity over GF(2^{12})), giving $n = 35$, $k = 24$ and $R = 68.5\%$.

References

[1] Bahl, L. R. and R. T. Chien: Single and multi- burst correcting properties of a class of cyclic product codes, IEEE Trans., IT-17 (1971), 594–600.

[2] Farrell, P. G.: An introduction to array error control codes, in: CISM Courses and Lectures 313—Geometries, Codes and Cryptography (Ed. G. Longo *et al*), Springer-Verlag, Wien, 1990, 101–128.

[3] Gilbert, E. N.: A problem in binary encoding, Proc. of Symposia in Applied Maths., 10—Combinatorial Theory, American. Math. Soc., (1960), 291–297.

[4] Olcayto, E. and T. Lesz: Class of linear cyclic block codes for burst errors occuring in one-,two- and three-dimensional channels, IEE Proc. Part F, 130(5) (1983), 468–475.

[5] Olcayto, E. and T. Lesz: Simple cyclic codes for burst errors, IEE Proc. Part F, 131(2) (1984), 153–155.

[6] Olcayto, E. and A. R. Birse: Split syndrome burst error correcting codes for one-dimensional channels, IEE Proc. Part F, 134(4) (1987), pp 373–376.

[7] Olcayto, E. and A. R. Birse: New class of cyclic, random and burst error correcting codes, Proceedings of AAECC-5, Springer-Verlag, 1988, 357–368.

DECODING 2D CYCLIC CODES
BY PARALLELING THE 2D BERLEKAMP-MASSEY ALGORITHM

S. Sakata
Toyohashi University of Technology, Toyohashi, Japan

ABSTRACT

In this paper † we show that it is possible to decode several binary 2D cyclic codes by applying the 2D B.M. algorithm in parallel with respect to some possible total orders, where we need no trial and error. The principle is that the 2D B.M. algorithm has many choices of total orders as the Gröbner basis algorithm and that some elements of a minimal polynomial set of a given 2D array are uniquely obtained from part of the given data, i.e., a certain fragment of the 2D array.

1. Introduction

A two-dimensional (2D) cyclic code of area $N_1 \times N_2$ is defined to be an ideal of the bivariate polynomial residue ring $K[x,y]/(x^{N_1}-1, y^{N_2}-1)$ [1], where $K := GF(q)$ is the Galois field of q elements which is identified with the symbol alphabet. Each codeword is represented either as a bivariate polynomial $a(x,y) = \sum a_{ij}x^iy^j \in K[x,y]$ of degree less than (N_1, N_2) or as a 2D array $a = (a_{ij})$ of size $N_1 \times N_2$. Every cyclic permutation of the array either along the x-direction or along the y-direction also gives a codeword of the same code. Thus, the 2D cyclic codes are a 2D generalization of well-known and important cyclic codes and belong to a subclass of Abelian codes [2]. In this paper we treat only semi-simple binary (i.e., $q = 2$) codes with odd integers N_1 and N_2. If $gcd(N_1, N_2) = 1$, any 2D cyclic code of area $N_1 \times N_2$ is equivalent to a cyclic code of length N_1N_2 (called a cyclic product code). In case of $gcd(N_1, N_2) > 1$, we have in general 2D cyclic codes of area $N_1 \times N_2$ which are not equivalent to any cyclic codes. Each 2D cyclic code is defined either by the zeros or by the nonzeros, where the zeros are common to all the polynomials (codewords) belonging to the code.

Although it has been shown that there are several good 2D cyclic codes [3], the problem of decoding these codes has been solved only partly, i.e., only a few of them have been shown to be correctable by applying the 2D Berlekamp-Massey (B.M.) algorithm (with or without trial and error) to the given 2D syndrome array [4]. For a binary code we have only to find the error locator of a given received word. The error locations of a received word define the error locator ideal I_e of the

† This work was partly supported by the International Communication Foundation (ICF).

polynomial ring $\tilde{K}[x,y]$, where $\tilde{K} = GF(2^M)$ is an extension of the symbol field $K = GF(2)$. Naturally the independent point set $\Delta(F)$ of the Gröbner basis F of the ideal I_e depends on the error pattern, where $\Delta(F)$ is defined to be a subset $\{(i,j) \in Z_+^2 | \neg(\exists f \in F)(deg(f) \leq (i,j))\}$ of the 2D integral lattice identified with the set Z_+^2 of all pairs of nonnegative integers. Therefore, we must take into consideration several forms of independent point sets for possible error patterns composed of the same number of errors. In many cases the known syndrome values are not enough to find the Gröbner bases of all the possible error locator ideals and to obtain the error pattern perfectly. Thus, we assume several possible values of the unknow syndromes necessary for the decoding procedure and to apply the B.M. algorithm to each of the compensated 2D syndrome arrays by trial and error. The computational complexity of this procedure tends to become larger for codes with bigger area $N_1 \times N_2$ [4].

In this paper we show that it is possible to decode several binary 2D cyclic codes by applying the 2D B.M. algorithm in parallel with respect to some possible total orders, where we need no trial and error. The principle is that the 2D B.M. algorithm has many choices of total orders as the Gröbner basis algorithm and that some elements of a minimal polynomial set of a given 2D array are uniquely obtained from part of the given data, i.e., a certain fragment of the 2D array. Since we have no knowledge of the form of the error pattern except for the number of correctable errors, we must execute the above procedure in parallel with respect to the total orders which correspond to all the possible error patterns.

2. Preliminaries

In this section, since we use almost the same terminology as in [4], we make only a brief sketch of our problem and the relevant concepts, omitting the details. Furthermore, we give a theorem and lemmas which will be useful to decode some binary 2D cyclic codes. In addition, we mention several choices of total orders over the 2D integral lattice Z_+^2.

For odd integers N_1, N_2, a binary (i.e., over $K = GF(2) = \{0,1\}$) 2D cyclic code C of area $N_1 \times N_2$ is defined by a bipartition of the set $D_{N_1N_2}$ of all representatives of cyclotomic cosets in $Z_{N_1N_2}$ $(:= \{(i,j) \in Z_+^2 \,|\, 0 \leq i \leq N_1 - 1, 0 \leq j \leq N_2 - 1\})$: $D_{N_1N_2} = D \cup \bar{D}$, where D and $\bar{D}$ correspond to the nonzeros and the zeros of the code C, respectively. For a prescribed couple of a primitive N_1-th root α and a primitive N_2-th root β, a bivariate polynomial $a(x,y)$ is a codeword of C if and only if $a(\alpha^i, \beta^j) = 0$ for $(i,j) \in \bar{D}$, or equivalently, for $(i,j) \in U$, where $\alpha, \beta \in \tilde{K} := GF(2^M)$ for the least integer M such that $lcm(N_1, N_2)$ divides $2^M - 1$, and U $(\subset Z_{N_1N_2})$ is the union of all cyclotomic cosets containing $(i,j) \in \bar{D}$.

Given a received word $r(x,y) = c(x,y) + e(x,y)$, we have the syndrome values

$s_{ij} := e(\alpha^i, \beta^j) = r(\alpha^i, \beta^j), (i,j) \in U$, where $c(x,y)$ is the sent codeword and $e(x,y) = \sum_{k=1}^{t} e_{i_k j_k} x^{i_k} y^{j_k}$ is the error pattern of size t with $e_{i_k j_k} \neq 0, 1 \leq k \leq t$. We will assume that t is less than the half of the minimum distance d of the code C. While we have no knowledge of the values $s_{ij} = e(\alpha^i, \beta^j)$, $(i,j) \in Z_{N_1 N_2} \setminus U$, we call the whole array $s = (s_{ij})$ defined by $s_{ij} := e(\alpha^i, \beta^j)$, $(i,j) \in Z_{N_1 N_2}$, by the name of the (2D) syndrome array. By the way, we remark the conjugacy of syndrome values, i.e., $s_{2i,2j} = (s_{ij})^2$. From the known syndrome values, we must determine a Gröbner basis F_e of the error locator ideal defined by

$$I_e := \prod_{k=1}^{t} (x - \alpha^{i_k}, y - \beta^{j_k}) \subset \tilde{K}[x,y], \tag{1}$$

the set $V(I_e)$ of zeros of which is just the error locators $\{(\alpha^{i_k}, \beta^{j_k}) | 1 \leq k \leq t\}$.

The syndrome array s over $Z_{N_1 N_2}$ can be extended to a doubly periodic (DP) array over Z_+^2, while several values are unknown. From now on, we denote any element $(i,j) \in Z_+^2$ as a point $p = (p_1, p_2)$. For a DP array $u = (u_p)$ defined over Z_+^2, a Gröbner basis F_u of the characteristic ideal $I(u)$ of u [5, 6] coincides with a minimal polynomial set of u, where both concepts are based on a total order $\leq_T$ as well as on the usual partial order $\leq_P$ defined over Z_+^2. A minimal polynomial set F $(\subset \tilde{K}[x,y])$ of a 2D array u (over $\tilde{K}$) is a set of polynomials f with a minimal (with respect to the partial order $\leq_P$) degree $deg(f)$ which are *valid* for u, in the sense that u satisfies the 2D linear recurring relation represented by f:

$$f[u]_p := \sum_{q \in Supp(f)} f_q u_{q+p-s} = 0, s \leq_T p, \tag{2}$$

where $f = \sum_{q \in Supp(f)} f_q x^{q_1} y^{q_2} \in \tilde{K}[x,y]$, f_q $(\in \tilde{K}) \neq 0$ for $q = (q_1, q_2) \in Supp(f)$ $(\subset Z_+^2)$, and $s = deg(f)$ is the degree of f which is the maximum (with respect to the total order $\leq_T$) element of the support $Supp(f)$ of f. For the 2D syndrome array s, the error locator ideal I_e is the characteristic ideal $I(s)$ of s, and a minimal polynomial set of s is a Gröbner basis of I_e.

A minimal polynomial set of a finite or DP 2D array can be obtained by the 2D B.M. algorithm [7]. During the process of iterations in the 2D B.M. algorithm, an array u is scanned according to the total order $\leq_T$ and a minimal polynomial set F of the checked part $u^p := (u_q | q <_T p)$ of the array is obtained at each point $p \in Z_+^2$. While a minimal polynomial set F of u^p is not necessarily unique, the independent point set $\Delta(F)$ of u^p is unique and denoted as $\Delta(u^p)$. $\Delta(F)$ is of echelon form as follows

$$\Delta(F) = \Gamma_C = Z_+^2 \setminus \Sigma_S, \tag{3}$$

where $\Gamma_C := \cup_{c \in C} \Gamma_c$, $\Sigma_S := \cup_{s \in S} \Sigma_s$ for a certain pair of *nondegenerate* subsets C, S of Z_+^2, and $\Gamma_c := \{q \in Z_+^2 | q \leq_P c\}$, $\Sigma_s := \{q \in Z_+^2 | q \geq_P s\}$. (If there exists no pair of distinct elements s, t of a subset T ($\subset Z_+^2$) such that $s \leq_P t$, T is said to be nondegenerate.) In fact, $S = \{deg(f) | f \in F\}$.

There is a point p such that a minimal polynomial set F of u^q is unique and the same for any $q \geq_T p$, which coincides with a Gröbner basis of $I(u)$. More precisely, let F be a minimal polynomial set of a finite 2D array u^p with support $Supp(u^p) = \Sigma^p := \{q \in Z_+^2 \,|\, q <_T p\}$, and assume that F is *reduced* in the sense that every $f \in F$ satisfies the condition:

(*) f has the leading coefficient $lc(f) = 1$ and the support $Supp(f) \subseteq \Delta(u^p) \cup \{deg(f)\}$.

Then, if Σ^p contains

$$\tilde{\Delta}(F) := \Delta(F) \cup (\cup_{f \in F}(deg(f) + \Delta(F))), \tag{4}$$

F is unique in the sense that there exists no other polynomial f' with $deg(f') = deg(f)$ for some $f \in F$ which is valid for u^p and satisfies the condition (*), where $q + \Delta := \{q + r | r \in \Delta\}$ for $q \in Z_+^2$ and $\Delta \subset Z_+^2$. Also for any point set of echelon form $\Delta = \Gamma_C = Z_+^2 \setminus \Sigma_S$ specified by either of a pair of nondegenerate subsets S, C, we define $\tilde{\Delta} := \Delta \cup (\cup_{q \in S}(q + \Delta))$ and call it the *ensuring set* of Δ. Let $\Delta(t)$ be the union of $\tilde{\Delta}$ for all possible sets Δ of echelon form with $\#\Delta = t$. Then, $\Delta(t)$ contains the ensuring set $\tilde{\Delta}(F_e)$ of the independent point set $\Delta(F_e)$ for every Gröbner basis F_e which depends not only on a particular error locator ideal I_e of a t-error pattern but also on a selection of a particular total order $\leq_T$. Therefore, if $\Delta(t) \subseteq U$, then we can find the Gröbner basis F_e of the error locator ideal I_e as a minimal polynomial set of a subarray s^p which is obtained according to either of the different total orders as a fragment of the syndrome array. Thus, $\Delta(t)$ corresponds to the perfect information for any t-error-correction [4]. On the other hand, if $\Delta(t) \setminus U \neq \emptyset$, we must have trial and error for the error-correction procedure of [4], where the 2D B.M. algorithm is applied once for each trial of assuming one of the all possible values (up to conjugacy) at each of the relevant points $p \in \Delta(t) \setminus U$.

As the present starting point, we give the following uniqueness theorem for any element of a minimal polynomial set. (The argument in the proof of the uniqueness theorem (Th. 4 of [7]) for the whole minimal polynomial set is applicable even to this more specific theorem.)

Theorem 1: Let F be a reduced minimal polynomial set of a finite 2D array u^p with respect to a total order $\leq_T$. $f \in F$ is unique, i.e., there exists no other polynomial f' with $deg(f') = deg(f)$ which is valid for u^p and satisfies the condition (*) if and only if $deg(f) + \Delta(u^p)$ $(= \{deg(f) + q \,|\, q \in \Delta(u^p)\}) \subseteq \Sigma^p$. In particular, for $f \in F$ which has the minimum (with respect to the total order $\leq_T$) degree among F, i.e., $deg(f) <_T deg(f')$ for any other $f' \in F$, f is unique if and only if the inequality $2deg(f) \leq_T p$ holds.

This theorem implies that we can obtain unambiguously an element or a subset of the reduced Gröbner basis from a certain fragment of the whole period of the DP array. Thus, from the known part of the 2D syndrome array s, we happen to be able to find some elements of the Gröbner basis F_e of the error locator ideal I_e. Furthermore, these polynomials ($\in F_e$) happen to give some information of the unknown syndrome values. To get greater possibility of obtaining the Gröbner basis totally, we need the following two lemmas.

Lemma 1: Let f be an element of the reduced Gröbner basis F_u of the characteristic ideal $I(u)$ of a DP array u. For a certain point $r \in Z_+^2$, if a finite subarray u' of u has the support of the form $Supp(u') \subset r + \Sigma^p$ which is so large as to contain the subset $r + deg(f) + \Delta(u)$, then it is possible to find the element f by starting from the point r and applying the 2D B.M. algorithm to u' with respect to the total order $\leq_T$ restricted within $r + \Sigma^p$. (The proof is based on the shift invariancy of any 2D cyclic code.)

Lemma 2: Let $\tilde{u}$ be the reciprocal of a DP array u, and F (resp. $\tilde{F}$) be a Gröbner basis of $I(u)$ (resp. $I(\tilde{u})$) with respect to $\leq_T$. (Remark: $\tilde{u}_{(N_1,N_2)-p} = u_p, p \in Z_{N_1 N_2}$.) Then, $\tilde{F}$ has the independent point set $\Delta(\tilde{F})$ which is reflexively symmetric to the independent point set $\Delta(F)$ of F with respect to the total order $\leq_T$. (The proof is based on the relationship with the reciprocal 2D cyclic code whose zeros are inverses of the zeros of the original code.)

Both lemmas combined with Theorem 1 often are useful to obtain the Gröbner basis of the characteristic ideal of a DP array containing some unknown values.

In the next section we will describe the details of our method for decoding some binary 2D cyclic codes in parallel with respect to several total orders, and show examples of its application. For the method it is important to recall that there are several choices of total orders $\leq_T$ which define different Gröbner bases of the same ideal [8]. Substantially, we have only to consider a class of total orders defined by a couple of relatively prime integers (a, b). More precisely we have four kinds of total orders for each couple (a, b), i.e., $(a, b)_x$-, $(a, b)_y$-, the inverse $(a, b)_x$-, and the inverse $(a, b)_y$-orders in the context of 2D cyclic codes. The following are the list of these orders:

(a) $(i, j) <_T (k, l)$ iff $ai + bj < ak + bl \vee (ai + bj = ak + bl \wedge i < k)$;
(b) $(i, j) <_T (k, l)$ iff $ai + bj < ak + bl \vee (ai + bj = ak + bl \wedge j .< l)$;
(c) $(i, j) <_T (k, l)$ iff $ai + bj > ak + bl \vee (ai + bj = ak + bl \wedge i > k)$;
(d) $(i, j) <_T (k, l)$ iff $ai + bj > ak + bl \vee (ai + bj = ak + bl \wedge j > l)$.

The usual total degree order is $(1, 1)_y$-order and the lexicographic order is $(0, 1)_x$- or $(1, 0)_y$-order.

Furthermore, the Gröbner basis $\Delta(\tilde{F})$ mentioned in Lemma 2 can be obtained by applying the 2D B.M. algorithm to an appropriate part of the original array u with respect to the reflexive inverse of the total order $\leq_T$, where, e.g., the reflexive inverse of $(a,b)_x$-order is the inverse $(a,b)_y$-order.

3. Parallel decoding method

Based on our observations given in the previous section, we have the possibility of handling all the cases of error patterns. That is, if we apply the 2D B.M. algorithm to several fragments of the syndrome array in parallel with respect to several kinds of total orders, we can possibly have the Gröbner basis F_e of the desired error locator ideal I_e among these results, though some of them might not give information of I_e.

For the number t of correctable errors of the code C under consideration, let $\Delta = Z_+^2 \setminus \Sigma_S$ be a point set of echelon form with $\#\Delta = t$, and $\tilde{\Delta}$ be the ensuring set of Δ. For each Δ, let $\leq_T$ and $p \in Z_+^2$ be a couple of a total order and a point such that, for the minimum (with respect to $\leq_T$) element q of S, $q + \Delta \subseteq \Sigma^p \cap U$. We call this $\leq_T$ the total order corresponding to Δ. Henceforth we assume that, for every type of Δ with $\#\Delta = t$, we can choose such a couple of $\leq_T$ and p. Otherwise, we cannot use the following parallel decoding method. If there exists a Gröbner basis F_e of I_e with $\Delta(F_e) = \Delta$ (with respect to $\leq_T$), we can find a minimal polynomial set F of s^p with $\Delta(F) \subseteq \Delta$ by applying the 2D B.M. algorithm to s^p (with respect to $\leq_T$). Now, assuming that, for a certain Δ, we have found a minimal polynomial set F of s^p with $\Delta(F) = \Delta$ by applying the 2D B.M. algorithm to s^p with respect to the corresponding total order $\leq_T$, we take the following procedure of calculating the other relevant unknown syndrome values, where let

U' be the subset of Z_+^2 on which the syndrome values are known before now;

$F'(\neq \emptyset)$ be a subset of F whose elements f satisfy the condition: $deg(f) + \Delta \subseteq U'$ (in other words, f can be obtained unambiguously by applying the 2D B.M. algorithm on a fragment of U');

U'' be the subset of $Z_+^2 \setminus U'$ on which the syndrome values are obtained by using the polynomials F' from the values on U'.

Procedure:

Step 1 (Initiation): $U' := U$;

Step 2: if $\#F' = \#F$ then stop; [Success]

Step 3: By using the polynomials of F', calculate the unknown syndrome values s_p, $p \in U''$;

Step 4: $F'' := \{f \in F \mid deg(f) + \Delta(F) \subseteq U' \cup U''\}$;

Step 5: if $F'' = F'$ then stop; [Failure]

Step 6: $F' := F''; U' := U' \cup U''$; go to Step 2.

In the following, we must assume that the above procedure with respect to every relevant total order $\leq_T$ will succeed. In Step 3, whether we can find some unknown syndrome values s_p, $p \notin U'$, depends on the support $Supp(f)$ of $f \in F'$. Furthermore, we remark in Step 3 that, in the process of applying the 2D B.M. algorithm successively (with respect to $\leq_T$) at every point $q \geq_T p$, we can skip the point $q \notin U' \cup \tilde{\Delta}$, i.e., we can calculate the unknown value s_q by using the current F of s^q and go to the next point $q \oplus 1$. (Because, if $f \in F$ failed to be valid at $q \notin \tilde{\Delta}$, the degree change would occur in view of Lemma 4 of [7], which implies that $\#\Delta(F_e) > t$.)

In fact, we do not know whether $\Delta(F_e) = \Delta$ for the unknown F_e. At most we can assume that, if $\#\Delta(F) \leq t$ and $F \subset I(s)$ for the whole syndrome array s whose unknown values are estimated by using F from the known values, F coincides with a Gröbner basis F_e of I_e. Consequently, we have the following alternative cases of F at a certain point $q \geq_T p$:

(1) $\#\Delta(F) > t$ with respect to a total order $\leq_T$: more than t errors occurred.

(2) $\Delta(F) = \Delta$ ($\#\Delta(F) = t$) with respect to the corresponding total order $\leq_T$: t errors occurred with the error pattern determined from F provided that $F \subset I(s)$, i.e., F is valid for s. (In this case, with respect to some other total orders $\leq_{T'}$ different from $\leq_T$, we might have another F such that $\#\Delta(F) = t$ but $\Delta(F) \neq \Delta$ with respect to $\leq_{T'}$, from which we get no information of the error pattern, since the minimal polynomial set F might not be unique or a Gröbner basis.) If $F \not\subset I(s)$, we must conclude that more than t errors occurred.

(3) $\#\Delta(F) < t$ with respect to every relevant total order $\leq_T$: From a minimal polynomial set F with respect to a total order $\leq_T$ for which $\#\Delta(F)$ is the maximum and $\Delta(F) \subset \Delta$, we can obtain the error pattern of size less than t in a similar way with (2).

The following examples will make our procedure clearer.

Example 1: We consider the binary 2D cyclic code of area $N_1 \times N_2 = 5 \times 5$ defined by $D = \{(1,2),(1,4)\}$ which is on the last row of Table IV of [4]. (Remark: In the description of [4], $D = \{(1,2)\}$ is wrong.) This is a linear $(n,k,d) = (25,8,8)$ code, which has the same code parameters as one of Jensen's codes [3], and is 3-error-correcting. The positions of known syndrome values are as shown in Fig. 1. (In Fig. 1, a period of the DP array is shown, where $\circ$ represents a point on which the syndrome value is known and $\diamond$ represents a point on which the syndrome value is not known.) By the method of [4], three errors can be corrected with trial and error, i.e., by assuming 2^4 possible values of unknown syndromes. At the most, 16 executions of the 2D B.M. algorithm are required. However, the following considerations will show that we can correct three errors without any assumption of the unknown syndrome values.

```
o o o o o
o o ◇ o ◇
o o o ◇ ◇
o ◇ ◇ o o
o ◇ o ◇ o
```

Fig. 1: The positions of known and unknown syndrome values.

In this case we have three kinds of point sets Δ of echelon form with $\#\Delta = 3$ (See Fig. 2a, 2b, and 2c, respectively. In these Figures, a period of each DP array (with some excess part in Figs. 2b, 2c) is shown, where $*$ represents a point of Δ, $\star$ represents a point of $\tilde{\Delta} \setminus \Delta$ on which the syndrome value is known, and $\diamond$ represents a point of $\tilde{\Delta} \setminus \Delta$ on which the syndrome value is unknown.)

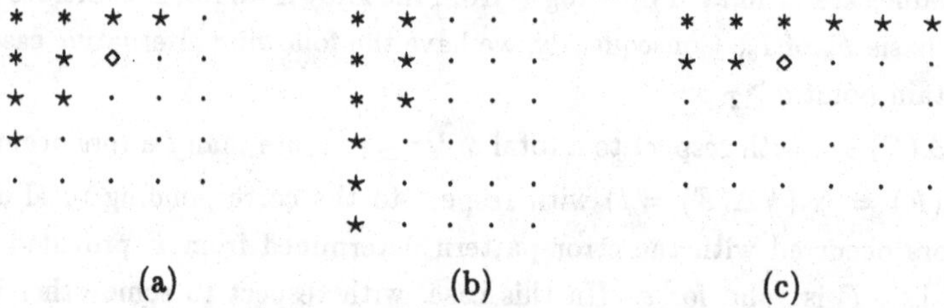

Fig. 2: Points sets Δ of echelon form and their ensuring sets $\tilde{\Delta}$.

(a) For the first type of Δ (Fig. 2a), we have the corresponding $(1,1)_y$-order and the point $p = (1,2)$. Among a minimal polynomial set $F = \{f, f', f''\}$ of s^p with $\Delta(F) = \Delta$, if it exists, we can obtain f unambiguously by applying the 2D B.M. algorithm to the syndrome subarray s^p, where $f = x^2 + f_{01}y + f_{10}x + f_{00}$ for certain $f_{01}, f_{10}, f_{00} \in \tilde{K} = GF(2^4)$. Thus, we initially have $F' = \{f\}$ in the above procedure. By f, we can calculate $s_{12} (= (s_{31})^2))$ as follows:

In case of $f_{01} \neq 0$, we obtain s_{31} from $\{s_{50} (= s_{00}), s_{40}, s_{30}\}$;

Otherwise, we obtain s_{31} from $\{s_{11}, s_{21}\}$.

(To obtain this type of Gröbner basis, we have the following alternative method: By using f, we can calculate the unknown value s_{23} from the known values $\{s_{03}, s_{13}, s_{04}\}$, and also s_{14} from $\{s_{-1,4} (= s_{44}), s_{-1,5}\ (= s_{40}), s_{04}\}$, and then s_{24} from $\{(s_{04}, s_{05} (= s_{00}), s_{14}$ (just found)$\}$. Therefore, we know the values $(s_p | p \in U')$ for $U' := (0,3) + \Sigma^{(4,0)}$. In view of Lemma 1, by starting at $(0,3)$ and applying the 2D B.M. algorithm to the part $(s_p | p \in U')$ with respect to $(1,1)_y$-order restricted within the region U', we can possibly find F.)

Thus, we can possibly find a unique minimal polynomial set F of s^q, $q = (4,0)$, by starting again the 2D B.M. algorithm at the halting point $(1,2)$ and stopping at $(4,0)$. Consequently, we obtain the Gröbner basis of the error locator ideal perfectly, provided that the 3-error pattern has this type of Gröbner basis with respect to $(1,1)_y$-order.

(b) For the second type of Δ (Fig. 2b), we have $(0,1)_x$-order and $p = (3,1)$. Since $\tilde{\Delta}(F) \subseteq U$, we can obtain F from the given known syndrome values, provided that the 3-error pattern has this type of Gröbner basis with respect to $(0,1)_x$-order.

(c) For the third type of Δ (Fig. 2c), we have $(1,0)_y$-order and $p = (0,6)$. Among $F = \{f, f'\}$, which we can possibly obtain by applying the 2D B.M. algorithm to the syndrome subarray s^p. Thus, we initially have $F' = \{f\}$ in the above procedure, where $f = y^3 + f_{02}y^2 + f_{01}y + f_{00}$ for certain $f_{02}, f_{01}, f_{00} \in \tilde{K}$, and f is unambiguous. If $f_{02} \neq 0$, then we can get the value s_{12} from $\{s_{10}, s_{11}, s_{13}\}$. Otherwise, we get the value s_{12} form $\{s_{13}, s_{15}(= s_{10})\}$ since $f_{00} \neq 0$. (Remark: if $f_{00} = 0$, then f has a zero of type $(0,*)$, which contradicts the fact that the error locator ideal I_e has zeros only of type (α^i, α^j), where α $(\in \tilde{K})$ is a primitive fifth root of 1.) Therefore, we know the values $(s_p | p \in U')$ for $U' := \Sigma^{(1,3)}$. By starting at $(0,0)$ and applying the 2D B.M. algorithm to the part $(s_p | p \in U')$ with respect to $(1,0)_y$-order, we can obtain the Gröbner basis of the error locator ideal, provided that the 3-error pattern has this type of Gröbner basis with respect to $(1,0)_y$-order.

In the above, for example, if the error locator ideal I_e has a Gröbner basis F_e with respect to $(1,1)_y$-order such that $\Delta(F_e)$ is not of the first type (as in Fig. 2a.), then the minimal polynomial set F of s^p, $p = (4,0)$, has $\tilde{\Delta}(F) \not\subseteq \Sigma^p$ so that F is not unique and cannot be regarded as F_e. Any such irrelevant F which has not been found to be unique yet is discarded. At any way, with respect to at least one of the three total orders, we necessarily have the relevant $F = F_e$ uniquely and unambiguously, provided that a certain 3-error pattern occurred. Therefore, we can correct any three errors by applying the 2D B.M. algorithm in parallel with respect to the above three kinds of total orders.

Remark: In general, if every t-error pattern can be corrected, any error of size less than t is also correctable, because the error locator ideal of a smaller error pattern has an independent point set of smaller size.

$$
\begin{array}{ccccc}
1 & \theta^{12} & \theta^{9} & \theta^{6} & \theta^{3} \\
\theta^{10} & \theta^{5} & \diamond & \theta^{3} & \diamond \\
\theta^{5} & \theta^{6} & \theta^{10} & \diamond & \diamond \\
\theta^{5} & \diamond & \diamond & \theta^{10} & \theta^{9} \\
\theta^{10} & \diamond & \theta^{12} & \diamond & \theta^{5}
\end{array}
$$

Fig. 3: The known syndrome values.

Example 2 (Example 1 revisited): For the code of Example 1, we assume the error pattern of size 3: $e(x,y) = y^4 + x^2y^3 + x^3y^3$. Then we have the known syndrome values as shown in Fig. 3, where θ is a primitive element of $GF(2^4)$ such that $\theta^4 + \theta + 1 = 0$. By applying the 2D B.M. algorithm to the syndrome subarray s^p, $p = (1,2)$, with respect to $(1,1)_y$-order we obtain a minimal polynomial set

$F = \{x^2 + \theta^{12}y + \theta^5 x + \theta^{13}, xy + y + \theta^{14}, y^2 + \theta^{12}y\}$. By using $f = x^2 + \theta^{12}y + \theta^5 x + \theta^{13} \in F'$ we calculate $s_{12} = \theta^{12}$. Then, we proceed to apply the 2D B.M. algorithm up to $q = (4,0)$, we obtain the unique minimal polynomial set $F = \{x^2 + \theta^{12}y + \theta^5 x + \theta^{13}, xy + y + \theta^9 x + \theta^9, y^2 + \theta^8 y + \theta^6\}$ of s^q. (By the way, we can invoke Lemma 1 instead to obtain the same F, as described in the remark.) As this F is valid for the whole s, F should be a Gröbner basis of I_e. In fact, from F we can find the correct error locations $\{(1,\alpha^4),(\alpha^2,\alpha^3),(\alpha^3,\alpha^3)\}$ as the common zeros of the polynomials of F, where $\alpha = \theta^3$. Furthermore, we can find the Gröbner basis $\{x^3 + \theta^{10}x^2 + \theta^{10}x + 1, y + \theta^3 x^2 + \theta^8 x + \theta\}$ of I_e by applying the 2D B.M. algorithm with respect to $(0,1)_x$-order, from which we obtain the same error pattern. On the other hand, with respect to $(1,0)_y$-order we find a minimal polynomial set F of s^p, $p = (1,2)$, such that $\Delta(F) \neq \Delta$, from which we get no information of the error pattern.

Example 3: Consider the binary 2D cyclic codes of area $N_1 \times N_2 = 3 \times 15$ defined by $D = \{(1,7),(2,3),(2,5),(2,7)\}$, which is a linear $(45,15,14)$ code and 6-error-correcting. (This is superior to a linear (cyclic) $(45,14,10)$ code which can be decoded by the method of Feng et al. [9].) The positions of known and unknown syndrome values are shown in Fig. 4.

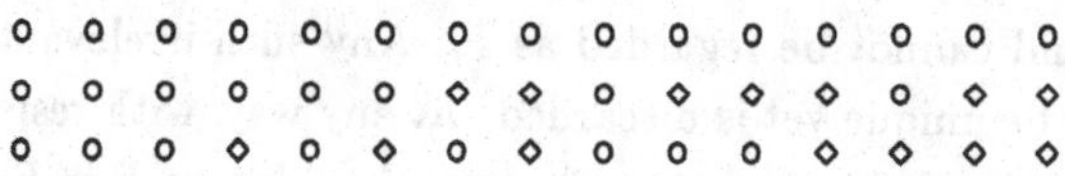

```
o o o o o o o o o o o o o o o
o o o o o o ◇ ◇ o ◇ ◇ ◇ o ◇ ◇
o o o ◇ o ◇ o ◇ o o o ◇ ◇ ◇ ◇
```

Fig. 4: The positions of known and unknown syndrome values.

For $t = 5$, which is by 1 less than the error correction capability, we have four kinds of point sets Δ of echelon form with $\#\Delta = t$ as shown in Figs. 5a, 5b, 5c, and 5d.

(a) For the first type of Δ (Fig. 5a), since $\tilde{\Delta}(F) \subset U$, we can possibly obtain F unambiguously by applying the 2D B.M. algorithm to the syndrome subarray s^p, $p = (0,11)$, with respect to $(1,0)_y$-order.

(b) For the second type of Δ (Fig. 5b), among $F = \{f, f', f''\}$, we possibly obtain $f = y^4 + f_{10}x + \sum_{i=0}^{3} f_{0i}y^i$ unambiguously by applying the 2D B.M. algorithm to the syndrome subarray s^p, $p = (0,8)$, with respect to $(4,1)_y$-order. By using f, we calculate s_{16} and then $s_{23} (= (s_{16})^8)$. Consequently, we can obtain F unambiguously by applying the 2D B.M. algorithm up to $q = (1,7)$.

(c) For the third type of Δ (Fig. 5c), among $F = \{f, f', f''\}$, we possibly obtain $f = y^3 + f_{11}xy + f_{10}x + \sum_{i=0}^{2} f_{0i}y^i$ unambiguously by applying the 2D B.M. algorithm to the syndrome subarray s^p, $p = (0,6)$, with respect to $(2,1)_y$-order. By using f, we calculate s_{23} and then obtain F unambiguously by applying the 2D B.M. algorithm up to $q = (1,5)$.

(d) For the fourth type of Δ (Fig. 5d), among $F = \{f, f', f''\}$, we possibly obtain $f' = xy + f_{20}x^2 + f_{01}y + f_{10}x + f_{00}$ unambiguously by applying the 2D B.M. algorithm to the syndrome subarray s^p, $p = (0,4)$, with respect to $(1,1)_y$-order. By using f', we calculate s_{23} and then obtain F unambiguously by applying the 2D B.M. algorithm up to $q = (6,0)$.

(e) For the fifth type of Δ (Fig. 5e), since $\tilde{\Delta}(F) \subset U$, we can possibly obtain F unambiguously by applying the 2D B.M. algorithm to the syndrome subarray s^p, $p = (0,4)$, with respect to $(1,1)_y$-order, while we can skip the points $\notin \tilde{\Delta}$ on which the syndrome values were unknown.

In this case, we can correct any error pattern of size 5 or less by applying the 2D B.M. algorithm in parallel with respect to the four kinds of total orders, i.e., $(1,0)_y$-, $(4,1)_y$-, $(2,1)_y$- and $(1,1)_y$-orders. (In fact, most of 6-error-patterns can be corrected.)

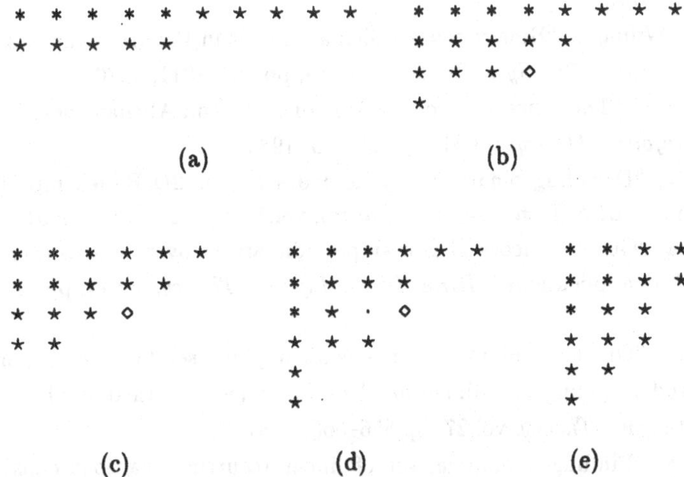

Fig. 5: Point sets Δ of echelon form and thier ensuring sets $\tilde{\Delta}$.

From this approach we can decode several binary 2D cyclic codes treated by Jensen [3] and some other similar codes.

4. Conclusion

We have proposed a method of decoding 2D cyclic codes efficiently by using the 2D B.M. algorithm in parallel. By this method we can decode some 2D cyclic codes which cannot be decoded by the method of [4] without trial and error and which cannot be decoded efficiently by any other previous method. Up to now we have not finished to investigate the extent of applicability of our method yet. Furthermore, at present, we have no exact estimate of computational complexity corresponding to several types of Δ with $\#\Delta = t$. In such a case, we can get some reduction of complexity, which is proportional to the number of relevant total orders. Comparison of complexity with the decoding method for generalized concatenated codes is also left open, while the latter is $\mathcal{O}(n^2 \log n)$ which seems to be much larger.

Acknowledgment

The author's thanks are due to Mr. Norifumi Kamiya of the NEC Corporation, who supplied him with an implementation of the 2D B.M. algorithm, and also to his student Mr. Yasuhiro Honda, who calculated examples of decoding some 2D cyclic codes.

REFERENCES

[1] H. Imai, "A theory of two-dimensional cyclic codes," *Inform. Control*, vol.34, pp.1–21, 1977.

[2] F.J. MacWilliams, "Binary codes which are ideals in the group algebra of an Abelian group," *Bell Syst. Tech. J.*, vol.49, pp.987–1011, 1970.

[3] J.M. Jensen, "The concatenated structure of cyclic and Abelian codes," *IEEE Trans. Inform. Theory*, vol.31, pp.788-793, 1985.

[4] S. Sakata, "Decoding binary 2D cyclic codes by the 2D Berlekamp-Massey algorithm," *IEEE Trans. Inform. Theory*, vol.37, pp.1200-1203, 1991.

[5] S. Sakata, "General theory of doubly periodic arrays over an arbitrary finite field and its applications," *IEEE Trans. Inform. Theory*, vol.24, pp.719–730, 1978.

[6] S. Sakata, "On determining the independent point set for doubly periodic arrays and encoding two-dimensional cyclic codes and their duals," *IEEE Trans. Inform. Theory*, vol.27, pp.556–565, 1981.

[7] S. Sakata, "Finding a minimal set of linear recurring realtions capable of generating a given finite two-dimensional array," *J. Symbolic Comp.*, vol.5, pp.321–337, 1988.

[8] K.P. Schemmel, "An extension of Buchberger's algorithm to compute all reduced Gröbner bases of a polynomial ideal," *Proc. EUROCAL'87*, pp.300-310, 1987.

[9] G.L. Feng and K.K. Tzeng, "Decoding cyclic codes and BCH codes up to actual minimum distance using nonrecurrent syndrome dependence relations," *IEEE Trans. Inform. Theory*, vol.37, pp.1716–1723, 1991.

APPENDIX

A short list of notation

$\tilde{K} := GF(2^M)$: the extension of the binary field $K := GF(2)$.

$D, \bar{D}$: the representatives of cyclotomic cosets for the nonzeros and zeros of a 2D cyclic code, respectively.

U: the positions of all zeros of C.

I_e: the error locator ideal of an error pattern.

F_e: a Gröbner basis of the error locator ideal I_e.

$I(u)$: the characteristic ideal of a DP array u.

$\leq_T$: the total order over the 2D integral lattice Z_+^2.

$\leq_P$: the partial order over the 2D integral lattice Z_+^2.

$f[u]_p = 0$: the linear recurring relation represented by a bivariate polynomial $f \in K[x]$ (resp., $\in \tilde{K}[x]$) for a 2D array u over K (resp., over $\tilde{K}$).

$deg(f)$: the degree of a polynomial f, i.e., the maximum (with respect to $\leq_T$) exponent pair of the nonzero terms of f.

$\Sigma^p := \{q \in Z_+^2 | q <_T p\}$.

$\Gamma_c := \{q \in Z_+^2 | q \leq_P c\}$.

$\Sigma_s := \{q \in Z_+^2 | q \geq_P s\}$.

S, C: a pair of nondegenerate subsets of Z_+^2.

$\Gamma_C := \cup_{c \in C} \Gamma_c$, $\Sigma_S := \cup_{s \in S} \Sigma_s$: subsets of Z_+^2 such that $\Gamma_C \cup \Sigma_S = Z_+^2$ and $\Gamma_C \cap \Sigma_S = Z_+^2$ for a pair of nondegenerate subsets S and C.

$\Delta = \Gamma_C = Z_+^2 \setminus \Sigma_S$: a point set of echelon form.

$s + \Delta := \{s + q | q \in \Delta\}$.

$\tilde{\Delta} = \Delta \cup (\cup_{s \in S}(s + \Delta))$: the ensuring set of Δ.

u^p: the restriction of an array u within Σ^p.

F: a minimal polynomial set of u^p.

$\Delta(F)$: the independent point set of u^p.

$\tilde{\Delta}(F)$: the ensuring set of the independent point set $\Delta(F)$.

THE POLYNOMIAL OF CORRECTABLE PATTERNS

N. Sendrier
INRIA, Rocquencourt, Le Chesnay, France

Abstract

We present here some new combinatoric tools to evaluate the performances of a decoding algorithm correcting either errors alone, either errors and erasures simultaneously. We show how the newly defined *polynomials of correctable patterns* and *polynomials of miscorrected patterns* can be connected with the correction an miscorrection probabilities in a given symmetric memoryless channel with or without erasures.

We propose a definition of a *decoding algorithm* that fits to symmetric channels with errors and erasures. In particular we define the closed algorithms: for instance the Berleykamp-Massey algorithm for BCH codes. For this class of algorithm, we are able to compute the polynomials of correctable and miscorrected patterns. Thus we have a short and easily computed formula for the probabilities of correction and miscorrection of any code using a closed decoding algorithm.

1 Introduction

Our purpose is to design some combinatoric tools that could help the evaluation of a coding design through a noisy channel whose characteristics are known. Mostly we will consider a q-ary symmetric memoryless error channel with or without erasures.

We define polynomials of correctable, uncorrectable and miscorrected patterns for both errors alone and errors and erasures together.

The polynomial of correctable patterns first appeared in [2] and the polynomials of uncorrectable and miscorrected patterns in [12]. However their definition is underlying in many papers [6, 14, 9, 8, 7, 11, 3]. They are defined, for a given decoding algorithm, as the Hamming weight enumerators of all three possible class of error patterns: those who are correctable, those leading to a failure and, the most grievious for applications: those leading to a miscorrection (i.e. the decoded word is different from the transmitted one).

In §2 we define the extended Hamming distance that takes in account the erasures, and which allows a definition of a decoding algorithm that fits to a symmetric error and erasure channel. Using the erasure weight and the error weight, we define in §3, with two variables, the polynomials of correctable, uncorrectable and miscorrected patterns. In this context, the case of error correction can be viewed as a particular case of error and erasure correction.

In the case of a transmission channel such that all patterns of same weight are equiprobable, the knowledge of the three polynomials described above gives a sufficient information for evaluating the performances of the algorithm (for an error and erasure channel "same weight" means same error weight and same erasure weigth). In particular for a q-ary memoryless symmetric error channel with or without erasures, we show in §4 the connection between the polynomials of correctable, uncorrectable and miscorrected patterns and the probabilities of correction, failure and miscorrection.

As an application, we consider in §5 the case of a closed decoding algorithm (Definition 3) for a code whose weight distribution is known. In this case, using generating function techniques, we give a new proof of the exact formula for the probability of miscorrection through an error channel [6, 8, 3], and the exact formula for the probability of miscorrection through an error and erasure channel.

2 Decoding algorithms

In all this paper $C[q; n, M, d]$ will denote a code of length n, cardinality M and minimum distance d over an alphabet A of size q. Let d_H and w_H denote Hamming distance and Hamming weight over A^n.

2.1 Transmission channel

A transmission channel is entirely defined by the knowledge of the triple $(A, B, P_{B|A})$ where A is the input alphabet, B the output alphabet and $P_{B|A}$ the conditional probability law of B knowing A. The channel is discrete if A and B are finite.

A q-ary symmetric channel is a discrete channel with $B = A$, $|A| = q$ and such that $P(a|a)$ has the same value for all a in A, and $P(a|a')$, $a \neq a'$ has the same value for all a, a' in A. We will denote $\mathcal{T}_A$ this channel.

A q-ary symmetric error and erasure channel is a discrete channel with $B = \tilde{A} = A \cup \{\infty\}$, $|A| = q$, having the same properties as the q-ary symmetric channel with additionally: $P(\infty|a)$ has the same value for all a in A. We will denote $\mathcal{T}_{\tilde{A}}$ this channel.

2.2 Extended Hamming distance

In an error and erasure channel, the output alphabet $\tilde{A}$ is equal to A plus an additionnal letter, denoted ∞, which represents an erasure. We wish to define a distance over $\tilde{A}^n$ such that its restriction to A^n is the Hamming distance.

Definition 1 *The* **extended Hamming distance** *is defined for all* $x = (x_1, \ldots, x_n)$ *and all* $y = (y_1, \ldots, y_n)$ *in* $\tilde{A}^n$ *by:*

$$\Delta_H(x, y) = \sum_{i=1}^{n} \delta_H(x_i, y_i),$$

where for all (a, b) *in* $\tilde{A}^2$: $\delta_H(a, b) = \begin{cases} 0 & \text{if } a = b \\ \frac{1}{2} & \text{if } a \neq b \text{ and } a = \infty \text{ or } b = \infty \\ 1 & \text{if } a \neq b,\ a \neq \infty \text{ and } b \neq \infty \end{cases}$

The **extended Hamming weight** *of an element* x *of* $\tilde{A}^n$*, denoted* $\Omega_H(x)$ *is defined to be its extended Hamming distance to the all zero vector.*

Of course Δ_H is a distance.

Remark 1 The metric Δ_H restricted to A^n coincide with the Hamming metric. In particular, the minimum distance of a code over A is the same for both usual and extended Hamming distance.

The weight $1/2$ given to the symbol ∞ is the smallest such that Δ_H remains a distance. Furthermore, it corresponds to the intuitive idea that an erasure correction "costs" half the price of an error correction.

Definition 2 *The* **erasure weight** *of a word* x *in* $\tilde{A}^n$*, denoted* $\rho_H(x)$*, is the number of coordinates of* x *equal to* ∞*. The* **error weight** *of* x*, denoted* $\nu_H(x)$ *is the number of coordinates of* x *different from zero and* ∞*.*

Remark 2 For all x in $\tilde{A}^n$, $\Omega_H(x) = \nu_H(x) + \dfrac{1}{2}\rho_H(x)$.

2.3 Decoding algorithm

Let $\mathcal{T}$ be an $(A, B, P_{B|A})$ transmission channel, we assume that $A \subset B$. We propose here a definition of a decoding algorithm for a given transmission channel as well as its most desirable properties. The code that we consider is not necessarely linear.

Definition 3 *A B-decoding* **algorithm** *for C is a mapping γ from B^n to $C \cup \{\infty\}$ such that for all x in C, $\gamma(x) = x$.*

Let D be a distance defined on B^n. The B-decoding algorithm γ for C is said to be **bounded** *by an integer e for the distance D if for all y in $\tilde{A}^n$ and all x in C,*

$$D(y,x) < \frac{e}{2} \Rightarrow \gamma(y) = x.$$

If also $\gamma(y) = x \neq \infty \Rightarrow D(y,x) < e/2$, then γ is said to be **closed** *by e (or e-closed).*

We use the same symbol ∞ for both decoding failure and erasure. Generaly the context will be sufficient to suppress this ambiguity.

2.4 Additive alphabet

We suppose now that $(A,+)$ is an abelian group. We will also denote by "+" the component-wise addition over A^n. We assume that the additive law can be extended in an internal law for the output alphabet B. This will obviously be possible if $B = A$. It is also possible for $B = \tilde{A} = A \cup \{\infty\}$, by saying that ∞ is an absorbing element.

Definition 4 *The B-decoding algorithm γ is said to be C-***additive** *if for all y in B^n and all x in C, $\gamma(y+x) = \gamma(y) + x$ (with the convention $\infty + x = \infty$).*

A pattern y in B^n is said to be **correctable** *by γ if for all x in C, $\gamma(y+x) = x$.*

Remark 3 Let $T_{\tilde{A}}$ be the q-ary symmetric error and erasure channel. Let γ be a $\tilde{A}$-decoding algorithm for C bounded by e for the extended Hamming distance:

1. e is lower or equal to the minimum distance of C,
2. any pattern y in $\tilde{A}^n$ such that $2\nu_H(y) + \rho_H(y) < e$ is correctable by γ,
3. if γ is closed, then it is C-additive.

Remark 4 If γ is C-additive and $0 \in C$, then the pattern y is correctable by γ if and only if $\gamma(y) = 0$, where 0 denotes the all-zero vector. The notion of C-additiveness for a decoding algorithm will fit to a linear code or a regular [4] non-linear code: when the code looks the same around any codeword, we want the decoding algorithm to behave accordingly. Viewing this property as an additive one is restrictive but convenient. Furthermore, for most practical codes we have an additive law for the alphabet.

3 Polynomial of correctable patterns

Let $T_{\tilde{A}}$ be the q-ary symmetric error and erasure channel, and let γ be a C-additive $\tilde{A}$-decoding algorithm for $C[q;n,M,d] \subset A^n$. We suppose $0 \in C$.

The definition of the extended weight enumerator $\mathcal{W}_E(X,Y)$ of a subset E of $\tilde{A}^n$ is given by Definition A.4 in the Appendix. Each monomial of degree i in X and j in Y "counts" the number of elements of E of error weight i and erasure weight j.

Definition 5 *The* **polynomial of correctable patterns** *of γ is the extended weight enumerator of the set $\left\{y \in \tilde{A}^n \mid \gamma(y) = 0\right\}$. The* **polynomial of uncorrectable patterns** *of γ is the extended weight enumerator of the set $\left\{y \in \tilde{A}^n \mid \gamma(y) = \infty\right\}$. The* **polynomial of miscorrected patterns** *of γ is the extended weight enumerator of the set $\left\{y \in \tilde{A}^n \mid \gamma(y) \in C \setminus \{0\}\right\}$.*

Proposition 1 *Let $P_0(X,Y)$, $P_1(X,Y)$ and $P_2(X,Y)$ denote the polynomials of correctable, uncorrectable and miscorrected patterns of γ. We have:*

$$P_0(X,Y) + P_1(X,Y) + P_2(X,Y) = (1 + Y + (q-1)X)^n. \tag{1}$$

proof: The extended weight enumerator of $\tilde{A}$ is $(1+Y+(q-1)X)$, thus the extended weight enumerator of $\tilde{A}^n$ is $(1+Y+(q-1)X)^n$. And since any pattern of $\tilde{A}^n$ is either correctable, uncorrectable or "miscorrectable" by γ, we have (1) □

A particular case: errors alone. If γ is a A-decoding algorithm for C, we define the polynomials of correctable, uncorrectable and miscorrected patterns respectively as the weight enumerators (Definition A.3 in Appendix) of $\{y \in A^n \mid \gamma(y) = 0\}$, $\{y \in A^n \mid \gamma(y) = \infty\}$ and $\{y \in A^n \mid \gamma(y) \in C \setminus \{0\}\}$. These polynomials are univariate, and the Proposition 1 applies with $Y = 0$.

4 Relationship with the error probabilities

We consider the q-ary symmetric error and erasure channel $\mathcal{T}_{\tilde{A}}$. The transition probability from an element of A to another is denoted p_t, and to an erasure is denoted p_∞. In such a channel the probability of correct transmission of a symbol of A is $p_c = 1 - (q-1)p_t - p_\infty$. In the case of errors alone we have $p_\infty = 0$.

Definition 6 *Let $P(X,Y)$ be the extended weight enumerator of a subset of $\tilde{A}^n$. We call* **homogenized polynomial** *of $P(X,Y)$, denoted $P^*(X,Y,Z)$, the polynomial with three variables $P^*(X,Y,Z) = Z^n P(X/Z, Y/Z)$.*

We consider γ a C-additive $\tilde{A}$-decoding algorithm for C. Let $P_0(X,Y)$, $P_1(X,Y)$ and $P_2(X,Y)$ denote its polynomials of correctable, uncorrectable and miscorrected patterns, and let $P_0^*(X,Y,Z)$, $P_1^*(X,Y,Z)$ and $P_2^*(X,Y,Z)$ be their homogenized polynomials.

Proposition 2 *Any codeword of C transmitted through the channel $\mathcal{T}_{\tilde{A}}$ will be decoded correctly by γ with probability $P_{cor}(\mathcal{T}_{\tilde{A}}) = P_0^*(p_t, p_\infty, p_c)$, it will be miscorrected with probablity $P_{err}(\mathcal{T}_{\tilde{A}}) = P_2^*(p_t, p_\infty, p_c)$, and the decoding will fail with probablity $P_{fail}(\mathcal{T}_{\tilde{A}}) = P_1^*(p_t, p_\infty, p_c)$.*

proof: In channel $T_{\tilde{A}}$, the probablity for a given pattern $y \in \tilde{A}^n$ to occur is equal to:

$$p_t^{\nu_H(y)} p_\infty^{\rho_H(y)} p_c^{n-\nu_H(y)-\rho_H(y)} = p_c^n \left(\frac{p_t}{p_c}\right)^{\nu_H(y)} \left(\frac{p_\infty}{p_c}\right)^{\rho_H(y)}.$$

By summing these probabilities over all patterns correctable, uncorrectable or miscorrected by γ, we obtain the results. □

Of course similar results can be derived for errors alone. Mainly the results above apply with $p_\infty = 0$.

5 Closed algorithms

5.1 Error correction

When a A-decoding algorithm γ is e-closed, with $e = 2t+1$, then obviously the polynomial of correctable patterns is $P_0(X) = \sum_{i=0}^{t} \binom{n}{i}(q-1)^i X^i$. We will show that if the weight distribution of the code is known then the polynomial of miscorrected patterns is known as well.

First we give the following lemma which is proved in Appendix.

Lemma 1 (P. Camion [1]) *The weight enumerator of a sphere of radius r centered on a word of weight w is equal to:*

$$S_{w,r}(X) = \sum_{0 \le i+j \le r} \binom{w}{i}\binom{n-w}{j}(1+(q-2)X)^i (q-1)^j X^{w-i+j}.$$

Proposition 3 *Let γ be a $(2t+1)$-closed A-decoding algorithm. Let $P_0(X)$ and $P_2(X)$ be its polynomials of correctable and miscorrected patterns.*

We have $P_0(X) = \sum_{i=0}^{t} \binom{n}{i}(q-1)^i X^i$ and $P_2(X) = \sum_{w>0} A_w S_{w,t}(X)$, where A_w is the number of words of weight w in C, and $S_{w,t}(X)$ the weight enumerator of a sphere of radius t centered on a word of weight w.

proof: Let $\mathcal{S}(c,t)$ be the sphere of radius t centered on the codeword c, we have $\mathcal{S}(c,t) = \{m \in A^n \mid \gamma(m) = c\}$ since γ is $(2t+1)$-closed. Thus

$$\{m \in A^n \mid \gamma(m) \in C \setminus \{0\}\} = \bigcup_{c \in C\setminus\{0\}} \mathcal{S}(c,t).$$

From the first point of Remark 3, we have $2t+1 \le d$. So the spheres of radius t centered on all codewords are disjoint. From Proposition A.1 in the Appendix and the definition of the polynomial of miscorrected patterns we have that

$$P_2(X) = \sum_{c \in C\setminus\{0\}} W_{\mathcal{S}(c,t)}(X) = \sum_{w>0} A_w S_{w,t}(X).$$

□

5.2 Error and erasure correction

Let γ be a $\tilde{A}$-decoding algorithm. As for error correction, when γ is closed, then the polynomial of correctable patterns is know, and furthermore we will show that if the weight distribution of the code is known then the polynomial of miscorrected patterns can be computed.

First we give the following lemma which is proved in Appendix.

Lemma 2 *The extended weight enumerator of a sphere of radius $r \in \frac{1}{2}\mathbb{Z}$ centered on a word of A^n of weight w is equal to:*

$$S_{w,r}(X,Y) = \sum_{i=0}^{2r}[Z^i]Q(X,Y,Z),$$

where $Q(X,Y,Z) = (X + YZ + Z^2 + (q-2)XZ^2)^w(1 + YZ + (q-1)XZ^2)^{n-w}$, and $[Z^i]Q(X,Y,Z)$ is the coefficient of Z^i in $Q(X,Y,Z)$ (i.e. a polynomial with variables X and Y).

The proof is given in Appendix as well as the Proposition A.3 which gives an explicit expression of the coefficients $[Z^i]Q(X,Y,Z)$.

Proposition 4 *We assume that γ is an e-closed $\tilde{A}$-decoding algorithm for C, with $e = 2t+1$, $t \in \frac{1}{2}\mathbb{Z}$. Let $P_0(X,Y)$ and $P_2(X,Y)$ be its polynomials of correctable and miscorrected patterns. We have:*

$$P_0(X,Y) = \sum_{2i+j<e}\binom{n}{i}\binom{n-i}{j}(q-1)^i X^i Y^j, \quad \text{and} \quad P_2(X,Y) = \sum_{w>0} A_w S_{w,t}(X,Y),$$

where A_w is the number of words of weight w in C, and $S_{w,t}(X,Y)$ is the extended weight enumerator of a sphere of radius t centered on a word of weight w.

proof: Similar to the proof of Proposition 3 □

Example 1 : $C = H(8,4,4)$ the binary extended Hamming code. Let $\tilde{\mathbb{F}}_2 = \mathbb{F}_2 \cup \{\infty\}$. We choose a $\tilde{\mathbb{F}}_2$-decoding algorithm closed by the minimum distance 4, by a table lookup for instance. For such an algorithm we have

$$\begin{aligned}
P_0(X,Y) &= 56\,XY + 8\,X + 56\,Y^3 + 28\,Y^2 + 8\,Y + 1\\
P_1(X,Y) &= 28\,X^6 + 168\,X^5Y^2 + 70\,X^4Y^4 + 224\,X^4Y^3 + 336\,X^4Y^2 + 56\,X^4 + 56\,X^3Y^5\\
&\quad + 28\,X^2Y^6 + 168\,X^2Y^5 + 420\,X^2Y^4 + 280\,X^3Y^4 + 224\,X^3Y^3 + 336\,X^3Y^2\\
&\quad + 224\,X^2Y^3 + 336\,X^2Y^2 + 28\,X^2 + 8\,XY^7 + 56\,XY^6 + 168\,XY^5 + 280\,XY^4\\
&\quad + 224\,XY^3 + 168\,XY^2 + 168\,XY^2 + Y^8 + 8\,Y^7 + 28\,Y^6 + 56\,Y^5 + 70\,Y^4\\
P_2(X,Y) &= X^8 + 8\,X^7Y + 8\,X^7 + 28\,X^6Y^2 + 56\,X^6Y + 56\,X^5Y^3 + 168\,X^5Y + 56\,X^5\\
&\quad + 56\,X^4Y^3 + 84\,X^4Y^2 + 280\,X^4Y + 14\,X^4 + 336\,X^3Y^3 + 224\,X^3Y^2\\
&\quad + 280\,X^3Y + 56\,X^3 + 336\,X^2Y^3 + 84\,X^2Y^2 + 168\,X^2Y + 56\,XY^3
\end{aligned}$$

Example 2 : $C = RS(15,8,8)$ the Reed-Solomon code of length 15, dimension 8 and minimum distance 8 over $\mathbb{F}_{16}$. C is a MDS code, so its weight enumerator is known [10, p. 319] and we consider a $\tilde{\mathbb{F}}_{16}$-decoding algorithm closed by 8 (exactly all patterns of extended Hamming weigth $3\frac{1}{2}$ or lower are corrected), as the Berlekamp-Massey algorithm for errors and erasures. For such an algorithm we have

$$\begin{aligned} P_0(X,Y) = \; & 1 + 15\,Y + 105\,Y^2 + 225\,X + 455\,Y^3 + 3150\,XY + 1365\,Y^4 \\ & + 20475\,XY^2 + 23625\,X^2 + 3003\,Y^5 + 81900\,XY^3 + 307125\,X^2Y \\ & + 5005\,Y^6 + 225225\,XY^4 + 1842750\,X^2Y^2 + 1535625\,X^3 \\ & + 6435\,Y^7 + 450450\,XY^5 + 6756750\,X^2Y^3 + 18427500\,X^3Y \end{aligned}$$

The polynomials of uncorrectable and miscorrected patterns can be computed as well, their expression are too large to be relevant here. However they can be computed in a relatively short time: using a program written in MAPLE langage on a Sun SparcStation2, the three polynomials of the above code are computed in about one second, and for the code RS(63,47,17) over $\mathbb{F}_{64}$ in about 70 seconds.

Appendix

A.1 Ordinary generating functions

Most of the combinatorial material used here can be found in [5, Chap. 2]. The presentation of is derived from [13, §1].

Definition A.1 *Let $\mathcal{U}$ be a finite or countable set, and let s be an integer valued function over $\mathcal{U}$, called the weight function, such that for each $n \geq 0$ the number U_n of elements in $\mathcal{U}$ of weight n is finite.* The ordinary generating function *(OGF)* $U(X)$ *of* $(\mathcal{U}, s)$ *is defined by*

$$U(X) = \sum_{n \geq 0} U_n X^n = \sum_{u \in \mathcal{U}} X^{s(u)}.$$

We can give a slightly more general definition, although it is conceptualy equivalent.

Definition A.2 *Let $\mathcal{U}$ be a finite or countable set, and let s_i, for $1 \leq i \leq k$, be k integer valued functions over $\mathcal{U}$, called weight functions, such that for each $\bar{n} = (n_1, \ldots, n_k) \in \mathbf{N}^k$ the number $U_{\bar{n}}$ of elements in $\mathcal{U}$ of weight $\bar{n}$ is finite, where the weight of an element u in $\mathcal{U}$ is the k-tuple $\bar{s}(u) = (s_1(u), \ldots, s_k(u))$. The k* variables ordinary generating function $U(\bar{X})$ *of* $(\mathcal{U}, \bar{s})$ *is defined by*

$$U(\bar{X}) = \sum_{\bar{n} \in \mathbf{N}^k} U_{\bar{n}} \bar{X}^{\bar{n}} = \sum_{u \in \mathcal{U}} \bar{X}^{\bar{s}(u)},$$

where $\bar{X} = (X_1, \ldots, X_k)$ and $\bar{X}^{\bar{n}} = X_1^{n_1} \ldots X_k^{n_k}$.

In the following theorem we refer to the more general definition of an OGF: a set with an arbitrary (finite) number of weight functions.

Furthermore, we assume that, for any of these weight functions, the weight of an element of a disjoint union is inherited from its weight in its original domain, and the weight of an element of a cartesian product is the sum of the weights of its component.

Theorem A.1 *The OGF of a disjoint union is equal to the sum of the OGFs of its operands, and the OGF of a cartesian product is equal to the product of the OGFs of its operands:*

$$\mathcal{W} = \mathcal{U} \cup \mathcal{V} \text{ with } \mathcal{U} \cap \mathcal{V} = \emptyset \Rightarrow W(\bar{X}) = U(\bar{X}) + V(\bar{X}) \quad (2)$$

$$\mathcal{W} = \mathcal{U} \times \mathcal{V} \Rightarrow W(\bar{X}) = U(\bar{X})V(\bar{X}) \quad (3)$$

A.2 Weight enumerators

Let w_H be the Hamming weight over A^n, where n is a positive integer.

Definition A.3 *For all subset E of A^n, we call* **weight enumerator** *of E the ordinary generating function of (E, w_H): $\mathcal{W}_E(X) = \sum_{v \in E} X^{w_H(v)} \in \mathbb{Z}[X]$.*

Let ν_H and ρ_H be the error and erasure weights over $\tilde{A}^n$, with $\tilde{A} = A \cup \{\infty\}$.

Definition A.4 *For all subset E of $\tilde{A}^n$, we call* **extended weight enumerator** *of E the ordinary generating function of $(E, (\nu_H, \rho_H))$: $\mathcal{W}_E(X, Y) = \sum_{v \in E} X^{\nu_H(v)} Y^{\rho_H(v)} \in \mathbb{Z}[X, Y]$.*

Remark 5 Let E be a subset of A^n. The extended weight enumerator of E considered as a subset of $\tilde{A}^n$ is equal to the weight enumerator of E as a subset of A^n. Thus all the following results, true for extended weight enumerators are also true for weight enumerators.

We define the cartesian product $E \times E'$ of a subset E of $\tilde{A}^n$ and a subset E' of $\tilde{A}^{n'}$ as the subset of $\tilde{A}^{n+n'}$ composed of all words produced by concatenation of a word of E and a word of E'. Clearly the error and erasure weights are consistent with this definition, that is ν_H and ρ_H verifie the assumptions made in Theorem A.1 for the weight functions. The two following results are then a direct consequence of this theorem.

Proposition A.1 *Let E and E' be two subsets of $\tilde{A}^n$ such that $E \cap E' = \emptyset$. Then the extended weight enumerator of $E \cup E'$ is $\mathcal{W}_{E \cup E'}(X, Y) = \mathcal{W}_E(X, Y) + \mathcal{W}_{E'}(X, Y)$.*

Proposition A.2 *Let E be a subset of $\tilde{A}^n$ and E' be a subset of $\tilde{A}^{n'}$. The subset $E \times E'$ of $\tilde{A}^{n+n'}$ has extended weight enumerator $\mathcal{W}_{E \times E'}(X, Y) = \mathcal{W}_E(X, Y)\mathcal{W}_{E'}(X, Y)$.*

Proofs omited.

A.3 Proof of Lemma 1

Lemma 1 (P. Camion [1]) *The weight enumerator of a sphere of radius r centered on a word of weight w is equal to:*

$$S_{w,r}(X) = \sum_{0 \le i+j \le r} \binom{w}{i} \binom{n-w}{j} (1 + (q-2)X)^i (q-1)^j X^{w-i+j}.$$

proof: Let c be an element of A^n of weight w, and $\mathcal{S}_{c,r}$ be the sphere of radius r centered on c. Let $U(X,Y)$ be the two variables OGF of A^n with weight functions $y \mapsto w_H(y)$ and $y \mapsto d_H(c,y)$,

$$U(X,Y) = \sum_{y \in A^n} X^{w_H(y)} Y^{d_H(c,y)},$$

and let $S(X,Y)$ be the two variables OGF of $\mathcal{S}_{c,r}$ with the same two weight functions,

$$S(X,Y) = \sum_{y \in \mathcal{S}_{c,r}} X^{w_H(y)} Y^{d_H(c,y)}.$$

We have $\mathcal{W}_{\mathcal{S}_{c,r}}(X) = S(X,1)$, where $\mathcal{W}_{\mathcal{S}_{c,r}}(X)$ is the weight enumerator of $\mathcal{S}_{c,r}$, and since $\mathcal{S}_{c,r} = \{y \in A^n \mid d_H(c,y) \le r\}$, we have $S(X,Y) = U(X,Y) \bmod Y^{r+1}$. This implies that

$$\mathcal{W}_{\mathcal{S}_{c,r}}(X) = \sum_{k=0}^{r} [Y^k] U(X,Y), \tag{4}$$

where $[Y^k]U(X,Y)$ denotes the coefficient of Y^k in $U(X,Y)$.

1. We will first establish that

   $$U(X,Y) = (X + Y + (q-2)XY)^w (1 + (q-1)XY)^{n-w}. \tag{5}$$

   Let $c = (c_1, \ldots, c_n)$, for all i, $1 \le i \le n$, the OGF of A with the two weight functions $a \mapsto w_H(a)$ and $a \mapsto d_H(c_i, a)$ is equal to

   $$A(X,Y) = \begin{cases} X + Y + (q-2)XY & \text{if } c_i \neq 0 \\ 1 + (q-1)XY & \text{if } c_i = 0 \end{cases}.$$

   Since $d_H(c,y) = \sum_{i=1}^{n} d_H(c_i, y_i)$, for all $y = (y_1, \ldots, y_n)$ in A^n we can apply Theorem A.1 for the cartesian product $A^n = A \times \ldots \times A$, and since the Hamming weight of c is w, exactly w of the c_i's are $\neq 0$, thus we have (5).

2. For any ring R, if $f(Y)$, $g(Y)$ and $h(Y)$ are elements of $R[Y]$ such that $f(Y) = g(Y)h(Y)$, then $[Y^k]f(Y) = \sum_{i+j=k} [Y^i]g(Y)[Y^j]h(Y)$. With $R = \mathbb{Z}[X]$, we deduce from (4) and (5):

   $$\mathcal{W}_{\mathcal{S}_{c,r}}(X) = \sum_{i+j \le r} [Y^i]((X + Y + (q-2)XY)^w)[Y^j]((1 + (q-1)XY)^{n-w}). \tag{6}$$

   Since this weight enumerator only depends on r and w we will denote it $S_{w,r}(X)$.

3. From the binomial formula we have

$$[Y^i]((X+Y+(q-2)XY)^w) = \binom{w}{i} X^{w-i}(1+(q-2)X)^i, \tag{7}$$

and

$$[Y^j]((1+(q-1)XY)^{n-w}) = \binom{n-w}{j}(q-1)^j X^j. \tag{8}$$

Finally from (6), (7) and (8) we obtain

$$S_{w,r}(X) = \sum_{0 \le i+j \le r} \binom{w}{i}\binom{n-w}{j}(1+(q-2)X)^i (q-1)^j X^{w-i+j}.$$

□

A.4 Proof of Lemma 2

Lemma 2 *The extended weight enumerator of a sphere of radius $r \in \frac{1}{2}\mathbb{Z}$ centered on a word of A^n of weight w is equal to:*

$$S_{w,r}(X,Y) = \sum_{s=0}^{2r} [Z^s] Q(X,Y,Z),$$

where $Q(X,Y,Z) = (X+YZ+Z^2+(q-2)XZ^2)^w (1+YZ+(q-1)XZ^2)^{n-w}$.

proof: Let c be an element of A^n of weight w, and $\mathcal{S}_{w,r}$ be the sphere of radius r in $\tilde{A}^n$ centered on c. Let $Q(X,Y,Z)$ be the three variables generating OGF of $\tilde{A}^n$ with weight functions $y \mapsto \nu_H(y)$, $y \mapsto \rho_H(y)$ and $y \mapsto 2\Delta_H(y,c)$,

$$Q(X,Y,Z) = \sum_{y \in \tilde{A}^n} X^{\nu_H(y)} Y^{\rho_H(y)} Z^{2\Delta_H(y,c)}.$$

The OGF $S(X,Y,Z)$ of $\mathcal{S}_{w,r}$ with the same weight functions is equal to

$$S(X,Y,Z) = \sum_{y \in \mathcal{S}_{w,r}} X^{\nu_H(y)} Y^{\rho_H(y)} Z^{2\Delta_H(y,c)}.$$

We have $\mathcal{W}_{\mathcal{S}_{c,r}}(X,Y) = S(X,Y,1)$, where $\mathcal{W}_{\mathcal{S}_{c,r}}(X,Y)$ is the extended weight enumerator of $\mathcal{S}_{w,r}$, and since $\mathcal{S}_{w,r} = \{y \in \tilde{A}^n \mid \Delta_H(y,c) \le r\}$, we have

$$S(X,Y,Z) = \sum_{s=0}^{2r} [Z^s] Q(X,Y,Z).$$

We now need to establish that

$$Q(X,Y,Z) = (X+YZ+Z^2+(q-2)XZ^2)^w (1+YZ+(q-1)XZ^2)^{n-w}.$$

Let $c = (c_1, \ldots, c_n)$, for all i, $1 \leq i \leq n$, the OGF of $\tilde{A}$ with the three weight functions $a \mapsto \nu_H(a)$, $a \mapsto \rho_H(a)$ and $a \mapsto \delta_H(c_i, a)$ is equal to

$$A(X,Y,Z) = \begin{cases} X + YZ + (1 + (q-2)X)Z^2 & \text{if } c_i \neq 0 \\ 1 + YZ + (q-1)XZ^2 & \text{if } c_i = 0 \end{cases}.$$

Since $\Delta_H(c,y) = \sum_{i=1}^n \delta_H(c_i, y_i)$, for all $y = (y_1, \ldots, y_n)$ in $\tilde{A}^n$ we can apply Theorem A.1 for the cartesian product $\tilde{A}^n = \tilde{A} \times \ldots \times \tilde{A}$, and thus

$$Q(X,Y,Y) = (X + YZ + (1 + (q-2)X)Z^2)^w (1 + YZ + (q-1)XZ^2)^{n-w}.$$

□

The coefficients $[Z^s]Q(X,Y,Z)$ are relatively easy to compute, it is not necessary to expand the polynomial $Q(X,Y,Z)$.

Proposition A.3 *For all integer s, $0 \leq s \leq 2r$, the coefficient $[Z^s]Q(X,Y,Z)$ is equal to*

$$\sum_{u+v=s} \sum_{2i \leq u} \sum_{2j \leq v} \binom{w}{i}\binom{w-i}{u-2i}\binom{n-w}{j}\binom{n-w-j}{v-2j}(q-1)^j X^{w-u+i+j} Y^{u-2i+v-2j}(1+(q-2)X)^i,$$

with the convention that $\binom{a}{b} = 0$ if $b < 0$ or $b > a$.

proof: We have by definition of $Q(X,Y,Z)$, $Q = Q_1 Q_2$, with $Q_1 = {q_1}^w$, $Q_2 = {q_2}^{n-w}$ and

$$\begin{cases} q_1(X,Y,Z) &= X + YZ + (1 + (q-2)X)Z^2 \\ q_2(X,Y,Z) &= 1 + YZ + (q-1)XZ^2 \end{cases}.$$

For any integer s, $0 \leq s \leq 2r$,

$$[Z^s]Q = \sum_{u+v=s} [Z^u]Q_1 [Z^v]Q_2. \tag{9}$$

We have:

$$[Z^u]Q_1 = [Z^u](q_1)^w = \sum_{2i+i'=u} \binom{w}{i}\binom{w-i}{i'}([Z^2]q_1)^i([Z^1]q_1)^{i'}([Z^0]q_1)^{w-i-i'}, \tag{10}$$

and

$$[Z^v]Q_2 = [Z^v](q_2)^{n-w} = \sum_{2j+j'=v} \binom{n-w}{j}\binom{n-w-j}{j'}([Z^2]q_2)^j([Z^1]q_2)^{j'}([Z^0]q_2)^{n-w-j-j'}. \tag{11}$$

From (9), (10) and (11) we deduce the result. □

Acknowledgement

The author wish to thank H.F. Mattson, Jr for his helpful advice.

References

[1] P. Camion. Private communication.

[2] P. Camion and J.-L. Politano. Evaluation of a coding design for a very noisy chanel. In G. Cohen and J. Wolfmann, editors, *Coding Theory and Applications*, number 388 in LNCS. Springer-Verlag, 1988.

[3] K. Cheung. More on the decoder error probability for Reed-Solomon codes. *IEEE Transaction on Information Theory*, 35:895–900, July 1989.

[4] P. Delsarte. *An Algebraic Approach to the Association Schemes of Coding Theory.* Thèse, Université Catholique de Louvain, June 1973.

[5] I. Goulden and D. Jackson. *Combinatorial Enumeration.* John Wiley & Sons, 1983.

[6] Z. Huntoon and A. Michelson. On the computation of the probability of post-decoding error events for block codes. *IEEE Transaction on Information Theory*, 23:399–403, May 1977.

[7] S. Jennings. On computing the performance probabilities of Reed-Solomon codes. In J. Calmet, editor, *Proceedings of the 3rd International Conference, AAECC-3*, number 229 in LNCS, pages 59–68. Springer Verlag, July 1985.

[8] T. Kasami and S. Lin. On the probability of undetected error for the maximum distance separable codes. *IEEE Transaction on Communication*, 32:998–1006, Sept. 1984.

[9] T. Klove. The probability of undetected error when a code is used for error correction and detection. *IEEE Transaction on Information Theory*, 30:388–392, Mar. 1984.

[10] F. MacWilliams and N. Sloane. *The Theory of Error Correcting Codes.* North-Holland, 1977.

[11] R. McEliece and L. Swanson. On the decoder error probability for Reed-Solomon codes. *IEEE Transaction on Information Theory*, 32:701–703, Sept. 1986.

[12] N. Sendrier. *Codes Correcteurs d'Erreurs à Haut Pouvoir de Correction.* Thèse de doctorat, Université Paris 6, Dec. 1991.

[13] J. Vitter and P. Flajolet. Analysis of algorithms and data structures. In J. van Leeuwen, editor, *Handbook of Theoretical Computer Science*, volume A: Algorithms and Complexity, chapter 9, pages 431–524. North Holland, 1990.

[14] J. Wolf, A. Michelson, and A. Levesque. On the probability of undetected error for linear block codes. *IEEE Transaction on Communication*, 30:317–324, Feb. 1982.

ON MAXIMUN LIKELIHOOD SOFT DECODING OF BINARY BLOCK CODES

J. Snyders
Tel Aviv University, Ramat Aviv, Israel

Abstract

For accomplishing maximum likelihood soft decision decoding of a binary linear block code, the search through codewords is replacable (as in the case of hard decoding) by a search through a coset of the code, aimed at identifying the most likely error pattern. The way to perform an inverse switch, from coset-search to code-search, is described. For some codes, a rather efficient and simple decoder is obtainable with the aid of such procedure.

[1]E-mail: snyders@eng.tau.ac.il
This work was supported by the Basic Research Foundation administered by the Israel Academy of Sciences and Humanities.

1 Introduction

Let C be an (n,k,d) binary linear block code of length n, dimension k and minimum distance d. Assume that a codeword $\mathbf{c} = (c_1, c_2, \cdots, c_n)$ is transmitted through a binary-input memoryless channel with transition probability (or, probability density) $P(r|c)$; $r \in \mathcal{R}, c \in GF(2)$, where $\mathcal{R}$ stands for the real line. Let $\mathbf{r} = (r_1, r_2, \cdots, r_n) \in \mathcal{R}^n$ be the received word. Denote by

$$\mathbf{v} = (v_1, v_2, \cdots, v_n) \tag{1}$$

the n-tuple of log likelihood ratios, i.e.,

$$v_j = log\frac{P(r_j|c_j = 0)}{P(r_j|c_j = 1)}; \quad j = 1, 2, \cdots, n, \tag{2}$$

and let

$$\mathbf{b} = (b_1, b_2, \cdots, b_n) \in GF(2)^n \tag{3}$$

be the corresponding n-tuple of the bit-by-bit hard detected version of the received word, i.e., $b_i = \text{sgn}(v_i)$ for $i = 1, 2, \cdots, n$ where $\text{sgn}(v) = 0$ for $v \geq 0$ and $\text{sgn}(v) = 1$ for $v < 0$.

For an n-tuple $\mathbf{p} = (p_1, p_2, \cdots, p_n)$ over GF(2), $\mathbf{p}'$ will denote the n-tuple obtained by changing each 0 and 1 entry into 1 and -1, respectively, i.e., $p_j' = (-1)^{p_j}$. We shall write $< \cdot, \cdot >$ for the inner product over $\mathcal{R}$. A superscript t will stand for transposition.

It is well known that a most likely codeword attains the following maximum

$$\max_{\mathbf{c} \in C} < \mathbf{c}', \mathbf{v} > . \tag{4}$$

One may view the decoding procedure indicated by (4) as follows. Consider a generator matrix G of C, and equip each column $\mathbf{g}_i$ of G with the (signed) *weight* v_i given by (2). Generate all the codewords with the aid of G, compute the weight $< \mathbf{c}', \mathbf{v} >$ of each codeword $\mathbf{c}$ and then perform the maximization. Let us call such procedure, of searching through C, type CS (code-search).

Alternatively, maximum likelihood (ML) decoding may be accomplished [2], [3], [6] by first calculating the syndrome $\mathbf{s}$ corresponding to $\mathbf{b}$,

$$\mathbf{s} = H\mathbf{b}^t, \tag{5}$$

where H is a check matrix of C. Then, unless $\mathbf{s} = \mathbf{0}$, a search is carried out to identify a *least* weighing error pattern $\mathbf{e} \in GF(2)^n$, i.e., a vector that satisfies

$$\mathbf{s} = H\mathbf{e}^t, \tag{6}$$

with the *weight* of $\mathbf{e}$ defined to be the sum of the entries of

$$\mathbf{w} = (w_1, w_2, \cdots, w_n); \quad w_j = |v_j|$$

over the support of $\mathbf{e}$. The decoded word $\mathbf{c}$ is given by

$$\mathbf{c} = \mathbf{b} + \mathbf{e}. \tag{7}$$

The latter procedure may be viewed as follows: each column $\mathbf{h}_j$ of H is equipped with the weight w_j, among all the sets of columns of H that sum up to $\mathbf{s}$ the set $\mathcal{F}$ with least weight is identified, and then the bits of $\mathbf{r}$ corresponding to $\mathcal{F}$ are inverted. Such a procedure, of searching through a coset of C, will be named type ES (error-search).

There exist fast methods for performing each of the two aformentioned types of searches. More precisely, a fast implementation means replacement of the exhaustive search through the code, or one of its cosets, by a nonexhaustive or different kind of search which, nonetheless, may be regarded as an implementation of the basic procedure. Wolf's implementation [1] of the Viterbi algorithm belongs to type CS, despite that the trellis is described there with the aid of H. Fast implemementations of type ES are described in [2], [3]. The fast methods (hence, all ML soft decoding methods) have limited practical applicability. If the method is based on a trellis representaion then the number of the nodes of the best trellis available for the code should not be excessive. Utilization of reduced lists of error patterns [3] is confined to codes with few types of cosets.

An approach to ML decoding that recently gained much interest is coset decoding, i.e., separate decoding of all cosets of some subcode of C, followed by matching of all the decoded words. There are various implementations of coset decoding [4], [5], [6], differing one from another in the tools used rather than conceptually. In general terms, the implementation of [6] of coset decoding approaches the problem initially as a CS-type search, then, following a contraction of a submatrix G_s of G, adopts an explicit ES-type search. An essentially identical implementation is described in [5] for the case that the contracted code, the one spanned by the contracted version of G_s, is either a single check-bit code or a universe code. The approach of [4], that relies on trellis representation, addresses the same particular case. By the aforementioned contraction we mean that all but one of a family of repeated columns of G_s are discarded; the weight of each deleted column is added to the weight of the remaining representative column, thereby preserving the value of the total weight corresponding to any codeword of G_s.

Within the framework of coset decoding one may encounter a situation where the search for error patterns is apparently complicated, and a switch backward to codeword search is useful. Even several changes between these two types of searches can be desirable, particularly for large codes. In Section 2 the way to perform the change from ES to CS will be described. An example is provided in Section 3.

2 Replacing error search by code search

Two basic methods of ML decoding were described in the previous section. That description may be regarded as an outline for replacing CS by ES, since the starting point is the more familiar method represented by (4). Now we shall procede in the reverse direction.

Consider a situation where we aim at finding the least weighing error pattern for some nonzero syndrome $\mathbf{s}$ and a given vector $\mathbf{w} = (w_1, w_2, \cdots, w_n)$ of n *nonnegative* weights. Suppose that a direct search is rather complex in terms of time or memory, which is indeed the case for most codes, and a code-search appears to be simpler.

Let $\mathbf{e} = (e_1, e_2, \cdots, e_n)$ be *any* error pattern satisfying

$$\mathbf{s} = H\mathbf{e}^t. \tag{8}$$

(One such error pattern is obtainable by hard decision decoding, but for our applications in the sequel even this simple procedure will be skipped.) Let

$$\mathbf{v} = (v_1, v_2, \cdots, v_n) \tag{9}$$

where now, unlike for (1), v_j is determined by

$$v_j = (-1)^{e_j} w_j. \tag{10}$$

Let a code search, with respect to the vector of signed weights $\mathbf{v}$, yield the optimal (maximum weighing) codeword $\mathbf{c}_o$. Then the optimal (least weighing) error pattern $\mathbf{e}_o$ is given by

$$\mathbf{e}_o = \mathbf{e} + \mathbf{c}_o. \tag{11}$$

Indeed, with $\mathbf{e}$ of (8) taking the role of $\mathbf{b}$ given by (3) and in view of (10), $\mathbf{v}$ of (9) is equivalent to $\mathbf{v}$ of (1) with respect to the code search. Since $\mathbf{c}_o$ attains the maximum inner product (4), it follows by (11) that $\mathbf{e}_o$ is the error pattern corresponding to ML decoding.

We remark that the foregoing method for obtaining an optimal error pattern holds for *any* initial choice of $\mathbf{e}$, although different optimal codevectors $\mathbf{c}_o$ will correspond, in general, to different selections of $\mathbf{e}$.

3 An example

Consider the (18,7,5) code C generated by

$$G = \begin{pmatrix} 1&0&1&0&1&0&1&1&1&0&1&0&1&0&1&0&1&0 \\ 0&1&0&1&1&1&0&1&0&1&1&1&0&1&0&1&0&1 \\ 1&1&0&0&0&0&0&0&0&0&1&1&1&1&0&0&0 \\ 0&1&0&0&0&0&0&0&0&0&0&0&1&1&1&1&1 \\ 0&0&1&1&1&0&0&0&0&0&1&1&0&0&0&0&0 \\ 0&0&0&0&0&1&1&1&0&0&0&0&0&1&1&0&0&0 \\ 0&0&0&0&0&0&0&0&1&1&1&0&0&0&0&1&1&0 \end{pmatrix}. \quad (12)$$

Upon receiving $\mathbf{r}$, calculate $\mathbf{v}$ and determine $\mathbf{b}$ (see Section 1). Denote by G_1 the matrix cosisting of the last five rows of G. Pursuing the coset decoding approach, we shall decode each of the four cosets of the subcode C_1 of C generated by G_1. To accomplish these decodings we shall employ four versions [4], [6] of $\mathbf{v}$: the original weight vector, and three versions corresponding to a tranlate of C_1 by the first row, the second row, and the sum of the first two rows of G, respectively. For example, in case the shift is by the sum of the first two rows, the weight vector is given by

$$(-v_1, -v_2, -v_3, -v_4, v_5, -v_6, -v_7, v_8, -v_9, \cdots, -v_{18}).$$

For simplicity of exposition, we shall hereafter refer to *any* of these four vectors by $\mathbf{v} = (v_1, v_2, \cdots, v_{18})$.

In order to perform each of the four decodings, it is convenient to utilize a contracted version C_2 of C_1, spanned by the following generator matrix

$$G_2 = \begin{pmatrix} 1&1&0&0&0&1&1&0&0 \\ 0&1&0&0&0&0&1&1&1 \\ 0&0&1&0&0&1&0&0&0 \\ 0&0&0&1&0&0&1&0&0 \\ 0&0&0&0&1&0&0&1&0 \end{pmatrix}.$$

Denote by $\mathbf{v}_2 = (v_{21}, v_{22}, \cdots, v_{29})$ the corresponding weight vector. Then $v_{21} = v_1$, $v_{22} = v_2$, $v_{23} = v_3 + v_4 + v_5$, $v_{24} = v_6 + v_7 + v_8$, etc. Consider ES, rather than CS. For that purpose we use the following check matrix H_2 of the contracted code C_2

$$H_2 = \begin{pmatrix} 1&1&1&1&1&1&1&1&1 \\ 0&1&1&0&1&1&0&1&0 \\ 0&1&1&0&0&1&0&0&1 \\ 1&0&1&0&0&1&0&0&0 \end{pmatrix}.$$

together with the weights $\{w_{2i}\}$ given by $w_{2i} = |v_{2i}|$. The corresponding 'hard detected' vector $\mathbf{b}_2$, with entries $b_{2i} = \text{sgn}(v_{2i})$, determines the syndrome $\mathbf{s}_2$. Since H_2 has repeated columns, and the support of the winning error pattern has to correspond to a set of linearly independent columns ([2], [6]), we may employ for the decoding

$$H_3 = \begin{pmatrix} 1 & 1 & 1 & 1 & 1 & 1 \\ 0 & 1 & 1 & 0 & 1 & 0 \\ 0 & 1 & 1 & 0 & 0 & 1 \\ 1 & 0 & 1 & 0 & 0 & 0 \end{pmatrix}$$

with the appropriate weight vector $\mathbf{w}_3$ (for example, $w_{31} = w_{21}$, $w_{32} = w_{22}$ and $w_{33} = min\{w_{23}, w_{26}\}$) and syndrome $\mathbf{s}_3 = \mathbf{s}_2$.

Observe that the code C_3 specified by the check matrix H_3 is a shortened version of the (8,4,4) code, and for the latter an ES is conveniently performable due to its highly symmetric structure. However, such decoding of C_3 involves redundant operations. For an ES applied directly to C_3, on the other hand, one has to take into account several cases, i.e., the memory requirement is quite large (relative, of course, to the size of the code). Consequently, it seems to be preferable to convert the decoding procedure into a CS.

A generator matrix G_3 of C_3 is given by

$$G_3 = \begin{pmatrix} 1 & 0 & 1 & 0 & 1 & 1 \\ 0 & 1 & 0 & 1 & 1 & 1 \end{pmatrix}.$$

In compliance with the aforementioned condition $\mathbf{s}_3 = \mathbf{s}_2$, let $\mathbf{b}_3$ be obtained from $\mathbf{b}_2$ as follows: add (modulo 2) the entries of $\mathbf{b}_2$ over the positions corresponding to each family of identical columns of H_2, and place the sum in the location of the column of H_3 that represents the family. Obviously, $\mathbf{b}_3$ is qualified to be the initial error pattern (see Section 2) for applying CS to C_3, and then the weight vector $\mathbf{v}_3$ is formed accordingly (9), (10). Upon completion of the CS, the error pattern corresponding to the decoding of C_3, hence also the one corresponding to C_2, are readily available. In turn, the latter error pattern is applied for computing the optimal codevector in C_2, whose weight is equal to the weight of the optimal codevector in the coset of C_1 under consideration. (The latter weight is needed for the final stage of decoding, namely for comparing the winner codewords of the four cosets of C_1.)

For decoding C_3, let us contract it into C_4 generated by the following matrix

$$G_4 = \begin{pmatrix} 1 & 0 & 1 \\ 0 & 1 & 1 \end{pmatrix}.$$

It is rather easy to decode C_4 straightforwardly. Nonetheless, is is interesting to notice that we can proceed with our method: the check matrix H_4 of C_4 is given by

$$H_4 = \begin{pmatrix} 1 & 1 & 1 \end{pmatrix}$$

which, in turn, is replacable (by minimizing the three weights) by the following check matrix

$$H_5 = (1).$$

The combination of all the previous steps forms a rather efficient decoder for C generated by G of (12). Furthermore, the decoder has a simple structure: the only real number operations are summations (minimizations) of weights corresponding to identical columns of generator (check) matrices.

Another, obvious, strategy to decode C by coset decoding is to employ the subcode C^* generated by the last three rows of G, since C^* is contractible to the trivially decodable (3,3,1) code. However, C consists of 16 cosets of C^*, in contrast to only four cosets of C_1. Of course, there are many other options, but the one adopted here is distinguished for yielding a particularly simple and efficient decoding procedure.

To derive an efficent coset decoder by our approach for a code which is several times larger than C spanned by G of (12), we would quite likely have to consider a longer sequence CS-ES-CS- $\cdots$ -ES of changes.

References

[1] Wolf, J.K.: Efficient maximum likelihood decoding of linear block codes, *IEEE Trans. Inform. Theory*, vol. IT-24, pp. 76-80, 1978.

[2] Miyakawa, H. and T. Kaneko: Decoding algorithm of error-correcting codes by use of analog weights, em Electronics and Communications in Japan, vol. 58-A, pp. 18-27, 1975.

[3] Snyders, J.: Reduced lists of error patterns for maximum likelihood soft decoding, *IEEE Trans. Inform. Theory*, vol. IT-37, pp. 1194-1200, 1991.

[4] Conway, J.H. and N.J.A. Sloane: Decoding techniques for codes and lattices, including the Golay code and the Leech lattice, *IEEE Trans. Inform. Theory*, vol. IT-32, pp. 41-50, 1986.

[5] Forney, G.D.,Jr.: Coset codes II : Binary codes, lattices and related codes, *IEEE Trans. Inform. Theory*, vol. IT-34, pp. 1152-1187, 1988.

[6] Snyders, J. and Y. Be'ery: Maximum likelihood soft decoding of binary block codes and decoders for the Golay codes, *IEEE Trans. Inform. Theory*, vol. IT-35, pp. 963-975, 1989.

A UNIVERSAL UPPER BOUND ON THE MISCORRECTION PROBABILITY WITH BOUNDED DISTANCE DECODING FOR A CODE USED ON AN ERROR-VALUE SYMMETRIC CHANNEL

L.M.G.M. Tolhuizen
Phillips Research Laboratories, Eindhoven, The Netherlands

ABSTRACT

The well-known upper bound on the miscorrection probability with bounded distance decoding for a Reed-Solomon code is shown to hold for any code, assuming that error patterns are equiprobable if they have equal support, i.e. if they corrupt the same set of postions. In previous papers, it was assumed that error patterns of equal weight are equiprobable; this prevented application of the bound if the code is used on a bursty channel.
Moreover, it is shown that for the case of MDS codes, the number of error vectors with given support that yield a miscorrection, only depends on the size of the support.
A generalization to error-and-erasure decoding is included.

1 Introduction

If an error correcting code with a (fixed) decoding algorithm is to be used in practical situations, it is very important to know its performance for various channel characteristics. Such knowledge enables us to judge if the performance satisfies the system demands, and to do some fine tuning in the decoding algorithm for improving the performance if necessary.

One aspect of performance estimation is estimating the probability that the decoding algorithm retrieves the transmitted codeword. A more subtle, but also very important aspect is the estimation of the *failure* and *miscorrection* probabilities. If the received word is far away from the transmitted codeword, then it is either far away from all codewords, or it is close to a codeword different from the transmitted one. In the first case, an incomplete decoding algorithm reports a decoding failure: it cannot decode the received word. In the latter case, a miscorrection occurs: the decoding algorithm delivers a codeword different from the transmitted one.

In many applications, a miscorrection has a much more serious effect than a decoding failure. In a concatenated code, for example, a decoding failure in the decoding of the inner code gives rise to an erased position in the outer code, whereas a miscorrection possibly yields an error in a position of the outer code; a similar situation occurs for other kinds of coding schemes where error correcting codes are combined, such as product codes or the CIRC in the Compact Disc players [1]. Another example occurs in the situation that a powerful postprocessing is available for words that cannot be decoded; typically, one can think of concealment techniques if codewords represent source-coded audio samples or (parts of) a video scene. Incorrectly decoded codewords would yield a very annoying click in the audio or a flash in the video scene. Hence, determination of the miscorrection probability is of great practical interest.

In [1] and [2], an upper bound is given on the number of error patterns of (Hamming) weight r that are miscorrected by a t-error decoder for $t \leq \frac{1}{2}(d-1)$ for the case where the error correcting code is a (shortened) Reed-Solomon code with minimum distance d. In [3], an exact formula for this number is given, using the principle of inclusion and exclusion. The bounds from [1] and [2] imply an upper bound on the miscorrection probability of a t-error decoder for shortened Reed-Solomon code if the channel is such that error patterns of equal weight are equiprobable. They do not, however, imply a bound on the miscorrection probability if the channel is bursty, that is, if introduced errors tend to be close to each other.

In this paper, we show that the bounds from [1] and [2] hold under much more general conditions on the channel. It is sufficient that the channel is *error-value symmetric*, i.e., that error patterns are equiprobable if they have equal support (i.e., if they affect the same set of positions). So a burst error affecting b (fixed)

consecutive positions may be much more likely than the occurence of b errors in b (fixed) scattered positions; all bursts affecting exactly these b fixed consecutive positions, however, are required to be equiprobable. Moreover, we show that the bound holds for *all* codes, not just for Reed-Solomon codes.
Next, it is shown that for MDS codes on an error-value symmetric channel, the conditional probability that an error pattern induces a miscorrection given that it has support R, only depends on the the cardinality of R.
In the last section, we generalize the results to error-and-erasure decoding.

2 Counting words close to codewords.

Let C be an error correcting code of length n with minimum distance d over an alphabet Q with q elements. For each codeword **c**, each subset R of {1,...,n} and each integer s$\leq\frac{1}{2}$(d-1), let γ(**c**,R,s) denote the number of words that differ from **c** exactly in the positions of R and are at distance s from a codeword different from **c**. As fewer than d-s errors cannot move **c** within distance s of another codeword, γ(**c**,R,s)=0 if |R| < d-s. (Here and in the sequel, |X| denotes the cardinality of the set X.) It is our aim to prove the following theorem.

Theorem 1.

$$\text{(a) } \textit{If } \mid R \mid \geq d-s, \textit{ then } \gamma(\mathbf{c},R,s) \leq \binom{n}{s}(q-1)^{|R|+s-d+1}.$$

$$\text{(b) } \textit{If } d-s \leq \mid R \mid \leq d-1, \textit{ then } \gamma(\mathbf{c},R,s) \leq (q-1)^{|R|+s-d+1} \sum_{j=0}^{|R|-(d-s)} \binom{\mid R \mid}{j}\binom{n-\mid R \mid}{s-j}.$$

Note that the formula in (b) is a sharpening of the formula in (a): the sum equals the number of s-subsets of {1,...,n} with at most |R| -(d-s) elements in R, which surely is at most $\binom{n}{s}$, the number of s-subsets of {1, ...,n}.

In order to simplify the proof, we assume without loss of generality that 0$\in$Q and that **c**=**0**.

Definition. The *support* of a word **x**$\in$Q^n, denoted by supp(**x**), is the set of indices of its non-zero positions, that is supp(**x**) = {i$\in${1,...,n}|x_i $\neq$0}.

Definition. The number of vectors with support W at distance s from a fixed vector with support R is denoted by f(R,W,s).[1]

[1]It is easy to see that f(R,W,s) is independent of the specific vector with support R that is selected.

Lemma 1. *If R and W are two subsets of {1,...,n}, then*

$$(q-1)^{|R|}f(R,W,s) = (q-1)^{|W|}f(W,R,s).$$

Proof. The equality is obtained by counting in two ways the number of pairs of vectors $(\mathbf{r},\mathbf{w})$ satisfying $supp(\mathbf{r})=R$, $supp(\mathbf{w})=W$ and $d(\mathbf{r},\mathbf{w}) = s$. □

Definition. The number of codewords with support W is denoted by A_W.

With the above notations, $\gamma(\mathbf{0},R,s)$ can be expressed as follows:

$$\gamma(\mathbf{0},R,s) = \sum_{\mathbf{c}\in C\setminus\{\mathbf{0}\}} |\{\mathbf{x} \mid \mathrm{supp}(\mathbf{x}) = R, d(\mathbf{x},\mathbf{c}) = s\}| = \sum_{W:|W|\geq d} A_W f(W,R,s).$$

Using Lemma 1, we find

$$\gamma(\mathbf{0},R,s) = \sum_{W:|W|\geq d} A_W (q-1)^{|R|-|W|} f(R,W,s). \tag{1}$$

In order to find an upper bound to $\gamma(\mathbf{0},R,s)$, we first give an upperbound on A_W.

Lemma 2. *If* $|W| \geq d$, *then* A_W, *the number of codewords with support W, is at most* $(q-1)^{|W|-d+1}$.

Proof. Let C_W be the set of codewords with support W. By deleting the zeroes from the words of C_W, we obtain a code C_W^* of length $|W|$ over $Q\setminus\{0\}$ that has minimum distance at least d. According to the Singleton bound, we have: $|C_W| = |C_W^*| \leq (q-1)^{|W|-d+1}$. □

From Lemma 2 and Equation (1), we obtain

$$\gamma(\mathbf{0},R,s) \leq (q-1)^{|R|-d+1} \sum_{W:|W|\geq d} f(R,W,s). \tag{2}$$

All we are left with is bounding the sum in (2). Clearly, it is at most the number of *all* vectors at distance s from a fixed vector with support R; that is, it is at most $\binom{n}{s}(q-1)^s$. So we have proved part (a) of Theorem 1. For proving part (b), we need one more lemma.

Lemma 3. *If R is a subset of {1, ...,n}, then*

$$\sum_{W:|W|\geq d} f(R,W,s) \leq (q-1)^s \sum_{j=0}^{|R|-(d-s)} \binom{|R|}{j}\binom{n-|R|}{s-j}.$$

Proof. Let $\mathbf{r}$ be a vector with support R. By definition, we have

$$\sum_{W:|W|\geq d} f(R,W,s) = |\{\mathbf{w} \mid d(\mathbf{r},\mathbf{w}) = s, |\operatorname{supp}(\mathbf{w})| \geq d\}|.$$

Let $\mathbf{w}$ be a vector with support W at distance s from $\mathbf{r}$, and suppose that $|W| \geq d$. Let S denote the set in which $\mathbf{w}$ and $\mathbf{r}$ differ. If $i \in W \setminus (W \cap R)$, then $w_i \neq 0$ and $r_i = 0$, so $i \in S$. Hence we have

$$W \setminus (W \cap R) \subset S \setminus (S \cap R).$$

From this inclusion relation, we conclude that $|W| - |W \cap R| \leq |S| - |S \cap R|$. As $|S| = s$, $|W| \geq d$ and $|W \cap R| \leq |R|$, we find that $|S \cap R| \leq |R| - d + s$. Consequently, we have

$$\{\mathbf{w} \mid d(\mathbf{r},\mathbf{w}) = s, |\operatorname{supp}(\mathbf{w})| \geq d\} \subset \bigcup_{S:|S\cap R|\leq|R|-d+s} \{\mathbf{w} \mid \{i \mid w_i \neq r_i\} = S\}.$$

As each of the sets in the above union has $(q-1)^s$ elements, Lemma 3 holds. □

Part (b) of the Theorem now follows by combining Lemma 3 and (2).

3 Consequences for miscorrection probabilities

In this section, we discuss the relevance from Theorem 1 to the miscorrection probability for a bounded distance decoder for C if it is being used on a error-value symmetric channel. We will derive an upper bound on the probability that a received word is at distance s from a codeword different from the transmitted one. This upper bound is independent on the probability distribution of the supports of the error patterns on the channel; all we require is that the channel is error-value symmetric.

Suppose C is used on an error-value symmetric channel. Let $\mathbf{c} \in C$ be transmitted and let $R \subset \{1,\ldots,n\}$ have cardinality at least d-s. As the $(q-1)^{|R|}$ error patterns affecting $\mathbf{c}$ in the positions of R are equally likely, the *conditional* probability that the received word is at distance s from C, *given* that it differs from $\mathbf{c}$ in the positions of R, equals $\gamma(\mathbf{c},R,s) / (q-1)^{|R|}$. Using Theorem 1, we see that this last number is at most $\binom{n}{s} / (q-1)^{d-s-1}$. So we have the following theorem.

Theorem 2. *Suppose a code C of length n with minimum distance d is used on an error-value symmetric channel. Let* $s \leq \frac{1}{2}(d-1)$. *The probability that the received word is at distance s from a codeword different from the transmitted one is at most*

$$\frac{\binom{n}{s}}{(q-1)^{d-s-1}} P(|R| \geq d-s),$$

where $P(|R| \geq d-s)$ denotes the probability that the transmitted word is affected in at least $(d-s)$ positions.

Theorem 2 has the following corollary.

Corollary 1. *If a code of length n and minimum distance d is used on an error-value symmetric channel, then the probability that the received word is at distance $s \leq \frac{1}{2}(d-1)$ from a codeword different from the transmitted one is at most*

$$\binom{n}{s}/(q-1)^{d-s-1}.$$

Proof. This follows immediately from Theorem 2, as $P(|R| \geq d-s)$, being a probability, is at most 1. □

The upper bound from Theorem 2 will yield a much smaller value than the bound from Corollary 1 if the channel is such that error patterns affecting at most d-s-1 postions are very likely. A typical example of such a channel is a q-ary symmetric channel with small error probability. The bound from Corollary 1, however, has the advantage that it holds for *all* error probability distributions on an error-value symmetric channel.

4 Conditional miscorrection probabilities for MDS codes

In [3], exact formulas are given for the number of error patterns of weight $u \geq d-s$ at distance s from C for the case that C is an MDS code, i.e. , if $|C|=q^{n-d+1}$. If C is used on a channel for which error patterns of equal weight are equiprobable, then $p_s(u)$, the conditional probability that an error pattern is at distance s from C, given that it has weight u, is found by dividing this number by $\binom{n}{u}(q-1)^u$, the number of error patterns of weight u.
Now suppose C is used on an error-value symmetric channel. The conditional probability that a vector is at distance s from C, given that it has support R, is denoted by $p_s(R)$. The following theorem implies that $p_s(R)$ only depends on the cardinality of R.

Theorem 3. *Let C be an MDS code of length n and minimum distance d. If C is used on an error-value symmetric channel, $s \leq \frac{1}{2}(d-1)$, and R and T are two subsets of $\{1,\ldots,n\}$ of equal cardinality, then $p_s(R) = p_s(T)$.*

Proof. We explicitly compute $p_s(R)$. As C is used on an error-value symmetric channel, we have

$$p_s(R) = \gamma(0, s, R)/(q-1)^{|R|} = \sum_{W:|W|\geq d} A_W (q-1)^{-|W|} f(R, W, s),$$

where we used Formula 1 to obtain the second equality.
Now we use the basic fact that in an MDS code, A_W only depends on |W|. This is most easily seen by considering the MDS code C_W of length |W| and minimum distance d that is obtained by shortening C outside the positions of W. Clearly, A_W equals the number of weight |W| words in C_W; this number is exactly known ([4], Ch. 11).
Using A_w to denote the number of words in C with a given support of size w, we find

$$p_s(R) = \sum_{w=d}^{n} A_w (q-1)^{-w} \sum_{W:|W|=w} f(R, W, s).$$

The inner sum in this formula equals the number of vectors of weight w at distance s from a given vector with support R; clearly, this number only depends on |R| . □

As a consequence, if C is an MDS code, then $p_s(R) = p_s(|R|)$, so we can apply the formulas from [3] to find the exact conditional miscorrection probabilities.

5 Generalization to error-and erasure decoding

The foregoing results can easily be generalized to bounds on miscorrection probabilities for error-and-erasure decoding (cf [1]).
Suppose that upon reception, the positions of the set E are erased, i.e. marked as unreliable, and that $e=|E|\leq d-1$. If outside E at most $\frac{1}{2}(d-e-1)$ errors occured, then the transmitted codeword can be retrieved. For important classes of codes such as BCH codes and RS codes, efficient algorithms for error-and-erasure decoding exist [5, Sec. 9.2].

Definition. For $s\leq \frac{1}{2}(d-e-1)$, $c\in C$ and $R\subset\{1,\ldots,n\}\setminus E$, $\gamma_E(c,R,s)$ denotes the number of vectors that outside E differ from **c** in the positions of R and in s positions outside E from a codeword different from **c**.

Clearly, C_E, the code of length (n-e) obtained by puncturing C in the postions from E, has minimum distance at least d-e. Therefore $\gamma_E(c,R,s)$ is upper bounded by the expression for $\gamma(c,R,s)$ from Theorem 1, where n-e and d-e should be substi-

tuted for n and d, respectively.[2]
The generalization of Theorem 2 is obvious: if the channel is error-value symmetric outside E, then the probability that a received word differs in s positions outside E from a codeword different from the transmitted one, is at most

$$\frac{\binom{n-e}{s}}{(q-1)^{d-e-s-1}} P(|R| \geq d-e-s),$$

where $P(|R| \geq d-e-s)$ denotes the probability that outside E the transmitted word is affected in at least (d-e-s) positions.
Also the generalization of Corollary 1 is obvious.

Finally, if C is an MDS code, then the exact formula for γ_E(**c**,R,s) is obtained from the exact formula foı γ(**c**,R,s) from Section 4, substituting n-e and d-e for n and d, respectively. This is due to the fact that C_E is an MDS code of length n-e and minimum distance d-e. Also the obvious generalization of Theorem 3 is valid. That is, let **c** be transmitted and suppose the channel is error-value symmetric outside E. The probability that the received word differs in s positions outside E from a word from C\{**c**}, given that outside E it differs from **c** in the the positions of R, only depends on |R|.

References

[1] L.H.M.E. Driessen and L.B. Vries, "Performance Calculations of the Compact Disc Error Correcting Code", *Proc. Int. Conf. Video and Data Recording*, Univ. of Southampton, April 20-23, 1982, pp. 385-395.

[2] R.J. McEliece and L. Swanson, "On the decoder error probability for Reed-Solomon codes", *IEEE Trans. on Inform. Th.*, **IT-32**, 5, 1986, pp. 701-703.

[3] Kar-Ming Cheung, "More on the decoder error probability for Reed-Solomon codes", *IEEE Trans. on Inform. Th.*, **IT-35**, 4, 1989, pp. 895-900.

[4] F.J. MacWilliams and N.J.A. Sloane, *The Theory of Error-Correcting Codes*, North-Holland, 1977.

[5] R.E. Blahut, *Theory and Practice of Error Control Codes*, Addison-Wesley, 1983.

[2] The upper bounds from Theorem 1 are decreasing in d; therefore, the proposed bound for γ_E(**c**,R,s) remains valid if the minimum distance of C_E is larger than d-e.

5

INFORMATION THEORY

ON THE ZERO-ERROR CAPACITY
OF MULTIPLE ACCESS CHANNELS

A. Fioretto, E. Inglese and A. Sgarro
University of Trieste, Trieste, Italy

ABSTRACT

We consider the zero-error capacity of multiple access channels with two inputs. A simple criterion for the possible degeneracy of the capacity region is provided. The problem of coding is formulated in purely graph-theoretic terms. Examples are given.

1. INTRODUCTION

Stationary memoryless 2-MACs (MAC is the acronym of *multiple access channel*) are channels with two inputs and one output, which are described through a stochastic matrix W: $\mathcal{X}\times\mathcal{Y}\rightarrow\mathcal{Z}$, $\mathcal{X}$ and $\mathcal{Y}$ being the finite input alphabets, and $\mathcal{Z}$ being the finite output alphabet. W is extended to a matrix W^n: $\mathcal{X}^n\times\mathcal{Y}^n\rightarrow\mathcal{Z}^n$ in the usual way:

$$W^n(\underline{z}|\underline{x},\underline{y}) = \Pi_i\ W(z_i|x_i,y_i),\quad \underline{x}=x_1...x_n,\ \underline{y}=y_1...y_n$$

A *code* (more precisely: an n-code) for 2-MAC W and for (finite) message sets $\mathcal{M}_X$ and $\mathcal{M}_Y$ is a triple (f,g,ϕ) where mappings f and g are the *encoders*, and mapping ϕ is the *decoder*; in symbols, f: $\mathcal{M}_X\rightarrow\mathcal{X}^n$, g: $\mathcal{M}_Y\rightarrow\mathcal{Y}^n$, ϕ: $\mathcal{Z}^n\rightarrow\mathcal{M}_X\times\mathcal{M}_Y$. Correspondingly, $R_X = n^{-1}\log|\mathcal{M}_X|$, and $R_Y = n^{-1}\log|\mathcal{M}_Y|$ are the transmission *rates*, and

$$P_{err} = \max W^n(\mathcal{Z}^n - \phi^{-1}(m_X,m_Y)|f(m_X),g(m_Y))$$

is the *error probability* (the max is taken with respect to all pairs of messages m_X in $\mathcal{M}_X$ and m_Y in $\mathcal{M}_Y$; since we are dealing below with the zero-error case, it would make no difference to define P_{err} in terms of average error, rather than maximal error; logarithms are to the base 2). A rate-pair (R_X,R_Y) is ε-*achievable* if it is possible to "asymptotically transmit information at speeds R_X,R_Y in a reliable way, that is while keeping the decoding error probability below ε". More formally, for each positive δ and for all n high enough, there must exist n-codes (f,g,ϕ) with error probability at most ε, and rates at least R_X-δ, R_Y-δ, respectively. The problem is determining (through a single-letter description, i.e. through a computable, non-asymptotical, formula) the ε-*capacity region*, that is the region of ε-achievable rates. (For a deep discussion of these notions, model and formalization, we refer to [1,2]; we are adopting the notation of [1]). Much attention has been paid in the literature to the case $\varepsilon>0$ (cf [1,2]), while to our knowledge the case $\varepsilon=0$ has never been tackled before, save for very special cases as the arithmetic adder MAC (cf. the examples below). A reason for this is that the zero-error case appears to be hopeless already in the one-dimensional (one-input) case, which is known to be equivalent to a notoriously intractable problem of graph theory, namely the determination of *graph capacity* [1]. Consequently, only "framework" results are to be hoped for in the case of two-dimensional channels such as 2-MACs and our paper is devoted to these more accessible results.

2. RESULTS

Let the matrix W be given, W: $\mathcal{X}\times\mathcal{Y}\rightarrow\mathcal{Z}$, and let $\mathcal{C}_0$ be the zero-capacity region of the corresponding 2-MAC, as defined above ($\varepsilon=0$). Let $\mathcal{R}$ be the rectangle in the (R_X,R_Y)-plane with vertices at points $(0,0)$, $(\log|\mathcal{X}|,0)$, $(0,\log|\mathcal{Y}|)$ and $(\log|\mathcal{X}|,\log|\mathcal{Y}|)$. One has:

<u>Proposition</u>. $\mathcal{C}_0$ is a closed convex corner entirely contained in $\mathcal{R}$.

The proof is omitted, as it is a straightforward variation of arguments used in the case $\varepsilon>0$ (e.g. to prove convexity one can resort to the standard technique of time-sharing, which, given two codes, consists of using one code for a slot of the transmission time, and the other code for the remaining slot; cf[1,2]; we recall that if a point (a,b), a>0, b>0, belongs to a convex corner of the positive quadrant, then the rectangle with vertices at (0,0), (a,0), (0,b), (a,b) is entirely contained in the convex corner).

Let S_X and S_Y be the highest rates which are achievable when one of the two inputs is kept idle; in other words: $(S_X,0)$ and $(0,S_Y)$ are achievable; if $(R_X,0)$ and $(0,R_Y)$ are achievable, then $R_X \leq S_X$ and $R_Y \leq S_Y$. Let $\mathcal{T}$ be the triangle with vertices at (0,0), $(S_X,0)$ and $(0,S_Y)$, and let $\mathcal{R}'$ be the rectangle with three vertices as for $\mathcal{T}$ and the fourth vertex at (S_X,S_Y). The proposition immediately implies:

$$\mathcal{T} \subseteq \mathcal{C}_0 \subseteq \mathcal{R}' \subseteq \mathcal{R}.$$

The zero-capacity region $\mathcal{C}_0$ will be called *positive* when $S_X>0$ and $S_Y>0$, *degenerate* when $S_X=S_Y=0$, *semidegenerate* when $S_X>S_Y=0$, or when $S_Y>S_X=0$. We begin by establishing a simple criterion of degeneracy and of semi-degeneracy for 2-MACs.

Criterion. $S_X>0$ iff there are two distinct letters x and x' in $\mathcal{X}$ and a letter y in $\mathcal{Y}$, such that $W(z|x,y)\,W(z|x',y) = 0$ for every letter z in $\mathcal{Z}$.

Proof. If the condition on the right holds, the achievability of (1,0) soon follows by constraining one input to use only letter y, while the other input uses all the codewords in $\{x,x'\}^n$. Conversely, if the condition on the right does not hold, any code with two distinct codewords $\underline{x}$ and $\underline{x}'$ in $f(\mathcal{M}_X)$ gives a positive error probability, as soon as these codewords are sent jointly with any fixed codeword $\underline{y}$ in $g(\mathcal{M}_X)$.

A similar criterion holds to have $S_Y>0$; this soon gives a criterion for $\mathcal{C}_0$ to be positive (recall that $\mathcal{C}_0$ is positive iff $S_X>0$ and $S_Y>0$).

We now re-cast our problem into the language of graph-theory. We first recall what happens in the one-dimensional case of the "usual" channel (cf [1]). Formally, this case can be derived from ours when one of the two input alphabets, $\mathcal{Y}$, say, has only one letter and $\mathcal{M}_Y$ contains only one message: then we can as well drop y from our notation, and write $W(z|x)$ rather than $W(z|x,y)$. To W (to the corresponding channel) we can associate a *distinguishability* graph G=G(W) such that the vertices are the letters of $\mathcal{X}$ and two vertices x and x' are connected iff they are distinguishable beyond the channel, that is iff $W(z|x)\,W(z|x') = 0$ whatever the output letter z. (Note that to define distinguishability we need not know the precise numerical values of the entries of W, but only whether these entries are zero or non-zero). The distinguishability relation extends to product graphs G^n over $\mathcal{X}^n$, taking W^n rather than W: then two vertices of G^n (two n-tuples) are distinguishable (are connected by an edge) iff there is at least a position i, $1 \leq i \leq n$, where the corresponding letters are distinguishable according to G. The problem of finding the highest-rate code with zero error-probability (the problem of determining S_X when $|\mathcal{Y}|=1$) becomes then the problem of finding the size of the largest *clique* in G^n, or rather the problem of determining the *graph capacity* c(G), where $c(G) = \lim_n n^{-1} \log cl(G^n)$

($cl(G^n)$ is the *clique number* of G^n, that is the size of the largest clique in G^n; we recall that any two vertices of a clique are connected by an edge). The definition of graph capacity, however, is itself not a single-letter one, and so remains uncomputable. When G is a perfect graph, it is quite simple to give a single-letter characterization of C(G); namely, C(G) = log cl(G), while in general one has only c(G) ≥ log cl(G). Unfortunately, for even such an innocent-looking imperfect graph as the heptagon the problem of determining the numerical value of C(G) remains open (the case of the pentagon was solved by L. Lovász after having remained a famous open problem for two decades).

Let us go back to our 2-MAC W, where $\mathcal{Y}$, in general, contains several distinct letters. For a moment, we might forget about the two-dimensional nature of $\mathcal{X}\times\mathcal{Y}$, and consider the distinguishability graph G=G(W) defined as above, whose vertices are the pairs (x,y) in $\mathcal{X}\times\mathcal{Y}$. The number C(G) of course would be perfectly defined: from an engineering point of view, all of this would amount to allowing the encoders f and g to "co-operate", or, rather, to have a single encoder f, f: $\mathcal{M}_X\times\mathcal{M}_Y\rightarrow\mathcal{X}^n\times\mathcal{Y}^n$. Instead, dealing with a 2-MAC, whose encoders are *not* allowed to co-operate, obliges us to add a "rectangular" constraint: only "rectangular" cliques, that is cliques of the form $\mathcal{A}\times\mathcal{B}$, $\mathcal{A}\subseteq\mathcal{X}^n$, $\mathcal{B}\subseteq\mathcal{Y}^n$, are of interest to us. So, the problem of finding good codes for the 2-MAC becomes the combinatorial problem of finding rectangular cliques in G^n which have "large" sides.

Even if G(W) is useless for 2-MACs, we can use distinguishability graphs for the determination of S_X and S_Y. We assume that the Y-side, say, has length zero and try to maximize the length of the X-side. To each letter y in $\mathcal{Y}$, we can associate a distinguishability graph G_y over $\mathcal{X}$ by declaring x and x' distinguishable iff W(z|x,y) W(z|x',y) = 0 whatever the output letter z; in engineering terms, we are constraining the $\mathcal{Y}$-input to use only letter y, so that, in fact, we are dealing with a one-dimensional channel from $\mathcal{X}^n$ to $\mathcal{Z}^n$. Set C_y=C(G_y). Clearly:

<u>Bounds</u>: $\max_y C_y \leq S_X \leq \log |\mathcal{X}|$

We stress that, in general, the bound on the left is not a computable one. It is a surprising fact, in sharp contrast with engineering intuition and with the case ε>0, that S_X can be strictly larger than $\max_y C_y$. (To see this, use the example given by Haemers in [3] to answer some questions posed by Lovász).

3. EXAMPLES

Binary adder MAC. This is a 2-MAC with binary input alphabet and binary output alphabet, where each output digit is deterministically equal to the binary sum of the two corresponding input digits. Clearly, (1,0) and (0,1) are achievable (if y=0, say, is fixed throughout, the $\mathcal{X}$-input is perfectly reproduced at the ouput); therefore the triangle $\mathcal{T}'$ with vertices at those two points and at the origin is contained in $\mathcal{C}_0$; actually, one has $\mathcal{C}_0=\mathcal{T}'$, the bound $R_X+R_Y \le \log 2 = 1$ being obvious for a binary-output channel.

Noiseless binary MAC. The inputs are as above; the output is an ordered pair which deterministically reproduces the pair of the two binary inputs. Here, trivially, $\mathcal{C}_0=\mathcal{R}$. This example shows that time-sharing between S_X and S_Y can be beaten (one can have $\mathcal{C}_0$ strictly larger than the triangle $\mathcal{T}$). It is nicer, however, to exhibit clever combinatorial constructions that beat time-sharing for more appealing 2-MACs than the noiseless one, such as the aritmetic adder MAC.

Arithmetic adder MAC. The output is ternary ($R_X+R_Y \le \log 3$), and the channel performs the integer sum rather than the modulo-two sum. Trivially, $\mathcal{T}'$ is achievable; time-sharing between (0,1) and (1,0) can be beaten, e.g. by extending the following code due to Kasami and Lin ([4]): $|\mathcal{M}_X|=3$, $|\mathcal{M}_Y|=2$, n=2, $f(\mathcal{M}_X)=\{00,01,10\}$, $g(\mathcal{M}_Y)=\{00,11\}$.

A degenerate 2-MAC. This is not a deterministic example (W is not a zero-one stochastic matrix). The inputs are binary, the output is quaternary:

$$\begin{array}{cc} x\ y & W \\ \begin{bmatrix} 0 & 0 \\ 0 & 1 \\ 1 & 0 \\ 1 & 1 \end{bmatrix} & \begin{bmatrix} * & * & 0 & 0 \\ 0 & * & 0 & * \\ * & 0 & * & 0 \\ 0 & 0 & * & * \end{bmatrix} \end{array}$$

(we did not bother to specify the value of positive probabilities). Note that the pairs 00 and 11 are distinguishable, and so are the pairs 01 and 10: then, with co-operative encoders (with a single encoder f, $f: \mathcal{M}_X \times \mathcal{M}_Y \to \mathcal{X}^n \times \mathcal{Y}^n$) the zero-capacity of the channel would be positive. However, with non co-operative encoders, as is the case for 2-MACs, the zero-capacity region degenerates to the point (0,0), as shown by the degeneracy criterion.

Acknowledgment. Thanks are due to J. Körner and L. Tolhuizen for their very helpful advice.

REFERENCES

1. I. Csiszár, J. Körner: *Information theory* , Academic Press, 1981
2. Th. M. Cover, J. A. Thomas: *Elements of information theory*, Wiley, 1991
3. W. Haemers, On some problems of Lovász concerning the Shannon capacity of a graph, IEEE Transactions on Information Theory, IT-25, pp 231-232, 1979
4. T. Kasami, S. Lin: Coding for a multiple access channel, IEEE Transactions on Information Theory, IT-22, pp 129-137, 1976

6

CONVOLUTIONAL CODING, SYNCHRONISATION

NEW TECHNIQUES FOR WORD SYNCHRONISATION

M. Darnell and M. Grayson
University of Hull, Hull, UK

B. Honary
Lancaster University, Lancaster, UK

Abstract

This paper describes two techniques for identifying a known sequence, when embedded in random data and corrupted by additive noise. The first method is derived from the maximum likelihood algorithm. The original implementation of this ML algorithm uses amplitude-shift-keyed transmission and shows typical improvements of 3 dB over the simple correlator rule. An algorithm for using the maximum likelihood technique with frequency-shift-keyed transmission is derived which, in simulation trial, shows typical improvements of 2 dB at low signal-to-noise ratios, when compared with the simple correlator rule. The second technique presented improves the probability of sequence detection when soft decision information is not available in the detector. Instead of looking for a single peak, this approach searches for the entire auto-correlation function of the known sequence. Under certain conditions, the performance of such a scheme can approach that of the ML algorithm.

Introduction

Massey [1] has derived the optimal pattern identification rule for an amplitude-shift-keyed (ASK) sequence when embedded in random data and perturbed by additive white Gaussian noise. This decision rule is a maximum likelihood (ML) detector, which chooses the word synchronisation epoch to be the value of μ which maximises

$$\sum_{i=0}^{m-1}\left(s_i\rho_{i+\mu}\right)-\sum_{i=0}^{m-1}f\left(\rho_{i+\mu}\right) \tag{1}$$

where s_i is the ith digit of the m digit sequence (where s_i is +/-1), ρ_i is the ith digit of the actual received noisy data sequence prior to a hard decision and where

$$f(x)=\frac{N_o}{\sqrt{2E}}\ln\left[\cosh\left(\frac{2\sqrt{E}x}{N_o}\right)\right] \tag{2}$$

N_o is the one-sided nose spectral density and E corresponds to the signal energy in one digit of the received sequence.

The first term in (1) is the simple correlation (SC) rule used by Barker [2]. The second term represents a correction term required to account for the effect of the random data surrounding the sequence. Using a high signal-to-noise ratio (SNR) approximation to the non-linearity in (2) produces the following decision rule

$$\sum_{i=0}^{m-1}\left(s_i\rho_{i+\mu}\right)-\sum_{i=0}^{m-1}\left|\rho_{i+\mu}\right| \tag{3}$$

Both Massey [1] and Nielson [3] have analysed the above synchroniser using Monte Carlo simulation techniques. The simulated results indicate that the high SNR approximation of the ML detector yields virtually optimum performance for values of SNR of practical interest, showing approximately 3 dB gain over the SC rule.

It is apparent from the above description that soft decision information is required in the form of the magnitude of $\rho_{i+\mu}$ before the ML algorithm can be implemented. If only hard decision information is available, the ML algorithm simplifies to the SC rule proposed by Barker.

ML algorithm used with FSK transmission

The realisation of the ML algorithm is somewhat complicated when FSK transmission is used, with no clear measurement of the symbol modulus being present within the demodulator. If it is assumed that the received data is bipolar and corrupted by additive Gaussian noise, then, at the correct frame synchronisation epoch, the high SNR algorithm can be written as

$$\sum_{i=0}^{m} s_i(s_i+n_i)-\sum_{i=0}^{m-1}|s_i+n_i| \tag{5}$$

where n_i is sampled Gaussian noise with zero mean and standard deviation σ. As expected, the only difference between the ML and SC detectors is the second term in (5). If the ASK data is assumed to have unity magnitude, the mean of this second term can be expressed as

$$\int_{-1}^{\infty}\frac{1}{\sqrt{2\pi}\sigma}e^{-x^2/2\sigma^2}(1+x)\,dx - \int_{-\infty}^{-1}\frac{1}{\sqrt{2\pi}\sigma}e^{-x^2/2\sigma^2}(1+x)\,dx \tag{6}$$

This function has been plotted for various values of noise standard deviation and the results are shown in Fig. 1. The graph indicates that, if the received signal is a random variable with unity mean and standard deviation σ, the mean of the ML correction term can be approximated to by 0.8σ. Hence, all that is required to realise the ML algorithm in an FSK environment is an estimate of the instantaneous SNR, or more accurately an estimate of the variance of the additive noise, at each received symbol interval.

A common metric used within FSK demodulation is the ratio, r, of the largest matched filter output to the smallest. A mathematical model for the probability density function (PDF) for such a ratio when used with a non-coherent detector has been derived [4]. This PDF has been used to calculate the expected value of the ratio for various values of E_b/N_o. The results are

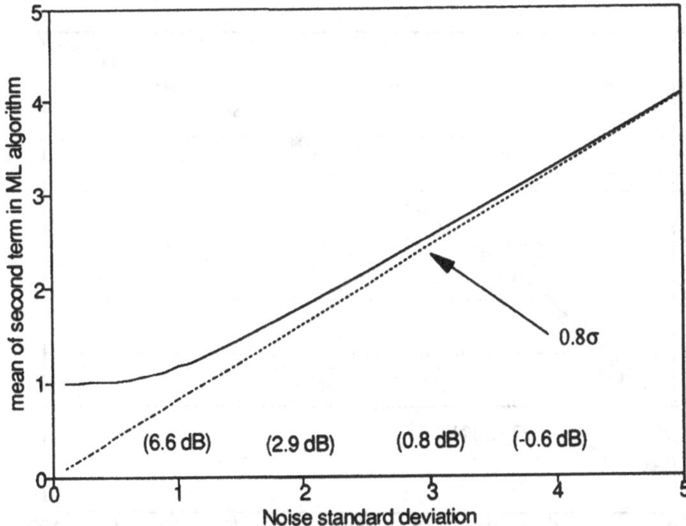

Figure 1 *Mean of the ML correction term as a function of additive noise standard deviation*

shown in Fig. 2 together with a plot of the function $1.77(E_b/N_o)^{1/2}$. The comparison clearly indicates that, except at low values of E_b/N_o, the mean of the ratio of the filter outputs can be approximated to by this function.

Now, assuming unity signal power, it can be shown that for non-coherent FSK [5]

$$E_b / N_o = 1/2\sigma_T^2 \tag{7}$$

and hence $\sigma_T = 1.25/r$ (8)

Consequently, the FSK implementation of the ML synchroniser chooses the value of μ which maximises

$$\sum_{i=0}^{m-1}\left(s_i\rho_{i+\mu}\right) - \sum_{i=0}^{m-1} r_{i+\mu}^{-1} \tag{9}$$

where now ρ_i is the ith received symbol after a hard decision has been made and r_i is the ratio of the largest and smallest ith matched filter outputs.

Improved sequence detection in the absence of soft decision information

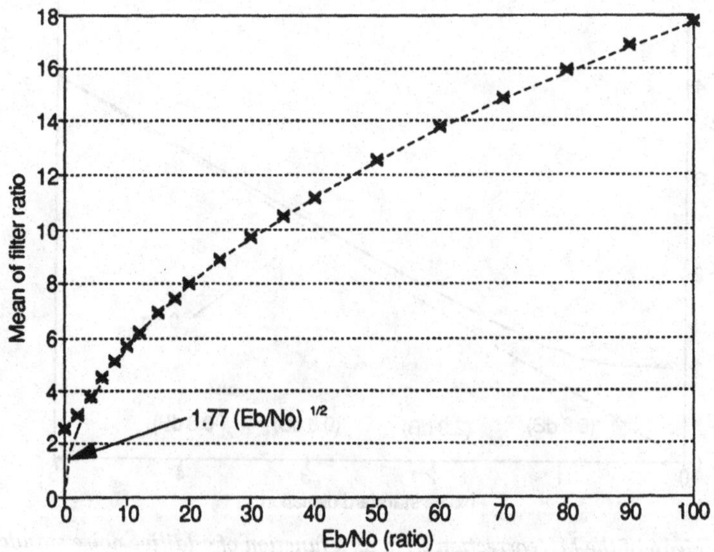

Figure 2 *Mean of ratio of matched filter ouputs for non-coherent reception*

The heart of the ML algorithm is the correlation rule which cross-correlates delayed versions of the received data stream with a replica of the transmitted sequence. Both the ML and SC techniques then attempt to identify the peak in the detector output, corresponding to the peak in the auto-correlation function (ACF) of the synchronising sequence. Furthermore, both schemes ignore all information in the sequence ACF except its value at zero time shift. This situation has arisen since an ideal synchronisation sequence must be impulse equivalent, i.e. have an ACF which is zero at all values of time shift except zero. Clearly in such a situation, the ML and SC algorithms cannot derive any further information from the actual result of the correlation of the transmitted sequence with the received sequence surrounded by random data.

It has been shown, however, that when no phase ambiguity exists in the receiver, improved frame synchronisation performance can be achieved by using a sequence with an ACF with a characteristic structure [6]. Specifically, negative sidelobes should be made as large as possible and positioned towards the extremes of the ACF. Conversely, all positive sidelobes should be made as small as possible and positioned towards the centre of the ACF. Using such sequences, which can only be described as approximately impulse equivalent, valid information concerning the correct synchronisation epoch can be obtained from the complete profile of the sequence ACF. This is the basis of a novel technique for improved synchronisation acquisition, which exploits the whole profile of the sequence's ACF, has been developed and has been termed "matched filter profile synchronisation" (MFPS) [7].

The MFPS estimator is essentially a double matched filter device as shown in Fig. 3. The first filter is matched to the transmitted sequence waveform and the second is matched to the undistorted sequence ACF. It should be noted that for truly impulse equivalent sequence, the above structure simplifies to the simple correlator.

Simulation results

The simulations have used Monte Carlo techniques to model the density function of both the matched filter outputs from a non-coherent FSK demodulator and the filter outputs from a bipolar ASK detector. The chosen sequence was embedded in an appropriate amount of random data and the composite symbol stream used as the input to the demodulator and sequence detector. Trials were repeated 1000 times for various values of SNR and the effective probability of synchroniser failure calculated.

The simulated performance of the FSK ML detector was found to be always superior to the simple correlator rule. Typical performance comparisons for the identification of Barker's 7 bit code are shown in Fig. 4. The graph shows the performance for the synchroniser when the sequence was embedded in various amounts of random data and indicate a 2 dB performance advantage at low SNR for differing amounts of surrounding data.

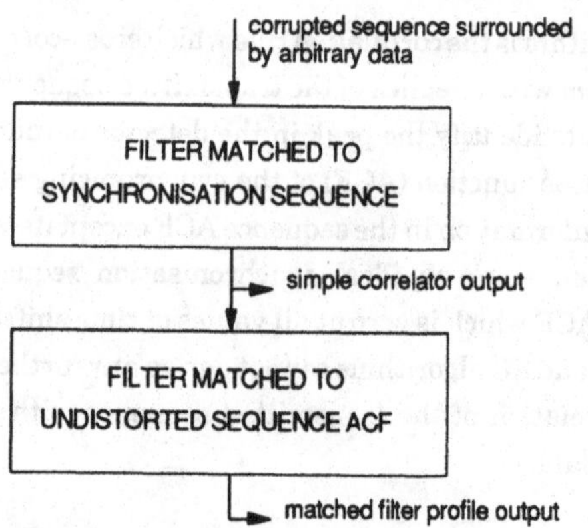

Figure 3 *Structure of a matched filter profile synchroniser*

The simulated results for the MFPS detector were obtained for ASK transmission allowing both the SC and true ML rule to be implemented. The results were obtained using the 11 bit Barker code. The results are shown in Fig. 5 and indicate that the MFPS synchroniser provides approximately half the performance gain of the ML estimator when compared with the SC rule, without the necessity for soft decision information.

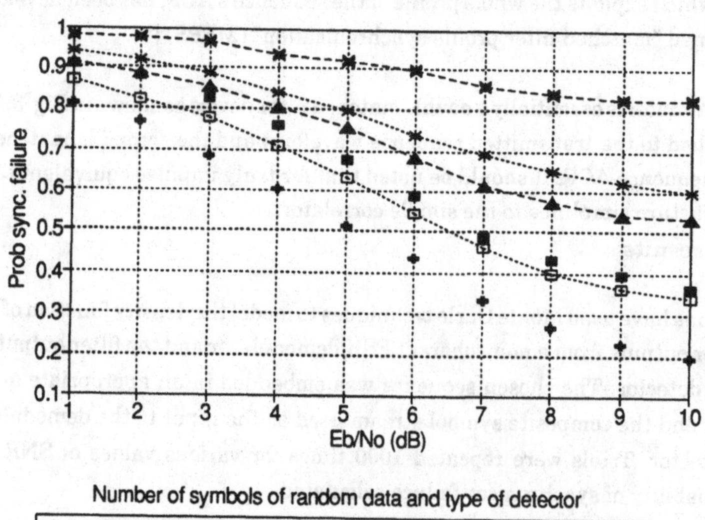

Figure 4 *Performance comparisons for the FSK detection algorithms using a 7 bit Barker code when surrounded by various amounts of random data*

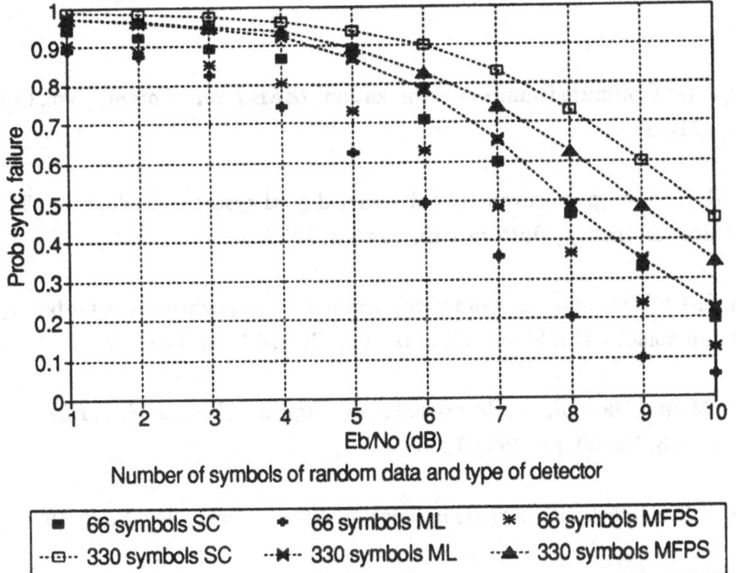

Figure 5 *Performance comparisons for the ASK detection algorithms using an 11 bit Barker code when surrounded by various amounts of random data*

Discussion and conclusions

By a rather empirical technique, an algorithm for FSK sequence detection has been derived from the original ML algorithm. Simulation trials have shown that this synchroniser offers a 2 dB performance improvement when compared with the simple correlator rule, as opposed to an observed 3 dB gain for the true ML algorithm with ASK transmission.

The double matched filter algorithm, whilst not employing soft decision information, utilises the complete profile of the sequence ACF. The results presented are for an 11 bit Barker code which has no positive sidelobes in its ACF. However, simulation trials have shown that with sequences with positive sidelobes, the ACF used in the search should not contain any positive peaks (except at zero time shift) so as to avoid falsely identifying the main peak in the correlator output as a possible sidelobe. The simulated performance improvements offered by using such a technique have been shown to be of the order of 1 dB.

In an attempt to try and mathematically predict this reported improvement, a mathematical model for the failure mechanism of the double correlator has been developed. This has been derived from a model for the simple correlator [6,8] which has been shown to produce an upper bound on the probability of failing to detect the sequence. Unfortunately, this model makes the assumption of statistical independence between separate estimator outputs. Hence, the second correlation used during the MFPS technique produces inaccurate results and valid comparisons between the two models cannot be made.

References

[1] Massey, J. L.: 'Optimum frame synchronization', *IEEE Trans. Comms.*, Vol. COM-20, No. 2, pp. 115-119

[2] Barker, R. H.: 'Group synchronizing of binary digital systems', in Jackson, W. (Ed.), 'Communication theory', Butterworth, London, 1953

[3] Nielsen, P. T.: 'Some optimum and suboptimum frame synchronizers for binary data in Gaussian noise', *IEEE Trans. Comms.*, Vol. COM-21, pp. 770-772

[4] Grayson, M. and Darnell, M.: 'Improved FSK frame synchronisation', *Electronics Letters*, Vol. 28, No. 20, pp. 1904-1905

[5] Whalen, A. D.: Detection of signals in noise, Academic Press, New York, 1971, p. 204

[6] Grayson, M. and Darnell, M.: 'Optimum synchronisation preamble design', *Electronics Letters*, Vol. 27, No. 1, pp. 36-38

[7] Grayson, M. and Darnell, M.: 'Adaptive synchronisation techniques for time-varying dispersive radio channels', Proceedings of the 5th International Conference on '*HF Systems and Techniques*', July 1991, pp. 261-264

[8] Grayson, M. and Darnell, M.: 'Synchronisation preamble design: New results', *Electronics Letters*, Vol. 26, No. 21, pp. 1775-1776

TRELLIS DECODING TECHNIQUE FOR ARRAY CODES

B. Honary and G.S. Markarian
Lancaster University, Lancaster, UK

M. Darnell
University of Hull, Hull, UK

L. Kaya
Lancaster University, Lancaster, UK

1. INTRODUCTION

Array codes were first introduced by Elias [1], and have been proposed for many burst and random-error control applications [2-5]. The essence of an array code is that the combination is based on a geometrical construction, the component codes are simple and decoding of array codes is relatively easy. The simplest array code is the row-and-column parity code, which also is called a two-coordinate, bidirectional, bit and block parity and has been widely used in data transmission systems and computer memories [2]. The code may be square or rectangular and has parameters (n_1n_2,k_1k_2,d_{min}), where (n_1,k_1) and (n_2,k_2) are row and column codes respectively, and minimum Hamming distance d_{min}=4. These codes are easy and flexible to design and relatively simple to decode. However these codes do not have the full power of block linear code of length $n=n_1n_2$, and the conventional decoding algorithms [2] for array codes, do not make use of maximum power of the code and are not maximum likelihood decoding algorithms.

In this paper we show that array codes can be decoded with a low complexity, fast, trellis decoding algorithm, which gives maximum likelihood performance. These codes can be used effectively in combined coding and modulation (CCM) schemes for band-limited channels to give higher transmission rates than the same complexity conventional CCM schemes based on convolutional codes. Section 2 introduces a simple procedure to construct the trellis diagram of any two-dimensional array code, and the algorithm is followed by a detailed example. The application of array codes in CCM schemes, and the bounds of the transmission rates for different types of CCM signals are presented in the Section 3. Section 4 presents computer simulation results for trellis decoding of array codes. Finally, conclusions and further work are mentioned.

2. TRELLIS REPRESENTATION OF ARRAY CODES

Trellis decoding was originally devised as a tool for decoding of convolutional codes [6], but it can also be used to decode block codes [7,8]. It has been shown by J.Wolf [7], that the trellis diagram of product code with symbols from GF(q) will have

$$N_s = \min\,[q^{k_1\cdot(n_2-k_2)},\ q^{k_2\cdot(n_1-k_1)}] \tag{1}$$

states and can be constructed using the parity check matrix. However, if the size of array increases, the number of calculations, required to design a trellis diagram increases dramatically.

We show that the trellis diagram of any array code with symbols from GF(q), where both row and column codes are single parity check codes, can be constructed easily. The number of states in this trellis diagram is given as:

$$N_s = q^{\min(n_1-1;\,n_2-1)} \tag{2}$$

The code can be square or rectangular in shape, and the trellis diagram can be obtained in the following manner.

1. Select the smaller of the two values n_1 and n_2 (say n_1).
2. Choose the number of states N_s, and the trellis depth (number of columns), N_c, in the trellis diagram as:

$$N_s = q^{(n_1-1)} \quad , \qquad N_c = n_2 \tag{3}$$

3. Identify each state at depth p, p = 0,1,...N_c-1, by a k_1-tuple q-ary vector, $S^p(A)=S^p(a_1,a_2,...a_{k1})$, where a_j = 0,1,...q-1, and j = 1,2,...k_1.
4. Identify the trellis branches which start and finish at depth p=0 and p = N_c-1, as S_0(00...0) and S_{n2}(00...0), respectively.
5. The trellis branches at each depth p, which connect two states $S_p(A)$ and $S_{p+1}(B)$, are labelled by a k_1-tuple q-ary vector $C^p{}_A$, which is determined in the following manner:

$$C_A^p = S_p(A) + S_{p+1}(B) \qquad \text{mod}\, q \tag{4}$$

where A and B are all possible k_1-tuple q-ary vectors, and addition is on modulo q.
6. Each branch on the trellis diagram is labelled also with the second value $D_p(A)=(a_1,a_2,...a_{n1})$, which represents the output vector.

There are $q^{(n1-1)(n2-1)}$ distinct paths through this trellis diagram, and each path corresponds to a unique codeword.

Example 1: Let the aim be to design the trellis diagram of the binary (9,4,4) array code, which consists of the (3,2) row code and the (3,2) column code.
1. Following the procedure, outlined above, we determine the number of states N_s = 4, and the trellis depth N_c = 4.
2. We identify each state at depth p (p=0,1,...4) by a 2-tuple binary vector $S_p(a_1,a_2)$, where a_i = 0,1, i=1,2.
3. At depth p=0 and p=4 the trellis has only one state, namely $S_0(00)$ and $S_3(00)$ respectively.
4. The trellis branches at depth p=0 are obtained as:

$$\begin{aligned} C_{00}^0(00) &= S_0(00) + S_1(00) = (00) \\ C_{00}^0(01) &= S_0(00) + S_1(01) = (01) \\ C_{00}^0(10) &= S_0(00) + S_1(10) = (10) \\ C_{00}^0(11) &= S_0(00) + S_1(11) = (11) \end{aligned} \tag{5}$$

and each state in second column, $S_1(a_1,a_2)$, is connected with the $S_0(00)$ by one line, labelled according to (5). The set of transition at next depth is obtained as follows:

$$
\begin{array}{ll}
C^1_{00}(00) = S_1(00) + S_2(00) & C^1_{01}(00) = S_1(01) + S_2(00) \\
C^1_{00}(01) = S_1(00) + S_2(01) & C^1_{01}(01) = S_1(01) + S_2(01) \\
C^1_{00}(10) = S_1(00) + S_2(10) & C^1_{01}(10) = S_1(01) + S_2(10) \\
C^1_{00}(11) = S_1(00) + S_2(11) & C^1_{01}(11) = S_1(01) + S_2(11) \\
C^1_{10}(00) = S_1(10) + S_2(00) & C^1_{11}(00) = S_1(11) + S_2(00) \\
C^1_{10}(01) = S_1(10) + S_2(01) & C^1_{11}(01) = S_1(11) + S_2(01) \\
C^1_{10}(10) = S_1(10) + S_2(10) & C^1_{11}(10) = S_1(11) + S_2(10) \\
C^1_{10}(11) = S_1(10) + S_2(11) & C^1_{11}(11) = S_1(11) + S_2(11)
\end{array}
\tag{6}
$$

and each state in the third column is connected to the previous set of states by 4 line, which are defined according to (6). Following a similar procedure, the trellis branches at the last step will be obtained as follows:

$$
\begin{array}{l}
C^2_{00}(00) = S_2(00) + S_3(00) = (00) \\
C^2_{01}(01) = S_2(01) + S_3(00) = (01) \\
C^2_{10}(10) = S_2(10) + S_3(00) = (10) \\
C^2_{11}(11) = S_2(11) + S_3(11) = (11)
\end{array}
\tag{7}
$$

and state $S_3(00)$ is connected to each state in the third column by one branch, which are labelled according to (7). The design of such a trellis diagram is very simple, and is shown in Fig.1.

The rate of array codes can be increased by adding columns (or rows) to the two-dimensional array. If the length of one dimension is kept constant (assume the number of rows is fixed), the number of states of the trellis diagram will be fixed, and the code rate can be increased by appending more columns to the trellis. When the number of trellis states is fixed the increase in the depth of trellis adds a little complexity to the decoding procedure. Figure 2 illustrates the trellis diagram of such a (3,2)(4,3) array code.

3. ARRAY CODES IN COMBINED CODING AND MODULATION SCHEMES.

The trellis representation of array codes allows us to design CCM with higher information digit per channel symbol rate in comparison with the same complexity conventional CCM technique, based on convolutional codes. In conventional CCM schemes using M-ary signals, the maximum transmission rate, R^c_{max} is:

$$R^c_{max} = M-1 \quad infor.bits/channel\ symbol \tag{8}$$

For example, using 4-PSK modulation with 1/2 rate convolutional code, $R^c_{max}=\log_2 4-1=1$ information bits per channel symbol. In the case of 8-PSK, using a 2/3 rate convolutional code, $R^c_{max} = 2$ info.bits/channel symbol (these rates could be increased by using concatenations, however the complexity of such a CCM scheme will be also increased).

The trellis structure of array codes enables us to increase the rate of information bits per channel symbol without increasing the CCM complexity. A sufficient condition for the design of CCM, based on M-ary signals and array code is as follows:

$$M = k_1, \qquad N_{ch} = n_2 \tag{9}$$

where, N_{ch} is the number of channel symbols per codeword. Each row of the array code has k_1 information bits, so there are q^{k1} possible combinations of length n_1, and we match each row with one of the M-ary signals. The transmission of the whole code requires n_2 channel symbols, so the transition rate for such a CCM scheme is $R_a=k/n_2$ info.digits/channel symbol. For instantce, if we employ the (9,4) array code in conjunction with M-ary (M=4) signals, we will need to transmit only three channel signals and R_a = 4/3 = 1.33 info.digits/channel symbol. This value is greater than R^c_{max} of the same complexity signalling set, using convolutional codes. If the depth (number of columns) of array increases, the R_a will increase respectively. For the (3,2)(10,9) array code with 4-ary signals, R_a = 18/10 = 1.8 info.digits/channel symbol. Table 1 shows information rates per channel symbol for most widely used signals in CCM employing array codes.

Table 1. Information rates for CCM, based on array codes.

	Row codes (3,2)	Row codes (4,3)
	Signalling set: M=4	Signalling set:M=8
Column codes	Information Rate	Information Rate
(3,2)	1.33	---
(4,3)	1.5	2.25
(5,4)	1.6	2.4
(6,5)	1.66	2.5
(7,6)	1.71	2.57
(8,7)	1.75	2.62
(9,8)	1.77	2.66
(10,9)	1.8	2.7

Lemma: In CCM based on array codes and M-ary signalling set, the maximum rate of information bits per channel symbol, R^a_{max}, is

$$R^a_{max} = \lim_{n_2 \to \infty} R^a = M \tag{10}$$

Proof:

$$R^a = \frac{Number\ of\ information\ digits}{Number\ of\ channel\ symbols} = \frac{k_1 k_2}{n_2} = \frac{k_1 k_2}{k_2 + 1};$$

$$\lim_{n_2,\, k_2 \to \infty} \frac{k_1 k_2}{k_2 + 1} = k_1 = M \tag{11}$$

Q.E.D.

Unlike the case of trellis coded modulation using convolutional codes

(TCM), the use of array codes in CCM is not based on set partitioning, which matches the coded bits to the signal points in a way that maximizes Euclidean distance between signal points. Thus the performance of an array code in CCM will be less than of a TCM achieved with the same number of trellis states.

4. COMPUTER SIMULATION RESULTS

The simulation test were carried out under Additive White Gaussian Noise (AWGN) channel conditions. The simulation results are plotted for the probability of bit error rate (BER) as a function of E_b/N_o, where E_b is the energy per information bit and N_o is equal to the noise variance. The results presented in this paper have been obtained for the binary unipolar signalling scheme, and for all cases perfect bit and block synchronisation are assumed.

The performance of unquantised trellis algorithm is compared with uncoded transmission and hard decision decoding. As is expected, the unquantised (or soft decision) trellis decoding has more than 2 dB better coding gain than hard decision, which is illustrated in Fig.3, for the (9,4,4) array code.

An interesting simulation result is that, as the array size increases the energy gain also rises respectively, see Fig.4. The simulation results suggest that the probability of bit error does not deteriorate with the increase in the array size for fixed number of trellis states. Figure 5 shows that the bit error rate of three different array sizes are nearly the same, while code rate shifts from 0.44 to 0.6.

The performance of the 4-state, half-rate (3,2)(4,3) array code has been compared with the half-rate convolutional code of constraint length three, which has d_{free}=5. Since convolutional codes have better minimum free distance properties than array codes, they can give additional performance improvement, and this has been confirmed with the simulation results in Fig.6. The performance of an array code is better for low E_b/N_o ratios, however, for high E_b/N_o values, the convolutional code provides about 1 dB more coding gain.

5. CONCLUSIONS

An easy trellis design procedure for array codes has been introduced in the paper. The trellis structure of array codes allows them to be used in combined coding and modulation schemes.

It has been shown that the combined coding and modulation schemes based on array codes provide higher information bit per channel symbol transmission rate, compared with the same complexity CCM based on convolutional codes.

The simulation results suggest that the probability of bit error does not deteriorate with increase of the array size for a fixed number of trellis states. Therefore array codes can be used in high rate applications, with a coding gain of about 2 dB. However, it is worth mentioning that asymptotically array codes give 3 dB coding gain in combined coding and modulation schemes.

REFERENCES

1. Elias P. Error free coding. -"IEEE Transactions", vol.IT-4, 1954, p.p.29-37.
2. Farrell P.G. Array codes. -"Algebraic Coding Theory and Applications". Ed. G.Longo, Springer-Verlag, 1979, p.p.231-242.
3. Burton H.O., Weldon E.J. Cyclic product codes. - "IEEE Transactions on Information Theory", vol. IT-11, July 1965, p.p.433-439.
4. Blaum M., Farrell P.G., Tilborg H.C.A. Multiple burst error correcting array codes. -"IEEE Transactions on Information Theory", vol.IT-34, September 1988, p.p.1061-1064.
5. Daniel J.S., Farrell P.G. Burst-error correcting array codes: Further Development. - "4-th International Conference on Digital Processing of Signals in Communications", Loughborough, April 1985, UK.
6. Viterbi A.J. Convolutional codes and their performance in communication systems. -"IEEE Transactions on Communications", col.COM-19, N:5, October 1971, p.p.751-772.
7. Wolf J.K. Efficient maximum likelihood decoding of linear block codes using a trellis.-"IEEE Transactions on Information Theory", vol.IT-28, No 2, 1982.
8. Forney G.D.Jr. Coset codes - Part 1: Introduction and Geometrical Classification. - "IEEE Transactions on Information Theory", vol.IT-34, No 5, 1988, p.p.1123-1151.
9. MacWilliams J.K., Sloane N.J. "The theory of error correcting codes." New York: North Holland, 1977.

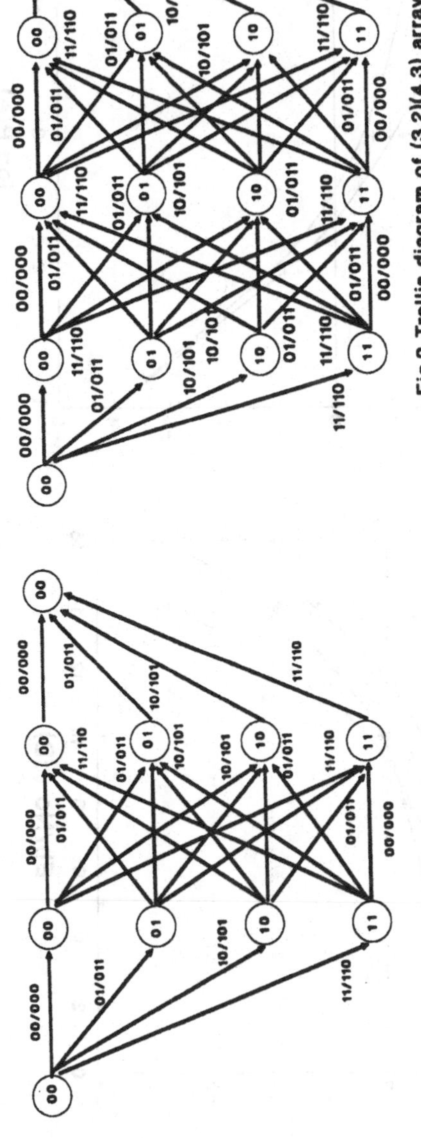

Fig.1.Trellis diagram of (3,2)(3,2) array code

Fig.2 Trellis diagram of (3,2)(4,3) array code

Fig.3. (9,4) array code
Trellis decoding

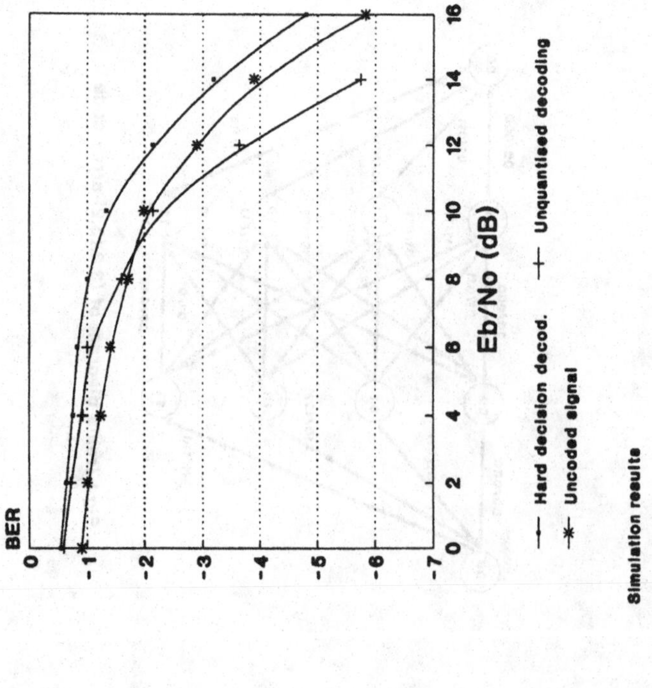
BER
0
-1
-2
-3
-4
-5
-6
-7
0
2
4
6
8
10
12
14
16
Eb/No (dB)
Hard decision decod.
Unquantised decoding
Uncoded signal
Simulation results

Fig.4. Energy Gains
Different array codes

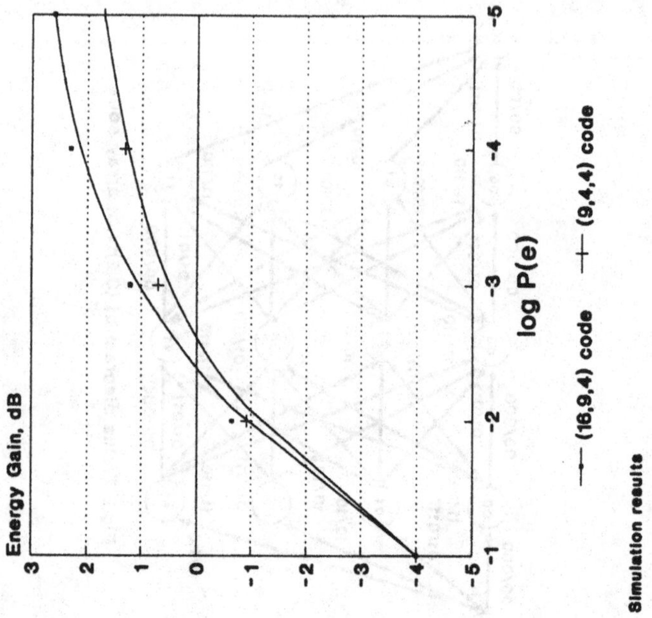
Energy Gain, dB
3
2
1
0
-1
-2
-3
-4
-5
-1
-2
-3
-4
-5
log P(e)
(16,9,4) code
(9,4,4) code
Simulation results

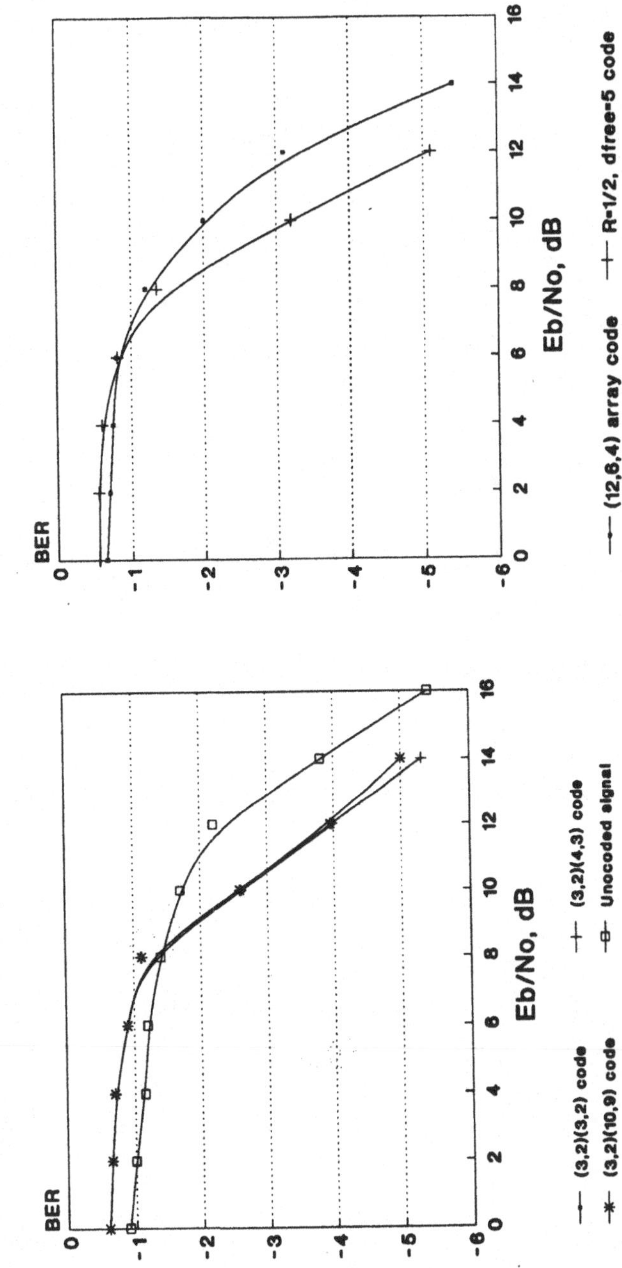
Fig.5. Trellis Decoding
Different array codes
BER
0
-1
-2
-3
-4
-5
-6
0
2
4
6
8
10
12
14
16
Eb/No, dB
(3,2)(3,2) code
(3,2)(10,9) code
(3,2)(4,3) code
Unecoded signal
Simulation Results
Fig.6. Trellis Decoding
Convolutional & array code
BER
0
-1
-2
-3
-4
-5
-6
0
2
4
6
8
10
12
14
16
Eb/No, dB
(12,6,4) array code
R=1/2, dfree=5 code
Simulation Results

7

CODING PERFORMANCE

WE CAN THINK OF GOOD CODES,
AND EVEN DECODE THEM

G. Battail
Ecole Nationale Supérieure des Télécommunications, Paris, France

ABSTRACT

"All codes are good, except those we can think of." This pessimistic opinion is widely shared among the coding theory community. We shall try instead to defend the optimistic statement of the above title. The word "good" is given in the quoted sentence the restricted meaning : "with a large minimum distance". From a communication engineering point of view, we were led to criticize this definition and to propose as a better criterion of goodness the proximity of the normalized distance distribution of a code with respect to that which results in the average from random coding, as measured for instance by the cross-entropy of these distributions. Looking for codes which are good in this sense for the Euclidean metric, as relevant to the additive white Gaussian noise channel, we show that combining a maximum distance separable code (e.g., a Reed-Solomon one) with an almost arbitrary one-to-one mapping of its q-ary symbols into a 2-dimensional constellation is a satisfactory solution provided q is large enough.

Near-capacity performance may be expected from such a *random-like* code, provided it can be decoded with reasonably low complexity. Using the Euclidean metric implies soft-decision decoding. We discuss two decoding algorithms for this purpose. The first one is sequential, as adapted to the finite context of a block code. It is potentially optimum, although practical considerations lead to limit the number of words to be tried. The second one is intrinsically nonoptimum and relies on an interpretation of a linear code as a kind of product of single-check codes which enables their cascaded weighted-output decoding. For large Reed-Solomon codes, none of these algorithms is fully satisfactory, but we suggest to combine them in order to obtain a near-optimum algorithm of acceptable complexity.

1. INTRODUCTION AND OUTLINE OF THE PAPER

Coffey and Goodman [1] recently gave a formal proof of a statement made by Wozencraft and Reiffen in 1961 namely, "Any code of which we cannot think is good". This paper is intended to show that some codes of which we can think are good. There is no logical contradiction, but we shall nevertheless criticize the rather technical meaning the word "good" is given in the above sentence. Section 2 is devoted to an attempt for answering the naive question : "What is a good code ?" We proposed an alternative definition of a good code which relies on the closeness of its distance distribution with respect to that of random coding, instead of the largeness of its minimum distance. Maximum distance separable (MDS) codes (e.g., Reed-Solomon ones) are good for this criterion, as applied to the Hamming distance. Considering the additive white Gaussian noise (AWGN) channel to which the Euclidean distance is relevant, we show in Section 3 that the system which results from MDS coding by one-to-one mapping of its q-ary alphabet into some almost arbitrary q-point 2-dimensional (say) constellation results in a Euclidean distance distribution close to that of random coding, too. Hence the combination of MDS coding and mapping is a good code for the AWGN channel according to our criterion.

Another point of fundamental importance is decoding. The availability of a good code is useless if there is no means for decoding it with reasonably low complexity. Optimal decoding for the Gaussian channel implies finding the codeword the closest to the received signal, according to the Euclidean metric, which is equivalent to optimal soft-decision decoding. In Section 4, we review a sequential soft-decision decoding algorithm for linear codes, hence which can be used for Reed-Solomon codes. It is optimal in principle, but its complexity forbids its use without some simplifications, which are possible at the expense of a slight performance impairment.

About decoding, we also proposed to change the game rule by extending its role to reassess the probabilities of the codewords, taking into account the code constraints, instead of merely determining the single codeword assumed to have been transmitted. Apparently, decoding as redefined becomes a heavier task. Since it is the most complex function to be performed in a communication link, it may seem unreasonable to still complicate it. Paradoxically, however, this is the key for alleviating the decoding process as a whole. The reason is that we may interpret any linear code as the intersection of several simple codes which contain it, as discussed in Section 5. Since decoding, as redefined, no longer incurs any information loss, it becomes possible to replace decoding of a single code by several successive decodings of its containing codes. Although the task of decoding a single containing code is actually complicated, the global complexity of decoding is widely diminished for a large-size code. Of course, strictly optimal decoding of the containing codes would still be exceedingly complex, but quasi-optimal decoding fortunately results in both a small loss of performance and a very large saving in complexity. Moreover, we show in Section 6 that the decoding algorithms discussed in Sections 4 and 5 can be combined in order to obtain a suboptimal algorithm which seems to provide an interesting compromise between performance and complexity.

The performance of random codes over the additive Gaussian channel was predicted by Shannon in the case of a finite number of dimensions [2]. Numerical computation of the error probability using the expressions he obtained results in a set of curves, for a given number of dimensions, as illustrated in Fig. 1. We may consider they provide a prediction of the performance which may be expected from the use of the combination of a Reed-Solomon code and some modulation, provided the average noise power is large with respect to the minimum squared distance between the constellation points. This condition is satisfied for a low enough signal-to-noise ratio as to make coding necessary.

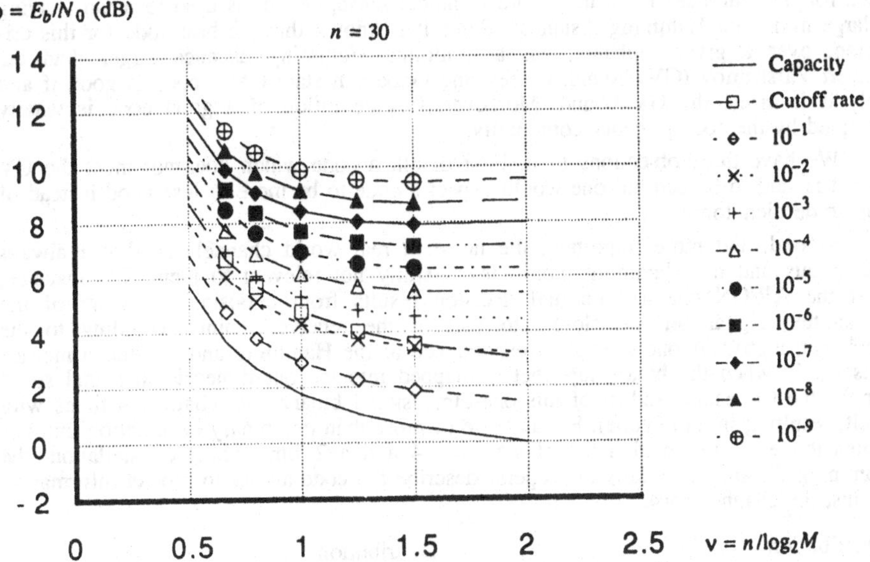

Fig. 1. Word error probability for constant energy random coding, computed according to Shannon [2].

The number of dimensions was assumed to be $n = 30$, in order to predict the attainable performance of a Reed-Solomon code of length $l = 15$ over GF(16) with a mapping of its symbols into a 2-dimensional 16-point constellation. The quantization effect due to the alphabet finiteness is neglected, a valid approximation only for a low enough signal-to-noise ratio.

The normalized signal-to-noise ratio $\rho = E_b/N_0$ (in decibels), where E_b is the received signal energy per binary information unit and N_0 the one-sided noise spectral density, was plotted in terms of the number of dimensions per binary information unit, $\nu = n/\log_2 M$, where M is the number of words belonging to the code. For a 2-dimensional constellation and a signal-to-noise ratio S/N, $\rho = \frac{\nu S}{2N}$. A family of curves in the plane (ν, ρ) results, each of them corresponding to a given word error probability, from 10^{-1} to 10^{-9} for integer values of the exponent. The channel capacity and its cutoff rate R_0 can be expressed in terms of the same parameters and the corresponding curves were also drawn in the figure.

2. WHAT IS A GOOD CODE ?

2.1. Criticism of the conventional criterion

According to the common sense, and from a communication engineering point of view, a good code is such that its use over a noisy channel results in an error probability reasonably close to the smallest possible one. Unfortunately, this definition is of no help for designing codes so it was necessary to replace it by a much simpler criterion concerning some code parameter. The minimum Hamming distance has been widely used for this purpose. Thus, in a more technical sense, "good" is used to mean "having a large minimum Hamming distance". Since it is known that the best code for this criterion, over a given alphabet, has a minimum Hamming distance larger than the Gilbert-Varshamov (GV) bound, Coffey and Goodman state that a code is good if and only if it meets the GV bound. Moreover, this definition of a good code is widely accepted by the coding theory community.

We have three objections to it. First of all, a code which does not meet the GV bound is said to be bad but one would expect a code to be more or less good instead of this crude dichotomy.

Second, and more important, the noise in real world channels is almost always continuous and the Hamming metric is generally not relevant to them. For instance, over the AWGN channel, optimal decision results from choosing the point of the transmitted signal set the closest to that of the received point, according to the Euclidean metric. A one-to-one correspondence of the Hamming and Euclidean metrics exists only when the binary alphabet is mapped into the one-dimensional signal space (or for the Cartesian product of this one-dimensional binary constellation m times with itself, resulting in an m-cube). For any rate of more than one binary information unit per dimension e.g., for an alphabet of size $q > 4$ and a 2-dimensional constellation, the Hamming distance thus fails to properly describe the code ability to protect information against the channel noise.

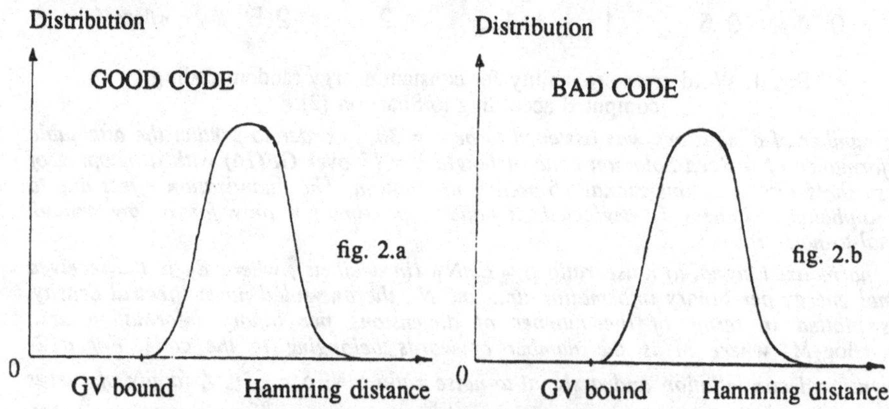

Fig. 2. Distance distribution of a good code (fig. 2.a), and of a bad code (fig. 2.b).

We plotted the normalized distribution of Hamming distances i.e., the ratio of the number of codeword pairs at a given distance to the total number of codeword pairs, in terms of this distance. We assume the code to be very large, so we may represent its distance distribution as a kind of continuous envelope. There is no visible difference (actually no difference at all) between the two drawings, but the distribution of the bad code has a peak of height 10^{-30}, *say, located at a distance below the Gilbert-Varshamov bound. Of course, this peak cannot be drawn and, if it could be, it could not be seen. There would be no noticeable difference in the performances of the "good" and the "bad" code, at least for reasonable error rates. This example depicts the situation experienced with the iterated product of single-parity check codes, for instance.*

Our third objection, too, is of basic importance. It concerns the choice of the *minimum* distance, whether Hamming or Euclidean, as the design parameter of a code. If a code has a large minimum distance, no doubt it is good according to the common sense i.e., results in a small error probability. The converse is not true, however, since large-size codes may have a poor minimum distance and still good performance (see Fig. 2). Actually, if only a very small fraction of the codeword pairs are at the minimum distance, it has no noticeable effect on the average error probability. An example of this case is given by the iterated product of simple codes. Such codes result in a vanishingly small error probability while keeping a nonzero rate as the length approaches infinity. As early as 1954, Elias noticed that the minimum distance criterion fails to apply to such codes [3]. We used such an iterated product of single-parity check codes over the Gaussian channel with weighted-output decoding at each stage, obtaining a vanishingly small error probability at a rate close to the channel cutoff rate [4]. As another example, (unexpurgated) random coding fails to have a good minimum distance since, strictly speaking, it has no minimum distance at all. The use of the minimum distance criterion is undisputable only if one assumes an almost noiseless channel. This led to the concept of "asymptotic coding gain" (especially in the field of coded modulations) where "asymptotic" means that the signal-to-noise ratio is assumed to increase indefinitely. In other words, the minimum distance criterion as well as the asymptotic coding gain are relevant only to the case where there is no need for coding. Very similar remarks are moreover valid for other codes specifically designed for the Euclidean space, like trellis-coded modulations and lattice codes.

2.2. Proposed criterion

Instead of the minimum distance, we proposed as a criterion a measure of similarity of the normalized distance distribution with respect to that obtained in the average by random coding using words of same length [4]. The Kullback divergence (or cross-entropy) of these distributions can conveniently be used for measuring the similarity of distributions. This criterion just recognizes that random coding is asymptotically (as the number of dimensions increases) the best and moreover answers the "more or less good" objection above. More precisely, we state that a family of codes is *asymptotically good* if the cross-entropy of its normalized distance distribution with respect to that of random coding approaches 0 as the code length tends to infinity. For a finite number of dimensions, we consider that a code whose normalized distance distribution is close to that of random coding is good, although better codes than the random one do exist in this case. Since we are interested in the AWGN channel, we apply this criterion to the Euclidean distance distribution. We shall in the following refer to this criterion, or to codes which satisfy it, as *random-like*.

Before going further, a remark should be made here. For an alphabet size q and a word length n, the uncoded case i.e., where $M = q^n$, results of course in the same distance distribution as obtained in the average by random coding. However, the "code" here is simply the Cartesian product of the alphabet n times with itself. This is a degenerate case where the benefit of coding i.e., of using an n-dimensional signal space, is lost since the signal space can be separated into n times the trivial 1-dimensional space. Thus, a necessary assumption is obviously that coding should be *redundant* i.e., the number M of codewords should be strictly smaller than q^n. In the most usual cases, the number of codewords is specified by the rate to be achieved. Thus, the alphabet size has to be chosen such that $\log_q M < n$. It will generally be assumed here that $M = q^k$, with k strictly less than n. We moreover assume that the code does not reduce to the Cartesian product of codes in signal spaces whose dimensionality is less than n.

In other words, we may think of coding as a process of selecting M words among the q^n n-tuples, to be referred to as a *decimation* process, and we assume that it cannot be described as separately operating in spaces of less than n dimensions. If it works independently, or almost independently, of the weight (or distance) of the codewords, then it keeps the overall shape of the weight (or distance) distribution of the set of all n-tuples, so it results in a good code according to the random-like criterion [4].

3. THE MDS CODES ARE ASYMPTOTICALLY GOOD FOR THE RANDOM-LIKE CRITERION

3.1. The MDS codes are good

Everybody is convinced that the Reed-Solomon codes are good due to their performance in actual operation. They are good according to the minimum Hamming distance criterion, since they belong to the MDS family, which means that they achieve the Singleton bound i.e., $d = n - k + 1$, where d, n and k are the minimum distance, the length and the dimensionality of the code, respectively [5]. This implies that they also exceed the Gilbert-Varshamov bound, although the near-equality of the code length n and the alphabet size q makes both increase when n increases (in contrast, the alphabet size is usually assumed to be a constant). We show that they are good according to our own criterion of proximity of distance distribution with respect to that of random coding, too. We shall assume they are used over the AWGN channel. This clearly implies that some modulation process is defined. We shall in the following restrict ourselves to the case where it consists of a one-to-one mapping of the q symbols of the alphabet into some q-point constellation (for instance, a 2-dimensional one). This constellation is almost arbitrary except that it should be symmetric in the sense that the set of distances between a given point and all others should not depend on the chosen point ; this condition is fulfilled by phase modulation, but it can actually be relaxed. The mapping itself is arbitrary since it is easily shown that the system performance does not depend on it.

If the alphabet size q and the number of dimensions k are large enough, then it is known that the Hamming weight distribution of the MDS codes is close to that obtained in the average by random coding [6]. If we assume that the q alphabet symbols are mapped into a q-point constellation, where the mapping is both random and variable with the symbol location, then the same is true for the Euclidean weight distribution (we define the Euclidean weight as the square of the Euclidean distance with respect to the point which represents the all-0 word). Moreover, we already noticed that an arbitrary deterministic mapping does not actually change the Euclidean weight distribution [7] (as illustrated in the example of Fig. 3). Since the distance distribution thus obtained approaches that of random coding as n goes to infinity, it becomes possible to approach the channel capacity by *purely deterministic and fully explicit means*.

3.2. Attempting an explanation

We must conclude that some codes having a very strong structure mimic random coding. We try to explain this paradox by the use of the algorithmic information theory of Kolmogorov and Chaitin [8] (interestingly, this theory was also invoked in the paper by Coffey and Goodman [1]). We may think of systematic coding as a means for associating a redundancy or check vector $\boldsymbol{s}$ with an information vector $\boldsymbol{u}$. This may be depicted according to Fig. 4, which shows a rectangular array of points where the rows are indexed by all the k-tuples and the columns by all the $(n-k)$-tuples. Any point corresponds to the n-tuple whose information vector is the row index and whose check vector is the column index, so this array represents the set of all n-tuples.

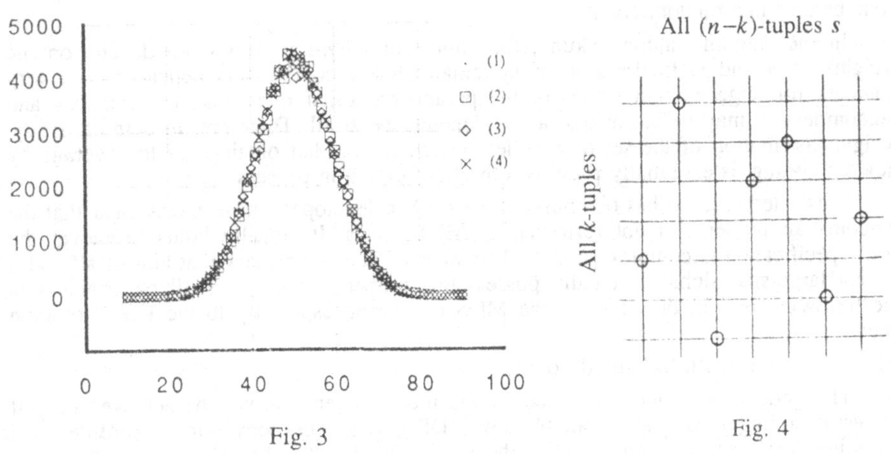

Fig. 3

Fig. 4

Fig. 3. Euclidean distance distributions.

Histograms of squared Euclidean distances obtained by MDS coding on the one hand, and in the average by random coding on the other hand. The code length was n = 30 and, in the former case, a (15,8) Reed-Solomon code over the 16-element field was used.

The mapping of the alphabet symbols into the 16 phases was : (1) random and varying with the symbol location ; (2) random but the same for all symbols in a given word ; (3) deterministic. For comparison purpose, the squared Euclidean distance distribution of randomly chosen 15-tuples was also drawn (4). The plots show the number of words found in each of 100 equally spaced intervals of the full possible range, for a total number of 100,000 words. No difference between the four histograms can be perceived. The cross-entropy of any of the corresponding normalized distributions with respect to that obtained in the average by random coding is about 10^{-3} *binary units.*

Fig. 4. Coding interpreted as a decimation process.

The set of all n-tuples is represented by an array of points where the rows are indexed by all the possible information k-vectors $\boldsymbol{u}$ *and the columns by all the check* $(n-k)$*-vectors s. A point of this array represents the n-tuple whose first k components are the vector which indexes its row and the last* $(n-k)$ *ones the vector which indexes its column. An (n,k) systematic coding results if a single point is chosen in each row, randomly or not (the circled points of the figure).*

Systematic (n,k) random coding results if one chooses at random a single point in each row (circled dots in the figure). In the case of linear systematic coding, a check vector s is associated with the information vector $\boldsymbol{u}$ according to the matrix equation $s = \boldsymbol{u}\mathbf{P}$, where $\boldsymbol{u}$ and s are row matrices, and $\mathbf{P}$ is a $k\times(n-k)$ matrix whose entries also belong to GF(q). For MDS codes, this matrix is "superregular" i.e., does not contain singular square submatrices of any order [5,9]. This property is clearly invariant by any column and row permutation, so a single superregular matrix actually belongs to a large set of matrices which also possess this property. Therefore, specifying a single matrix is complex enough, if the parameters n and k are large enough, to make the corresponding code behave like random coding.

In the case of random coding, the choice of an n-tuple does not depend on the weights of u and s. In the case of systematic linear coding, this dependence is very complex for large enough values of the parameters so, if one equates complexity and randomness, it may be considered as no dependence at all. Therefore, in both cases, the weight distribution of the set of n-tuples, which equals that obtained in the average by random coding, is essentially preserved by the decimation process.

This interpretation has obviously to be further developed but it makes clear that the random-like property is not restricted to MDS codes. It actually holds whenever the code specification is complex enough. This meets Cheung's remark that almost all codes over a large-size alphabet actually possess this property [10]. We shall nevertheless in the following restrict ourselves to the MDS codes and especially to the Reed-Solomon ones.

3.3. Necessity of soft decision decoding

The good performance promised by the previous remarks will be achieved only if proper decoding means are available for MDS codes. The conventional hard-decision decoding does not take into account the Euclidean distance between the signal which represents a codeword and the received signal, which may be thought of as an information loss. This loss can be avoided only by weighted, or *soft decision*, decoding, which takes into account the symbol *a priori* probabilities or, equivalently, uses the Euclidean metric. Therefore, we are led to look at weighted decoding algorithms which can be applied to MDS codes and especially to Reed-Solomon ones.

We may divide such algorithms into two broad classes. The first one contains the algorithms which exploit fine structural properties of the code. These algorithms are thus tailor-made for a specific code or a narrow family of codes. The second class is that of algorithms which rely only on the basic properties which are common to all members of a broad family of codes e.g., in the case of Reed-Solomon codes, on linearity and on the fact that the choice of k arbitrary symbols anywhere in the word always specifies a codeword. Such algorithms are the block-code equivalent of the sequential decoding and the Viterbi algorithm, which apply to any convolutional code.

Several algorithms belonging to the first class have already been reported. May be the second family of algorithms is less known, but we shall in the following restrict ourselves to it. Similarly to convolutional codes, block coding can be described by the trellis (or tree) diagram when properly extended to block codes [11-14]. Therefore, the sequential decoding and the Viterbi algorithm could in principle be used. If one attempts to apply it to Reed-Solomon codes, the Viterbi algorithm is of prohibitive complexity due to the number of states q^{n-k}, very large for reasonable values of the parameters. We shall first briefly consider a sequential decoding algorithm adapted to the case of block codes. This algorithm is optimal in its principle, although its practical implementation leads to use simplified versions of it. The volume of computation needed in order to decode a word is a random variable but it is possible, and in practice necessary, to fix an upper limit to the number of operations effected in order to decode a word. The algorithm is then no longer optimal, but the resulting impairment is small if the maximum number of trials is chosen large enough.

We shall then in Section 5 consider another algorithm which relies on a different principle namely, where the decoding process is decomposed into successive decodings of single-check symbol codes, each of which containing the original code. Such an algorithm is inherently nonoptimal but potentially simpler.

We already noticed that q-ary phase modulation fulfills the symmetry condition which, strictly speaking, is necessary for obtaining a random-like Euclidean distance distribution with an MDS code. Since for such a code a large codelength implies a large size alphabet, the minimum distance of the constellation is very small so the symbol error probability becomes very high. From a hard decision point of view, the decoding performance would be very poor. It is not necessarily so in a soft decision context although this fact may complicate the decoding process. We are currently studying means for coping with this difficulty. For instance, one can use a transmission alphabet of size $q' < q$ and actually transmit the n-tuple in this alphabet the closest to the "theoretical" codeword, i.e. which belongs to the MDS code over the q-ary alphabet. The difference between the theoretical codeword and the one actually transmitted (using the q'-ary alphabet) results in increasing the added noise but, for properly chosen parameters, this increase can be neglected at low signal-to-noise ratios for which the code is useful. The use of non-MDS codes, in order to avoid the dependence between the code length and the alphabet size, may also be contemplated.

4. SEQUENTIAL ALGORITHM

Let us first define the metric to be used. Let A and B denote the channel input and output alphabets, respectively. A is discrete but card(B) may be infinite e.g., the received symbols may be real numbers as, for instance, in the case of the unquantized additive Gaussian noise channel. The channel is assumed to be memoryless, so it is characterized by the transition probabilities that $b \in B$ is received when $a_i \in A$ is transmitted, namely $\Pr(b|a_i)$ (or by the corresponding probability densities). Let $\mathrm{h}(b)$ denote the hard decision on b i.e., the symbol of A such that

$$\Pr(b|\mathrm{h}(b)) = \max_{a \in A} \Pr(b|a) \tag{1}$$

and let us introduce the following nonnegative quantities, to be referred to as generalized weights :

$$\mathrm{v}(a_i,b) = \log[\Pr(b|\mathrm{h}(b))] - \log[\Pr(b|a_i)], \tag{2}$$

where the logarithms are to an arbitrary base. (Since only a ratio of probabilities is actually involved in this definition, the extension to the continuous case e.g., the unquantized one, is straightforward; it merely results from replacing the probabilities by probability densities.)

We may now associate with each received symbol $b \in B$ a q-vector (row matrix) $\mathrm{v}(b) = [\mathrm{v}(a_i,b)]$, $i = 0, 1, \cdots q-1$. Let $\boldsymbol{c}^m \in A^n = [c_1^m \cdots c_n^m]$ denote the m-th codeword and $\boldsymbol{r} \in B^n = [r_1 \cdots r_n]$ be a received vector, then the generalized distance between $\boldsymbol{r}$ and $\boldsymbol{c}^m$, or metric, is defined as :

$$\mathrm{Z}(\boldsymbol{r}, \boldsymbol{c}^m) = \sum_{j=1}^{n} \mathrm{v}(r_j, c_j^m). \tag{3}$$

Let us assume that the code is given in systematic form. Since the generalized distance is additive with respect to the word symbols, we may write (3) as the sum of two terms :

$$\mathrm{Z}(\boldsymbol{r}, \boldsymbol{c}^m) = \mathrm{Z}_u(\boldsymbol{r}, \boldsymbol{c}^m) + \mathrm{Z}_s(\boldsymbol{r}, \boldsymbol{c}^m), \tag{4}$$

where $\mathrm{Z}_u(\boldsymbol{r}, \boldsymbol{c}^m)$ and $\mathrm{Z}_s(\boldsymbol{r}, \boldsymbol{c}^m)$ are the contributions of the information and check symbols to the metric, respectively. Optimal decoding consists of determining $\hat{m}$ such that $Z(r, c^{\hat{m}}) < Z(r, c^{m})$ for any $m \neq \hat{m}$. $Z_u(r, c)$ is immediately determined once the information vector u which corresponds to a codeword c is given. On the other hand, $Z_s(r, c)$ can be obtained once the check symbols of c have been determined i.e., after u has been encoded. The algorithm is designed in order to minimize the average number of information vectors u which need be encoded in order to determine $\hat{m}$.

We are now able to briefly describe the sequential algorithm [15-17]. First of all, the received symbols are reordered by nonincreasing reliability of the corresponding hard decisions. The first k symbols after this reordering are taken as information symbols and the corresponding generator matrix in systematic form is computed, which is always possible for an MDS code. The vector $u^h = [h(r_1) \cdots h(r_k)]$, where $h(r_i)$ denotes the hard decision about the i-th received symbol r_i, determines a first codeword c^h such that $Z_u(r, c^h) = 0$. Encoding u^h enables to compute $Z(r, c^h)$ according to (4) and a threshold is set at this value. Then, other information vectors u, corresponding to vectors c which belong to the code, are tried in the order of non-decreasing $Z_u(r, c)$ and the generalized distance $Z(r, c)$ which corresponds to them is computed. If $Z(r, c)$ is found for some c less than the current threshold, then $Z(r, c)$ and c become the new

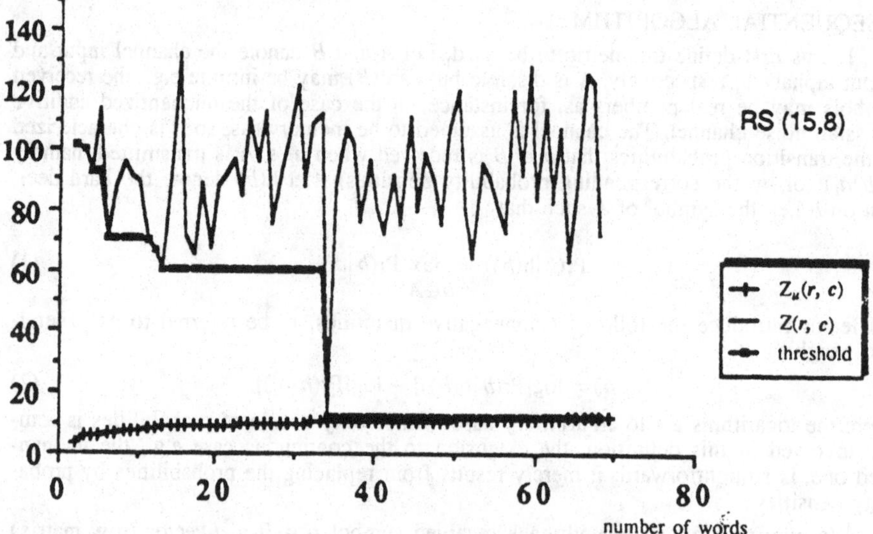

Fig. 5. Behaviour of the sequential algorithm.

The generalized distance $Z_u(r, c)$ associated with the information symbols, the one associated with the corresponding codeword $Z(r, c)$, and the threshold were plotted in terms of the number of words tried, for decoding a given received word r (belonging to a large sample of received RS(15,8) words whose transmission in the presence of additive Gaussian noise was simulated). The 34-th word tried is the best one, but $Z_u(r, c)$ exceeds the current threshold i.e., the previous word is known to be the best, only after the 69-th one has been tried. The difference between the indices of the best word and the last to be tried one is not very large here, but in most instances it is much larger, so the best codeword is typically found much before one becomes certain it is actually the best.

threshold and the last provisional candidate to the final decoding decision, respectively. The provisional candidate becomes the final decoding decision as soon as $\mathbf{Z}_u(\mathbf{r}, \mathbf{c})$ exceeds the current threshold.

The average complexity of this algorithm is small, but exactly optimal decoding may demand a high computation effort for certain (seldom met) received vectors. It is why suboptimal decoding should be preferred in practice. A straightforward simplification results from upper limiting the number of trials.

Interestingly, the average number of words which need be tried before the best one is found is much smaller than the average number of trials before the algorithm stops. In other words, it takes in the average much less time to find the best word than to make sure that the best one was found. It is why limiting the number of trials is possible with little performance impairment. Moreover, the later the best codeword is found, the more likely does it differ from the actually transmitted word. Fig. 5 shows the behaviour of the algorithm when decoding a particular received word.

5. CASCADED DECODING OF REED-SOLOMON CODES

We now describe the other algorithm. We begin with interpreting an arbitrary linear code as an intersection of "containing codes".

5.1 Interpreting a linear code as the intersection of simple containing codes

Let C be an arbitrary (n, k) linear code. Let $\mathbf{G}$ be a generator matrix of it in the systematic form $\mathbf{G} = [\mathbf{I}_k \mid \mathbf{P}]$, where $\mathbf{I}_k$ is the unit matrix of order k and $\mathbf{P}$ is a matrix having k rows and $n-k$ columns. Let $\boldsymbol{u}$ be the vector whose components are the k information symbols to be encoded, written as a row matrix. The result of encoding $\boldsymbol{u}$ according to C is the word $\boldsymbol{v} = [v_0\ v_1\ \cdots\ v_{n-1}]$ given by the matrix product $\boldsymbol{v} = \boldsymbol{u}\mathbf{G}$. Let us consider the matrix

$$\mathbf{T} = \begin{bmatrix} \mathbf{I}_k & \mathbf{P} \\ \mathbf{0} & -\mathbf{I}_{n-k} \end{bmatrix}, \tag{5}$$

whose first k rows are the generator matrix $\mathbf{G}$ of C while its $n-k$ last columns are a transposed check matrix of it, $\mathbf{H}^T$ [18]. One readily verifies that $\mathbf{T}$ is its own inverse and that its row space is the set E of all n-tuples. Let $\boldsymbol{u}'$ be the vector of dimension n which results from appending to $\boldsymbol{u} = [u_0\ u_1\ \cdots\ u_{k-1}]$ $n-k$ zero components : $u'_i = u_i$ if $0 \le i < k$, $u'_i = 0$ if $k \le i < n$. Using this definition, we have $\boldsymbol{v} = \boldsymbol{u}'\mathbf{T}$, hence

$$\boldsymbol{v}\mathbf{T} = \boldsymbol{u}'. \tag{7}$$

By expressing that the last $n-k$ components of $\boldsymbol{u}'$ are 0 in (7), $n-k$ scalar equalities written in vector form result :

$$\boldsymbol{v}\mathbf{H}^T = 0, \tag{8}$$

which are the check equations of the code. (Incidentally, this interpretation illustrates the remark of Section 2.2 about redundancy: $\boldsymbol{u}$ is trivially decomposed into k information symbols but, once $\boldsymbol{u}'$ has been right multiplied by $\mathbf{T}$, the information it bears is spread into the whole n-dimensional word which can no longer be decomposed.)

Let then C_1 be the set of words which result from right multiplying by $\mathbf{T}$ the vectors of dimension n whose symbol in a given position (k for instance) is 0, C_2 the set of words which result from right multiplying by $\mathbf{T}$ the vectors of dimension n whose symbols in two positions (e.g., k and $k+1$) are 0, etc. C_1 is also the set of words which satisfy one of the equalities (8) (the first one in the example), C_2 the set of words which satisfy 2 of them (e.g., the first 2 ones), ... C_{n-k} the set of words which satisfy all of them i.e., the code C itself. In other words, a check matrix of the set of words C_1 results from transposing a column of matrix **T** (e.g., the k-th one) , a check matrix of C_2 results from transposing two columns (both the k-th and (k+1)-th in the example), ... finally, a check matrix of C_{n-k} (= C) results from transposing its last $n-k$ columns. It is moreover clear that, E denoting the set of n-tuples over the q-ary alphabet, that

$$E \supset C_1 \supset C_2 \supset \cdots \supset C_{n-k} = C \tag{9}$$

since the same inclusion relations are true for the set of words before their multiplication by matrix **T**. The rightmost set in (9), C_{n-k}, is the code to be decoded. The order in which the zero components of u' are taken for defining the subsets is moreover arbitrary.

We proposed to extend the decoding role to reassess the *a posteriori* probabilities that the words are transmitted taking into account the code constraints, given the probabilities of each symbol separately (referred to as *a priori*) and thus ignoring these constraints. Then the decoding output is weighted as its input is. Optimal word by word decoding thus redefined, in its exact form, simply consists of renormalizing the *a priori* probabilities defined on the set E of the n-tuples, by restriction to the words which belong to the code [19]. Clearly, this restriction can be effected step by step, for instance from E to C_1, then from C_1 to C_2 and so on up to $C_{n-k} = C$, without any degradation. The interest of this step by step restriction will now become clear.

Each of the $n-k$ equalities (8) may be interpreted as defining a code (k+1, k) having a single check symbol (which moreover may be arbitrarily chosen due to the symmetry) computed in terms of the information symbols. Such a code, homologous to the single-parity check binary code, will be referred to in the following as a *simple code*. Then, restricting the set E to C_1 corresponds to word by word decoding of the simple code which has v_k (for instance) as check symbol, restricting C_1 to C_2 corresponds to decoding the result of decoding C_1 (the output of the previous decoder) with respect to the simple code having v_{k+1} as check symbol, etc, until the final decoding which uses v_{n-1} as check symbol.

Calculation of the *a posteriori* probabilities of the codewords when decoding a simple code would be of prohibitive complexity since they are as many as q^k. We proposed to perform it in the approximate form of a product of probabilities separately associated with each of the k information symbols [19]. In other words, the k parameters which specify a codeword are dealt with as independent. The number of probabilities which have to be computed is thus reduced to $(q-1)k$. Instead of word by word decoding, we have now to perform weighted-output *symbol by symbol* decoding, which is easy for a simple code. The decoding output of a given containing code is used as input for another one, until all are used. Moreover, the process just described is identical to the initialisation procedure of decoding by iteratively solving a system of implicit equations, as proposed in [20]. This initialisation procedure is in itself a decoding process, which is moderately improved by the final solution of the system of equations but obviously simpler.

To summarize, we decode each of the simple codes defined by equalities (8), symbol by symbol and with weighted output, the output of a particular decoder being used as the input of the following one. This process is continued until all these simple codes have been used. It would be optimal according to (9) if decoding would exactly reassess the word probabilities. Symbol by symbol decoding is a means for approximately performing it without prohibitive complexity. This approximation is the best possible if the code symbols are reordered by decreasing reliability, as in the sequential algorithm but for a different reason: in the sequential algorithm, this reordering resulted in decreasing the number of words to be tried before the optimal decoded word is found; here, it determines the quality of the approximation, hence of the result itself. We shall not give more details about this algorithm (of which many variants and refinements can be conceived) and we shall refer to it in the following as *cascaded decoding*.

5.2 Application of cascaded decoding to Reed-Solomon codes

Since the above decoding process pertains to any linear code, it can be used for Reed-Solomon codes. A generator matrix **G** of a Reed-Solomon code $C(n,k)$ has k rows and n columns and its entries are given by

$$g_i^j = \alpha^{ij},\ 0 \leq i < k,\ 0 \leq j < n, \tag{10}$$

where α is a primitive element of the field GF(q). The code length is $n = q - 1$.

In terms of the vectors $\boldsymbol{u}$ and $\boldsymbol{u}'$ already introduced, the codeword $\boldsymbol{v} = [v_0\ v_1\ \cdots\ v_{n-1}]$ which corresponds to $\boldsymbol{u}$ results from the matrix product $\boldsymbol{v} = \boldsymbol{u}\mathbf{G}$, and we have

$$\boldsymbol{v} = \boldsymbol{u}'\mathbf{W}, \tag{11}$$

where **W** is the $n \times n$ matrix whose entries are $w_i^j = \alpha^{ij}$, $0 \leq i, j < n$) since the first k rows of **W** are identical to matrix **G** defined in (10). The product by matrix **W** directly extends to the field GF(q) the *inverse Fourier transform* since α is an n-th root of unity in this field. One may thus interpret an (n, k) Reed-Solomon code as the set of vectors obtained by the inverse Fourier transform of vectors whose first k components are arbitrary while the last $n-k$ ones are 0 [21]. This last condition just expresses the code redundancy. Equality (11) results in

$$\boldsymbol{u}' = \boldsymbol{v}\,\mathbf{W}^{-1}, \tag{12}$$

and writing that the last $n-k$ components of $\boldsymbol{u}'$ are 0 results in the check equations of the code :

$$\sum_{m=0}^{n-1} v_m\, \alpha^{-(j+k)m} = 0, \quad 0 \leq j < n-k. \tag{13}$$

We notice that (12) results from (7) by substituting matrix $\mathbf{W}^{-1}$ for **T**. Equalities (7) and (8) apply to a Reed-Solomon code as well as to any linear code, but the generator matrix **G** should be expressed in the systematic form $\mathbf{G}_s = [\mathbf{I}_k \mid \mathbf{P}]$. The corresponding form of the check matrix is then

$$\mathbf{H}_s = [\mathbf{Q} \mid \mathbf{I}_{n-k}], \tag{14}$$

where $\mathbf{Q} = -\mathbf{P}^T$ is a matrix with $n-k$ rows and k columns whose entries belong to the field GF(q) and $\mathbf{I}_{n-k}$ denotes the unity matrix of order $n-k$. The last $n-k$ columns of **T**, according to (1), are then identical to $-\mathbf{H}_s^T$.

In the case of Reed-Solomon codes, where matrix **P** results from putting in systematic form the first k rows of matrix $\mathbf{W}^{-1}$, **P** and hence **Q** possess the important property of *superregularity* already referred to in Section 3. This property is actually shared by the family of linear maximum distance separable MDS codes to which the Reed-Solomon ones belong [5]. No square submatrix of **P** or **Q**, of any order, is singular, so in particular no entry in these matrices is 0. The check equations (8) of the code then read

$$\sum_{m=0}^{k-1} v_m\, h_j^m + v_{k+j-1} = \sum_{m=0}^{n-1} v_m\, h_j^m = 0, \quad 0 \leq j < n-k, \tag{15}$$

where h_j^m is an entry of $\mathbf{H}_s$. Due to the systematic form (14) of matrix $\mathbf{H}_s$, the first k symbols of a word, namely $v_0, v_1, \cdots v_{k-1}$, are its information symbols $u_0, u_1, \cdots u_{k-1}$.

Thus, each of the $n-k$ equalities (15) may be interpreted as defining a simple code $(k+1, k)$ whose single check symbol v_{k+j-1} is computed in terms of *all* the information symbols $\{v_i\}$, $0 \le i < k$, of the original Reed-Solomon code (for a non-MDS code, all the information symbols would not be simultaneously present in each check equation). We thus first decode the k information symbols of the simple code which is defined by one of the check equations (15). The result of this decoding is used as input for decoding another of the simple codes defined by (15) and this process is continued until all such simple codes were used. For weighted-output symbol-by-symbol decoding of the containing codes, we used replication decoding as extended to nonbinary codes [22,14].

6. COMBINING THE SEQUENTIAL AND CASCADED ALGORITHMS

Both the sequential and the cascaded decoding algorithms work satisfactorily when they are applied to short codes (e.g., Reed-Solomon codes over GF(8)) but some difficulties are experienced for larger codes :

- the sequential algorithm enables limiting the search among the codewords to a very small fraction of them. Only the metric of the codewords which belong to this fraction need be calculated. However, when the number of codewords $M = q^k$ is very large (e.g., $q = 32$ and $k = 20$ result in $M = 2^{100}$, over 10^{30}) even a small fraction of such a number is prohibitively large. If for the sake of simplification the search is limited to a too small number of codewords, performance impairment results ;
- the cascaded decoding algorithm does not suffer an exponential increase of complexity but it relies on the assumption that the decoded information symbols may be dealt with as statistically independent, in the sense that the *a posteriori* probability of a codeword can be estimated by the product of the *a posteriori* probabilities of its information symbols. It is a crude estimate but it appears difficult to avoid this assumption. When the number of check symbols, $n - k$, is large, many successive decodings are cascaded and the weighting of their outputs becomes less and less significant as the number of successive decodings increases. A less and less efficient use is thus made of the code redundancy.

Even though the cascaded decoding algorithm in its present form fails to reach a performance close to the optimal one as far as the choice of the best codeword is concerned, it nevertheless widely improves the codeword probability estimates with respect to those solely based on the *a priori* probabilities of the information symbols since it takes into account the code constraints. But trying successive words in the sequential algorithm according to increasing values of $Z_u(\mathbf{r}, \mathbf{c})$ precisely means that only the probability estimates based on the *a priori* probabilities of the information symbols are used for choosing these words. Therefore, the sequential and the cascaded algorithms may be combined by successively performing the following steps :

- the reordering step, which is common to both ;
- the cascaded decoding algorithm, resulting in estimates of the *a posteriori* probabilities of the information symbols ;
- finally, the sequential algorithm, but where the *a posteriori* probability estimates obtained at the previous step are used in order to choose the successive words, instead of the *a priori* ones. The metrics of these words remain however computed in terms of the *a priori* data.

Thanks to this combination, only the few words which are plausible candidates according to the cascaded algorithm need be tried, but the possible inaccuracy of its estimates of the *a posteriori* probabilities does not impair the final decision.

7. CONCLUSION

Finding good codes according to the conventional minimum distance criterion is indeed a very difficult task. From our communication engineering point of view, the very strong requirement of a large minimum distance appears as unnecessary. MDS and especially Reed-Solomon codes, which were discovered more than 30 years ago, are good also according to the proposed criterion of proximity of distance distribution with respect to that of random coding. For the AWGN channel and the Euclidean metric, this remains true when an almost arbitrary modulation is used in order to map the alphabet symbols into a two-dimensional constellation. The asymptotic Euclidean distance distribution of the code-and-modulation system thus obtained is therefore the same as that of random coding i.e., the channel capacity can in principle be achieved.

Not surprisingly, the challenging problem remains *decoding*. Maximum likelihood decoding of a linear code is known to be an NP-complete problem, so there is no hope to find an exactly optimal algorithm of reasonable complexity. We believe that the problem should rather be stated: "Find an *almost* optimal decoding algorithm" and that tremendous simplification is possible at the expense of little performance impairment. We briefly described above two algorithms intended to this aim and we hope the combination of both will provide a useful compromise between performance and complexity.

ACKNOWLEDGEMENTS

Some of the results mentioned in Section 4., including Fig. 5, are due to L. Cuvelier and R. Ryckaert of UCL, Louvain-la-Neuve, Belgium, who worked at Telecom Paris in 1991 on the sequential decoding of Reed-Solomon codes. The author expresses his thanks to the organizers of the Conference for giving him the opportunity to express his views in plenary session. He also thanks the attendees who participated in the discussion of his paper, especially Dr P. Piret for bringing to the audience attention his earlier work [23] which shows the feasibility of good codes in an algebraic context.

REFERENCES

[1] J.T. COFFEY and R.M. GOODMAN, Any Code of Which we Cannot Think is Good, IEEE Trans. Inf. Th., vol. **36**, n° 6, Nov. 1990, pp. 1453-1461

[2] C.E. SHANNON, Probability of Error for Optimal Codes in a Gaussian Channel, BSTJ, vol. **38**, n° 3, May 1959, pp. 611-656

[3] P. ELIAS, Error-Free Coding, IRE Trans. Inf. Th., 1954, pp. 29-37

[4] G. BATTAIL, Construction explicite de bons codes longs, Annales Télécommunic., vol. **44**, n° 7-8, July-Aug. 1989, pp. 392-404

[5] R.C. SINGLETON, Maximum Distance Q-nary Codes, IEEE Trans. Inf. Th., vol. **IT-10**, Apr. 1964, pp. 116-118

[6] K.M. CHEUNG, More on the Decoder Error Probability for Reed-Solomon Codes, IEEE Trans. Inf. Th., vol. **35**, n° 4, July 1989, pp. 895-900

[7] G. BATTAIL, H. MAGALHÃES de OLIVEIRA and ZHANG W., Coding and Modulation for the Gaussian Channel, in the Absence or in the Presence of Fluctuations, EUROCODE 90, Udine, Italy, 5-9 Nov. 1990 ; Lecture Notes in Computer Science, Springer Verlag, n° 514, pp. 337-349

[8] G.J. CHAITIN, Algorithmic Information Theory, Cambridge Univ. Press, 1987

[9] R.M. ROTH and A. LEMPEL, On MDS Codes via Cauchy Matrices, IEEE Trans. Inf. Th., vol. **IT-35**, n° 6, Nov. 1989, pp. 1314-1319

[10] K.-M. CHEUNG, Identities and approximations for the weight distribution of q-ary codes, IEEE Trans. on Inf. Th., vol. **36**, n° 5, sept. 1990, pp. 1149-1153

[11] L.R. BAHL, J. COCKE, F. JELINEK and J. RAVIV, Optimal Decoding of Linear Codes for Minimizing Symbol Error Rate, IEEE Trans. Inf. Th., vol. **IT-20**, n° 2, Mar. 1974, pp. 284-287

[12] G. BATTAIL and M.C. DECOUVELAERE, Décodage par répliques, Annales Télécommunic., vol. **31**, n° 11-12, nov.-déc. 1976, pp. 387-404

[13] J.K. WOLF, Efficient Maximum Likelihood Decoding of Linear Block Codes Using a Trellis, IEEE Trans. Inf. Th., vol. **IT-24**, n° 1, Jan. 1978, pp. 76-80

[14] G. BATTAIL, M. DECOUVELAERE and P. GODLEWSKI, Replication Decoding, IEEE Trans. Inf. Th., vol. **IT-25**, n° 3, May 1979, pp. 332-345

[15] G. BATTAIL, Décodage pondéré optimal des codes linéaires en blocs I.- Emploi simplifié du diagramme du treillis, Annales Télécommunic., vol. **38**, n° 11-12, Nov.-Dec. 1983, pp. 443-459

[16] G. BATTAIL and J. FANG, Décodage pondéré optimal des codes linéaires en blocs II. - Analyse et résultats de simulation, Annales Télécommunic., vol. **41**, n° 11-12, Nov.-Dec. 1986, pp. 580-604

[17] J. FANG, Décodage pondéré des codes en blocs et quelques sujets sur la complexité du décodage, ENST Doctor Thesis, Mar. 18, 1987

[18] G. BATTAIL, Description polynomiale des codes en blocs linéaires, Annales Télécommunic., vol. **38**, n° 1-2, Jan.-Feb. 1983, pp. 3-15

[19] G. BATTAIL, Le décodage pondéré en tant que procédé de réévaluation d'une distribution de probabilité, Annales Télécommunic., vol. **42**, n° 9-10, Sept.-Oct. 1987, pp. 499-509

[20] G. BATTAIL, Décodage par résolution d'un système d'équations implicites analogiques, Annales Télécommunic., vol. **45**, n° 7-8, July-Aug. 1990, pp. 393-409

[21] R.E. BLAHUT, A Universal Reed-Solomon Decoder, IBM J. Res. & Dev., vol. **28**, n° 2, Mar. 1984, pp. 150-158

[22] G. BATTAIL and P. GODLEWSKI, Emploi de représentations polynomiales à plusieurs indéterminées pour le décodage des codes redondants linéaires, Annales Télécommunic., vol. **33**, n° 3-4, Mar.-Apr. 1978, pp. 74-86

[23] P. DELSARTE and P. PIRET, Algebraic Constructions of Shannon Codes for Regular Channels, IEEE Trans. Inf. Th., vol. **IT-28**, n° 4, July 1982, pp. 593-599

SUBOPTIMAL DECODING OF LINEAR CODES

I.I. Dumer
Institute for Problems of Information Transmission, Moscow, Russia
and
Manchester University, Manchester, UK

Abstract

Suboptimal decoding algorithms of linear codes in an arbitrary "symmetric" memoryless channel are considered. The decoding error probability ϵ is upper bounded by twice the error probability ϵ_e of maximum likelihood (ML) decoding. For the q-ary codes of length $n \rightarrow \infty$ and code rate R the asymptotic equality $\epsilon \sim \epsilon_e$ holds, while the number of decoding operations is upper bounded by the value $q^{n(c+0(1))}$, where $0(1) \rightarrow 0$ and $c = \min (R(1-R), (1-R)/2)$. For channels with discrete (quantized) output the better estimate $c = R(1-R)/(1+R)$ is obtained. Suboptimal coverings with polynomial construction complexity are also considered.

* This work was supported in part by the Royal Society, UK.

1. Introduction

We consider the problem of constructing suboptimal decoding algorithms (SA) for linear codes with correcting capacity close to that of maximum likelihood (ML) decoding and smaller complexity. The problem of the construction of sufficiently simple SA is at present far from solution. The known SA [1-6] are obtained for the minimum distance (MD) decoding into the nearest codewood in the Hamming metric. These results are the following.

The SA discussed in [1] gives for virtually all q-ary linear codes (except for a part of codes going to zero) the exponent $c = c_1 = R(1-R)$ of the complexity $q^{n(c+0(1))}$, when $n \to \infty$. The algorithm in [2] gives exponent $c_2 = H(2\delta) - H(\delta)$ for binary codes, where $\delta \leq 1/2$ is the root of the equation $\delta = H^{-1}(1-R)$ and $H(x) = -x\log_2 x - (1-x)\log_2(1-x)$ is the binary entropy, $0 < x < 1$. Later the complexity exponents $c_3 = (1-R)/2$ [3] and $c_4 = (1-R(1-H(\delta/(1-R)))$ [4], [5] were also obtained for arbitrary q. The algorithm in [6] generalizes algorithms [1], [3], [4], and allows a minor decrase of the complexity exponent c_4 for all code rates.

Much less is known about SA for other channels. For an arbitrary additive q-ary channel the probability distribution is not defined by the Hamming weight wt(e) of vector $e \in E^n_q$. Moreover, in the case of "soft decision" decoding the output alphabet Y exceeds the input q-ary alphabet, giving either channels with discrete (quantized) output for finite Y, $|Y| > q$, or semicontinuous channels for infinite Y. For the important class of memoryless channels [7] Wolf's trellis algorithm [8] provides ML-decoding with complexity $q^{n(c+0(1))}$, where $n \to \infty$, $0(1) \to 0$ and $c = \min(R,1-R)$.

Below we consider "symmetric" channels with arbitrary output alphabet (section 2) and define suboptimal algorithms for these channels (sections 3 and 4). Then we generalize SA [1] and [3] of MD-decoding for "symmetric" memoryless channels (sections 5 and 6), obtaining new estimates of the suboptimal decoding complexity for linear codes. In section 7 we consider the complete MD-decoding algorithms and construct the corresponding suboptimal coverings with polynomial complexity.

2. Symmetric channel

Following [7], let us define at first a discrete symmetric (DS) channel with a discrete set X of Q inputs and an arbitrary output set Y ($|Y| \leq \infty$). Let $P_{Y|(y/x)}$ be a probability measure for each $x \in X$. For any finite output set $Y_\alpha \subset Y$, $|Y_\alpha| = J_\alpha$, consider the $(Q \times J_\alpha)$-matrix $P_\alpha = P(y/x)$, $x \in X$, $y \in Y_\alpha$, using inputs as rows and outputs as columns.

Definition 1: A channel with an arbitrary output set Y is called symmetric if Y can be partitioned into disjoint finite subsets Y_α, $Y = \cup_\alpha Y_\alpha$, in such a way that in any matrix P_α the following condition A1 holds.

A1 : Each row of this matrix is a permutation of another row and each column (if more than one) is a permutation of another column.

An important subclass of symmetric channels can be obtained by using the mappings of the sets X and Y. Let $F = \{f_1,...,f_m\}$, $m \geq Q$, be the a set of mappings, transforming the sets X and Y onto themselves and satisfying the following conditions X1, Y1, P1.

X1. For any input $x \in X$ the subset $F(x) = \{f_1(x),...,f_m(x)\}$ of its images forms the whole set $X:F(x) = X$.
Y1. For any output $y \in Y$ the subset $F(y) = \{f_i(y), i = 1,...,m\}$ of its images is generated by any image $y_i = f_i(y)$ under mappings from $F : F(y_i) = F(y), i = 1,...,m$.
P1. For any $x \in X$, $y \in Y$ and $f \in F$ the equality $P(y/x) = P(f(y)/f(x))$ holds.

Definition 2: The channel is called a "mapping" channel if a finite set F of mappings, satisfying conditions X1, Y1 and P1, exists.

Lemma 1 Any "mapping" channel is symmetric.

The proof is almost obvious. The output set Y is partitioned into a finite number of disjoint subsets F(y), $y \in Y$, under the mapping set F. Consider the probability matrix $P(z/x)$, $z \in F(y)$, $x \in X$. For any pair $x_1, x_2 \in X$ the mapping $f \in F$ exists, such that $f(x_1) = x_2$. Therefore condition A1 holds for the rows $P(z/x_1)$ and $P(z/x_2)$, $z \in F(y)$, under permutation $z \rightarrow f(z)$. The same holds for any pair of columns $P(z_1/x)$ and $P(z_2/x)$ under permutation $x \rightarrow f(x)$, where $f : z_2 = f(z_1)$.

Examples.

1. Consider the set $E = \{-1, +1\}$ of two antipodal input signals, being used n times in the memoryless channel with white Gaussian noise. The inputs form the set $X = E^n = \{-1,+1\}^n$ of all binary n-sequences and the ouputs form the set $Y = R^n$ of all real n-sequences. Consider any n-sequence $f \in E^n$, $f = f_1,...,f_n)$, and define the mapping $f(y) = fy = (f_1y_1,...,f_ny_n)$ for any point $y = y_1,...,y_n) \in R^n$. The set $F = E^n$ of 2^n mappings defines a partition of the ouput set Y into disjoint sets F(y). The cardinality of this position is $2^{w(y)}$, where w(y) is the number of non zero components of vector y. Obviously conditions X1 and Y1 hold. Moreover,

any mapping f preserves the Euclidean distance d(x,y) between any two points x,y ϵR^n, i.e., d(f(x),f(y)) = d(x,y). As the probability density P(y/x) is defined by the distance d(x,y), condition P1 also holds.

2. Consider q-PSK, being used n times in a memoryless channel with additive 2-dimensional white Gaussian noise. For any position the inputs form the set $E_q = \{e^{2\pi i/q}\ i = 0,...,q-1\}$, while the outputs form the set Z of complex numbers. Therefore $X = E^n_q$ and $Y = Z^n$. Define the mapping $f(y) = fy = (f_1y_n,...,f_ny_n)$ for any point $y = (y_1,...,y_n) \in Z^n$ and any $f = (f_1,...,f_n) \in E^n_q$. Similarly to example 1, the output set Y is partitioned into disjoint sets F(y) of cardinality $q^{w(y)}$, w(y) being the number of non-zero components of vector y, and conditions X1, Y1, P1 hold.

3. Consider a channel with additive noise and let Y and X form an Abelian group and subgroup respectively, $X \subseteq Y$. As before, the channel is defined by the condition P(y/x) = p(y-x). For any $y \in Y$ consider the set F of |X| mappings f(y) = y + f, $f \in X$. The set Y is partitioned by mappings into subgroup X and its cosets X + y of equal size |X|. Obviously conditions X1, Y1 and P1 hold.

3. Suboptimal decoding

For any output $y \in Y$ let us order in terms of the values P(y/x) all the Q elements $x \in X$ into the set

$$X_y = \{x(1),x(2),...,x(Q)/P(y/x(i)) \geq P(y/x(i)),i=1,...,Q-1\} \quad (1)$$

Note that the ordering is not unique iff $P(y/x_1) = P(y/x_2)$ at least for one pair $x_1,x_2 \in X$. Let N(x) denote the number of the vector $x \in X$ in the ordered set X_y (i.e., N(x) = i, iff x = x(i)) and let $X_y(M) = x(1),...,x(M)\} = x/N(x) \leq M\}$ denote the set of M most probable inputs for given output y, where $1 \leq M \leq Q$.

Consider now any channel (X,Y,P(y/x)) and let any subset A of S - |A| inputs among its Q = |X| inputs be used with equal probability 1/S in this channel. For any received output $y \in Y$ the ML-decoding algorithm D gives the most probably input x' in the subset or "code" A:

$$D(y) = x' \in A : N(x') < N(x), \forall x \in A, x \neq x'. \quad (2)$$

The decoding error probability $\epsilon_e = 1/S \sum_{y \in Y} \sum_{x \in A, x \neq x'} P(y/x)$ is known to be the minimal one.

For the given subset A of S inputs define the decoding algorithm D_M by the following rule:

$$D_M(y) = \begin{cases} x' \text{ if } N(x') \leq M \\ 0 \text{ otherwise.} \end{cases}$$

In other words D_M-decoding gives the most probably "code" input $x' \in A$ for any input y iff x' belongs to the list $X_y(M)$ of M most probably inputs. Let $\epsilon(m)$ be the error probability of D_M. Note that $\epsilon(M) = \epsilon_e + 1/S\sum_{y \in Y:N(x')>M} P(y/x')$.

Obviously, $\epsilon(Q - S + 1) = ... = \epsilon(Q) = \epsilon_e$ and $\epsilon(M + 1) \leq \epsilon(M)$ for all M = 1,...,Q-1. The following thoerems give upper bounds on $\epsilon(M)$ for $M \geq N, N = \lceil Q/S \rceil$ and are derived from Lemma 1 in [1].

Theorem 1 : For any symmetric channel the error probability of D_M-decoding can be upper bounded for any M = iN, i=2,...,S-1 as $\epsilon(M) \leq \epsilon_e + \epsilon_e/(i-1) = \epsilon_e [i/(i-1)]$.

The proof is based on the splitting of the output set Y into disjoint finite subsets Y_α, satisfying condition A1. For any subset the proof is based on Lemma 1 in [1].

Note that the ordered set $X_y(M)$ of M most probable inputs is not unique iff $p(M_1) = ... = p(M_2)$ for some $M_1 \leq M < M_2$ (see 2)) where $p(M) = p(y/x_M)$. Therefore any $M - M_1$ inputs can be included in $X_y(M)$ among M_2-M_1 inputs with the same p(M). In the worst case it implies that all "code" inputs $x' \in A$ are not included in the set $X_y(M)$, even in $P(y/x') = p(M)$ for some $x' \in A$. For the "mapping" channels a specific ordering in the sets $X_y(M)$ can be constructed, that matches the orderings for different $y \in Y_\alpha$ and provides slightly better estimates of the error probability $\epsilon(M)$.

Theorem 2 : For any symmetric "mapping" channel the error probability of D_M-decoding can be upper bounded for any M = iN, i = 1,...,S - 1 as $\epsilon(M) \leq \epsilon_e + \epsilon_e/i$.

4. Memoryless channel.

Define the input set X and the output set Y as Cartesian n-products of the input finite alphabet $U = \{a_1,...,a_q\}$ and the output alphabet V :

$$X = U^n = \{x = (e_1,...,e_n)/e_i \in U, i=1,...,n\},$$
$$Y = V^n = \{y = (v_1,...,v_n)/v_i \in V, i=1,...,n\}.$$

Without loss of generality we consider below that $U = E_q = \{0,...,q\text{-}1\}$ and $X = E^n_q$. Let $p(v/u)$ define a probability measure for $u \in U$, $v \in V$. Define a memoryless channel by the probability measure $P(y/x) = \prod_{i=1}^{n} p(v_i / e_i)$.

Let the channel $(U,V,p(v/u))$ be symmetric and let V_β be the corresponding partition of the set V into disjoint finite subsets, $V = \cup_\beta V_\beta$. This partition generates the "Cartesian" partition of the set Y into disjoint finite subsets Y_α, where $\alpha = (\beta_1,...,\beta_n)$ and $Y_\alpha = \{y=(v_1,...,v_n)/v_i \in V_{\beta i}, i=1,...,n\}$.

Lemma 2 : The channel (X,Y,P(y)/x) is symmetric, if the alphabet channel (U,V,p(v/u)) is symmetric. The channel (X,Y,P(y/x) is a "mapping" channel if the alphabet channel (U,V,p(v/u)) is a "mapping" channel.

The proof is based on the following arguments. For any subset Y_α the corresponding matrix $P(y/x)$, $y \in Y_\alpha$, $x \in X$ is the Cartesian n-product of alphabet matrices $(p(v_i/e_i), v_i \in V_{\beta i}, e_i \in U$, $i = 1,...,n$. Each alphabet matrix is a permutation one. Therefore the product matrix is also a permutation matrix, as the Cartesian product of matrices preserves their permutation property.

For the mapping alphabet channel (U,V) a finite set $G = \{g_1,...,g_h\}$ of mappings, $h \geq q$, satisfying conditions X1, Y1 and P1, exists. Define the set of mappings $F = \{f_1,...,f_m\}$, $m = h^n$, as the Cartesian n-product $F = G^n$. In this case conditions X1, Y1 and P1 hold for the product channel X, Y.

From now on we consider the problem of constructing the set $X(M) = \{x(1),...,x(M)\} = \{x/N(x) \leq M\}$ of M, $1 \leq M \leq q^n$, most probable inputs for given output y in the symmetric memoryless channel. For a given output $y = (v_1,...,v_n)$ let us calculate the (qxn)-matrix $P = \| p_j(i) \|$, $i=0,...,q\text{-}1; j=1,...n$, of a posteriori probabilities $p_j(i) = p(i/v_j)$ of any alphabet symbol i in position j, if the output symbol v_j is received. Finally define the corresponding (qxn)-matrix $W = \|w_j(i)\|$ of weights $w_j(i) = -\log_2 p_j(i)$ of any symbol $i \in E_q$ in position j and generalized weight

$$w(x) = \sum_{i=1}^{n} w_i(e_i) \tag{3}$$

of any vector $x = (e_1,...,e_n), e_i \in E_q$.

According to the definition (3) the set $X(M) = \{x(1),x(2),...,x(M)\}$ of M most probable inputs is the set of M lightest inputs in E^n_q, ordered according to their generalised weights:

$$w(x(1)) \leq w(x(2)) \leq ... \leq w(x(M)).$$

Therefore the problem of D_M-decoding is restricted to the problem of constructing the sublist of code vectors in the list of the M lightest vectors. Below we consider this problem for linear q-ary (n,k)-codes via constructing the corresponding lists on subblocks of smaller length $s < n$, including about $M^{s/n}$ lightest subvectors.

5. Basic lemmas

Let $I = \{0,...,n\text{-}1\}$ and let $L = L(j,s) = \{j,j+1(\text{mod}n),...,j + s - 1(\text{mod}n)\}$, $j = 0,...,n\text{-}1$, $s=1,...,n$, be the ordered subset of cardinality s. Let $e_L = (e_i,i \in L)$ denote the restriction of the vectors $x = (e_1,..,e_n) \in E^n_q$ on the set L and define the weight $w(e_L) = \Sigma_{i\in L}w_i(e_i)$.

Order all q^s vectors of the set $E_L = \{e_L\}$ according to their weights:

$$w(e_L(1)) \leq w(e_L(2)) \leq ... \leq w(e_L(q^s)). \quad (4)$$

Below for any j and s all vectors of E_L with equal weights are ordered lexicographically i.e., vector $a_L = (a_i ,i \in L)$ follows vector $c_L = (c_i,i \in L)$ if $\Sigma_{i\in L}a_iq^i > \Sigma_{i\in L}c_iq^i$. Let $N(e_L)$ be the number of the vector e_L in the set (4) and let $E_L(M) = \{e_L(1),...,e_L(M)\}$ be the set of M lights subvetors on the set L. For $M > q^s$ we define $E_L(M)$ as $E_L(q^s)$.

Lemma 3 : For all vectors $e \in E^n_q$ and all j, s the following inequality holds:

$$N(e) \geq N(e_L)N(e_{I/L}). \quad (5)$$

Proof: Consider the subset of vectors $A = \{a = (a_L,a_{I/L}) \in E^n_q/N(a_L),N(a_{I/L}) \leq N(e_{I/L})\}$. Obviously, $|A| = N(e_L)N(e_{I/L})$ and e is the last vector in A.

Definition 3 : The subblock E_L of a vector $e \in E^n_q$ is called good, if $N(e_L) \leq N(e)^{/L/n}$.

Below, $Z(L) = \{L_1,...,L_m\}$ denotes any decomposition of a subset L into m,m $\leq |L|$, disjoint consecutive subsets $L_1,...,L_m$.

Lemma 4: Any vector $e \in En_q$ contains at least one good subblock e_{Li} on any decomposition $Z(I) = \{L_1,...,L_m\}$ of the set I, $1 \leq i \leq m$.

Corollary 1 : If s/n, then any vector $e \in E^n_q$ contains at least one good subvector e_{Li} of length s for some $1 \leq i \leq m$.

This corollary can be modified for any s in the following way.

Definition 4 : The subblock e_L, $L = L(j,s)$, of a vector $e \in E^n_q$ is called decomposable, if there exists a decomposition $Z(L) = \{L_1,...,L_m\}$, which generates m good subblocks e_{Li}, $1 \leq m \leq s$:

$$N(e_{Li}) \leq N(e)^{|Li|}/n, \; i = 1,...,m. \tag{6}$$

We prove below that any vector $e \in E^n_q$ contains a decomposable subblock E_L, $L = L(j,s)$, of arbitrary length s, $1 \leq s \leq n$, at least for one $j = 0,...,n-1$ and construct the set of corresponding decompositions.

Let us consider any subset L(j,s) and represent s/n as a continued fraction:

$$s/n = \cfrac{1}{d_1 + \cfrac{1}{\dots + \cfrac{1}{d_m}}} \tag{7}$$

It is well known [9], that

$$m \leq \lfloor \log_2 s \rfloor + 1. \tag{8}$$

Define the sequence $n = d_1s_1 + s_2, s_1 = d_2s_2 + s_3,...,s_{m-1} = s_md_m, s_{m+1} = 0$, where $s_1 = s$, $s_i < s_{i-1}$ for all $i = 2,...,m$ and $s_m = gcd(n,s_1)$. For any L(j,s) let us denote by $Z = Z(p_1,t_1|p_2,t_2|...|p_f,t_f)$ the decomposition of the subset L(j,s) into $P = p_1 + ... + p_f$ subsets $L_1,...,L_p$ of f types, including p_i subsets of cardinality t_i for any type $i = 1,...,f$. To avoid ambiguity all subsets of type $h = 2,...,f$ are presumed to follow all subsets of type $i < h$. Consider the following set of decompositions $Z_i, i=0,...,l, l = \lfloor m/2 \rfloor$, into $P_i = 1 + \sum_{k=1} d_{2k}$ subsets:

$$Z_0 = Z(1,s), Z_1 = Z(d_2,s_2|1,s_3),...,Z_i = Z(d_2,s_2|...|d_{2i},s_{2i}|1,s_{2i+1}), i = 1,...,l. \quad (9)$$

Lemma 5: For any vector $e \in E^n_q$ and any length s, $1 \leq s \leq n$, there exists at least one pair (i,j), $0 \leq j \leq n-1$, $0 \leq i \leq l$, such that subblock e_L, $L = L(j,s)$, is decomposable by decomposition Z_i (9).

Idea of the proof: Consider at first the trivial decomposition Z_0 (9) for al j = 0,...,n-1. If some subblock e_L, $L = L(j,s)$ is good, then the lemma is proved. Otherwise, subblocks e_{L2}, $L_2 = L(j,s_2)$ are good for all j = 0,...,n-1 according to lemma 4. Consider now the decomposition Z_1 for all j = 0,...,n-1. If at least one subblock e_{L3}, $L_3 = L(j,s_3)$ is good for some j, then subblock e_L, $L = L(j,s)$, is decomposable by decomposition Z_1. Otherwise, subblocks e_{L4}, $L_4 = L(js_4)$, are good for all j = 0,...,n-1, since any good subblock e_{L2} contains at least one good subblock on the decomposition $Z(d_3,s_3|1,s_4)$. We continue with this procedure up to the step l. In this case subblocks e_{Lm}, $L_m = L(j,s_m)$ are good for any j = 0,...,n-1. Therefore any subblock $e_{L(j,s)}$, j = 0,...,n-1, is decomposable by decompositon Z_l.

Note. According to this lemma, at most $ln \sim n\lfloor \log_2 s \rfloor/2$ decompositions should be inspected in order to find at least one decomposable subblock e_L, $L = L(j,s)$ and its decomposition Z_i. This estimate can be improved. Namely, if a/b is the irreducible representation of the fraction s/n, then the number of decompositions can be upper estimated as b.

This lemma provides the generalization of algorithm [1]. The next lemma provides the generalization of algorithm [3]. Let $L = L(j,s)$, $L' = L\backslash j$, $L'' = I\backslash L$.

Lemma 6 : For any vector $e \in E^n_q$ the following statements hold.

1. There is a unique s, $1 < s < n$, such, that $N(e_{L'}) \leq N(e)^{1/2}$, $N(e_{L''}) \leq N(e)^{1/2}$, where $L = L(0,s+1)$.

2. There is at least one j, $0 \leq j \leq n-1$, such that $N(e_{L'}) \leq N(e)^{1/2}$, $N(e_{L''}) \leq N(e)^{1/2}$, where $L = L(j,\lceil n/2 \rceil)$.

Let $\theta(M)$ be any algorithm for sorting M real numbers with a complexity $n(M) \leq cM\log_2 M$. The following lemma describes an algorithm for constructing the list of lightest vectors on an arbitrary subset and defines its complexity.

Lemma 7: The set of M lightest vectors on any subset $L = L(j,s)$ can be constructed in $C(s,M) \le csr(q)M\log_2(r(q)M) = O(sM\log_2 M)$ operations with real numbers, where $r(q) = 1 + 1/2 + ... + 1/q$.

Proof: The procedure of constructing the set $E_L(M)$ of M lightest vectors on the set L is the following one. Consider the set of all vectors $E_{L(j,p)}$ on the length p $= \min (\lfloor \log q\, M \rfloor, s)$. If $p = s$, then $E_{L(j,p)} \equiv E_{L(j,p)}(M)$. If $p < s$, we construct the set $E_{L(j,t)}(M)$ recursively for all $t = p,...,s$. Let the set $E_{L(j,t)}(M)$ be given. Let $P = \{t + 1\}$. Order all the q symbols $a \in E_q$ in $(t + 1)$-position into the set $E_P(q) = \{a_P(1),...,a_P(q)\}$ (see (5)). Let $D_i(t + 1) = \{(e, a_P(i)) \mid e \in E_{L(j,t)}(\lfloor M/i \rfloor)\}$ and $D(t + 1) = \cup_{i=1m} D_i(t + 1)$. Note, that $E_{L(j,t+1)}(M) \subset D(t + 1)$. The set $D(t + 1)$ has cardinality $H_t \le Mr(q)$. Order all the vectors in the set $D(t + 1)$ according to (4) by sorting procedure $\theta(H_t)$. Then the first M vectors form $E_{L(j,t+1)}(M)$. □

Let the volume M, $1 \le M \le Q$, and the length s, $1 \le s \le n$, be given. For any $L = L(j,s)$ define the set of subvectors

$$E_L(i,j) = \{e_L \in E_L \mid N(e_{Lm}) \le M^{|Lm|/n}, m = 1,...,P_i\}, \quad (10)$$

where L_m is the m-th subset in the decomposition Z_i (9) of the set $L(j,s)$. The following corollary results from lemmas 5 and 7.

Corollary 2 : The restriction x_L of any vector $x \in X(M)$ belongs to the subset $E_L(i,j)$ at least for one pair (i,j). Each subset $E_L(i,j)$ has the volume $|E_L(i,j)| \le M^{s/n}$ and can be constructed in $\kappa(s,M) \le csr(q)\, M^{s/n} \log_2((r(q)M^{s/n}))$ operations for any pair (i,j).

Note that each set $E_L(i,j)$ is constucted as the concatenation $\{e_L = (e_{L1},...,e_{Lpi})\}$ of the lists $E_{Li}(M^{|Li|/n})$ for all $k = 1,...,P_i$.

Finally, the following well known lemma defines the conditions for information set decoding.

Lemma 8: For vitually all linear codes of length $n \to \infty$ and size $Q = q^k$, $0 < k/n < 1$, any subset $L = L(j,s)$ of s consecutive positions is an information subset, if $s - k - \log_2 n \to \infty$. Below we use $s = K + \lfloor 2\log_2 n \rfloor$ to obtain the information subsets. Note that for all cyclic codes $s = k$.

6. The algorithms

Below we generalize the algorithm [1] of minimum distance decoding of virtually all linear (n,q^k) codes for suboptimal decoding in an arbitrary symmetric memoryless channel. Let $M = tq^{n-k}$, $t \geq 2$, and $s = k + \lceil 2\log_2 n \rceil$. For given s and n calculate the parameters l, p_i and t_i of the decomposition Z_i (9) for all i = 0,...,l.

Algorithm 1.

1. For the received ouput $y = (v_1,...,v_n)$ define the (q x n) matrix $W = (w_j(i))$ for all j = 0,...,n-1 and $i \in E_q$.

2. For any pair (i,j), $0 \leq j \leq n - 1$, $0 \leq i \leq l$, construct the set $E_L(i,j)$ (10) of subvectors e_L on the set L = L(j,s). Perform the encoding of each vector e_L into the code vector $a \in C$. Calculate the weight w(a) (4) of each vector $a = a(e_L)$. Choose vector a if its weight is less than the weight w (a') for all the a' inspected before.

3. Repeat step 2 for all pairs (i,j) and choose the vector a with minimum weight.

According to the corollary 2, algorithm 1 provides suboptimal D_M-decoding. It is easy to check that the complexity $\kappa_1(s,M)$ of algorithm 1 satisfies the equality

$$\kappa_1(s,M) = 0(tq^{(n-k)s/n}n^3\log_2 n).$$

Therefore, choosing a slowly growing parameter t (for example, $t \sim \log_2 n$) we obtain:

Theorem 3: Algorithm 1 provides for the decoding, in the memoryless symmetric channel, of virtualy all q-ary linear codes and all cyclic codes of length $n \to \infty$ and code rate R, $0 < R < 1$, with error probability which is equivalent to the error probability of ML-decoding and complexity

$$\kappa_1 = q^{n(c+o(1))},$$

where $o(1) \to 0$ and $c = R(1 - R)$.

The following algorithm is based on the examination of n subsets L = L(j,s), j = 0,...,n - 1, of length $s = \lceil n/2 \rceil$ according to lemma 6 and generalises the algorithm in [3] for any symmetric memoryless channel. Let $H = (h_{ij})$, i = 1,...,n - k, j = 0,...,n - 1, be the parity check matrix of the code A. For any subset L = L(j,s) define the submatrix $H_L = (h_{ij})$, i = 1,...,n = k, $j \in L$. Let $h(e_L) = e_L(H_L)^T$ denotes the syndrome of the subvector e_L. As before $L' = L \setminus j$, $L'' = I \setminus L$. Let $K = \lfloor M^{1/2} \rfloor$.

Algorithm 2.

1. The same as step 1 of Algorithm 1.

2. Choose any j, $0 \le j \le n - 1$ and consider the sets $E' = \{(c_j,e_{L'}) \mid e_{L'} \in E_{L'}(K), c_j \in E_1\}$ on the subset L, and $E'' = E_{L''}(K)$ on the complementary subset L". For any $e' \in E'$ calculate the syndrome h(e') and form the set $H' = \{(e',K - N(e'),0,h(e')) \mid e' \in E'\}$. Similarly for any $e'' \in E''$ for the set $H'' = \{(e'', N(e''),1,-h(e'')) \mid e'' \in E''\}$. Consider the set $H^* = H' \cup H''$ and sort all its elements as natural q-ary numbers with right significant digits. Look through the ordered set H*, choosing all adjacent pairs of elements, equal in the most significant n - 1 digits and differing in the next position. For each pair calculate the weight $w = w(e') + w(e'')$ and choose the pair with minimal weight w.

3. Repeat step 2 for all j = 0,..., n - 1 and choose the pair e = (e',e") with minimal weight over all j.

Note: in other words, all adjacent pairs (e',e") of vectors, saitsfying the condition $h(e') = -h(e'')$, are looked through in H*. These pairs form the codewords, since $h(e') + h(e'') = 0$. Moreover, all vectors $\{e' \in E' | h(e') = a\}$ are ordered with decreasing numbers N(e') in the set H*, while all the elements e" with the same syndrome $h(e'') = b$ are ordered with increasing numbers N(e"). Therefore only adjacent pairs (e',e") of vectors should be checked in H*.

Theorem 4: Algorithm 2 provides for the decoding, in the memoryless symmetric channel, of all q-ary linear codes of length $n \to \infty$ and code rate R, $0 < R < 1$, with error probability, which is equivalent to the error probability of ML-decoding, and complexity

$$\kappa_1 = q^{n(c+o(1))},$$

where $o(1) \to 0$ and $c = (1 - R)/2$.

Suboptimal decoding can also decrease the complexity for short length codes. For example, the binary (24,12) Golay code can be decoded using the most reliable 64 trellis nodes on two information sets of the first 12 positions and the last 12 positions. Note that the complete trellis diagram includes 4096 nodes.

An improvement of estimates obtained for discrete memoryless channels is presented in the following statement, using the method in [6] of combining algorithms 1 and 2.

Theorem 5 Virtually all q-ary linear codes of length $n \to \infty$ and code rate R, 0 R 1, can be decoded in any discrete memoryless symmetric channel with an error probability which is equivalent to the error probability of ML-decoding, and a complexity $\kappa = q^{n(c+o(1))}$, where $o(1) \to 0$ and $c = R(1 - R)/(1 + R)$.

7. Coverings

Algorithms [4], [5], [6] for minimum distance (MD) decoding are based on suboptimal coverings in the Hamming metric. These coverings can be constructed by random search [10], providing decoding error probability ϵ, equivalent to the MD-decoding error probability ϵ_e, when $n \to \infty$. We consider methods of constructing suboptimal coverings with polynominal complexity, that give the complete MD-decoding ($\epsilon = \epsilon_e$).

Let S(n,t) be the set of vectors of Hamming weight t in F_2^n. A subset of S(n,t) is called a covering T(n,t,l), t > l, if any vector in S(n,l) is covered by some vector(s) from T(n,t,l). It is well known [10], that the minimal volume of the covering satisfies the inequality:

$$\binom{n}{l} / \binom{t}{l} \leq \min |T(n,t,l)| \leq \binom{n}{l} / \binom{t}{l} \qquad (11)$$

We call the covering T(n,t,l) suboptimal if it has the lowest exponential order, satisfying (11): $\log_2 | T(n,t,l) | \sim \log_2(\binom{n}{l} / \binom{n}{t})$, when $n \to \infty$. If the covering vector in T(n,t,l) can be constructed in polynominal time c(n) for any vector in S(n,l), we call T(n,t,l) a polynominal covering of complexity c(n).

Theorem 6: A suboptimal covering T(n,t,l) of complexity $O(n \log_2 n)$ can be constructed, when $n \to \infty$, $t = \alpha n$, $l = \beta n$, $0 < \beta < \alpha < 1$.

8. Conclusion

Two algorithms for suboptimal decoding in an arbitrary memoryless channel are presented. These algorithms provide asymptotically for codes of length $n \to \infty$ and code rate R the decoding complexity $q^{n(R(1-R)+o(1))}$ and $q^{n((1-R)/2+o(1))}$ respectively, where $o(1) \to 0$.

9. Acknowledgement

The author is grateful to P.G. Farrell and G.A. Kabatyanskii for valuable and helpful remarks.

References

[1] G.S. Evseev, On the complexity of decoding linear codes. *Problemy Peredachi Informatsii*, vol. 19, no. 1, pp. 3-8, 1983.

[2] L. Levitin and C. Hartmann, A new approach to the general minimum distance decoding problem: The zero-neighbours algorithm. *IEEE Trans. Inform. Theory*, vol. IT-31, pp. 379-384, 1985.

[3] I.I. Dumer, Two algorithms for linear codes decoding. *Problemy Peredachi Informatsii*, vol. 25, no.1, pp.24-32, 1989.

[4] E.A. Krouk, A bound on the decoding complexity of linear block codes. *Problemy Peredachi Informatsii*, vol. 25, no. 3, pp. 103-106, 1989.

[5] J.T. Coffey and R.M.F. Goodman, The complexity of information set decoding. *IEEE Trans. Inform. Theory*, vol. IT-36, pp. 1031-1037, 1990.

[6] I.I. Dumer, On minimum distance decoding of linear codes. *Int. Proc. of the Fifth Intern. Workshop on Information Theory. Convolutional Codes; Multiuser Communication*. Moscow, pp. 50-52, Jan. 1991.

[7] R.G. Gallager, *Information Theory and Reliable Communication*. New York: John Wiley & Sons, 1968.

[8] J.K. Wolf, Efficient maximum likelihood decoding of linear codes using a trellis. *IEEE Trans. Inform. Theory*, vol. IT-24, pp. 76-80, 1978.

[9] Donald E. Knuth, *The Art of Computer Programming, Vol. 2: Seminumerical Algorithms*. Reading, MA: Addison-Wesley, 1969.

[10] P. Erdos and J. Spencer, *Probabilistic methods in combinatorics*. Akademiai Kiado, Budapest, 1974.

SOFT DECISION DECODING OF REED SOLOMON CODES

P. Sweeney and S.K. Shin
University of Surrey, Guildford, UK

ABSTRACT

Reed Solomon codes form a well known family of multilevel cyclic block codes which meet the Singleton bound. As random symbol error correcting codes they are particularly useful in conditions where errors occur in bursts. For the AWGN channel, however, they suffer relative to other codes such as convolutional codes, at least at moderate bit error rates (around 10^{-5} to 10^{-6}). Much of this disadvantage results from the lack of a generally applicable method for soft decision decoding.

This paper reports on progress in applying soft decision decoding to Reed Solomon codes. Any algebraic approach to soft decision decoding must address a number of problems such as the relationship of the soft decision values to the field over which the code is defined, the calculation of Euclidean distances and the provision of an algorithm to find the maximum likelihood codeword. In this paper it is proposed that the algebraic problems may be eased by defining the code as a subfield subcode of a code defined over a larger field and by a specially devised mapping of symbol values onto detected bit values. In the absence of any known algorithm, the performance of soft decision decoded RS codes has been studied by the application of trellis decoding methods. Viterbi decoding has been used to limit the complexity of trellis decoding and the performance of a reduced search method has been assessed. It has been found that useful coding gains can be achieved at moderate bit error rates from (15, 13) and (15, 11) codes, using a method in which only the best B paths are updated at each incoming symbol, with B having values of 16 or 32.

1 INTRODUCTION

This paper contains a discussion of some of the problems of finding an algebraic approach to soft decision decoding of Reed Solomon codes and makes suggestions which may help lead to a solution. Since a complete algebraic solution is not currently available, a non-algebraic approach has also been studied to help establish the performance that could be expected. The approach adopted is one of trellis decoding, first suggested by Wolf [1]. Viterbi decoding has been implemented and the possibilities of reduced search methods considered in the hope that such methods might be feasible for implementation in certain cases.

2 ALGEBRAIC APPROACHES

2.1 Arithmetic in an Expanded Galois Field

In any algebraic approach to soft decision decoding, the inclusion of soft-decision information expands the size of the field that must be processed at the decoder and the operations on transmitted symbol values must be consistent between the two fields. The problem can be overcome by encoding in the field whicn will eventually be used for the decoding. If the code is over GF(q), then it is seen that encoding can be achieved as a sub-field subcode over GF(q^m), where m is any positive integer. For any integer value of q and m, $q^m-1 = (q-1)(q^{m-1}+q^{m-2}+ \ldots\ldots +1)$. Thus there will be some element β in GF(q^m) of order q-1. Any vector $v(\alpha)$ over GF(q) of length q-1 can therefore be replaced by $v(\beta)$ over GF(q^m). If we use the definition of the Fourier transform over a finite field as given by Blahut[1] then the properties of the Fourier transform of $v(\beta)$ over GF(q^m) will be identical to those of the Fourier transform of $v(\alpha)$ over GF(q). Since a Reed Solomon code is defined in terms of its Fourier transform, it can therefore be defined in the extension field.

From the above, it can be seen that there is no difficulty in principle in providing a suitable finite field arithmetic in which to do the algebra of soft decision decoding. The practical difficulty is that if we assume that q is an integer power of 2 and we wish to provide 2^m levels of soft decision on each bit of the code over GF(q), the algebra must be carried out in the much larger field of GF(q^m).

To take a specific example, we might wish to define a Reed Solomon code of length 7 over GF(8). If three transmitted bits are to be used for each code symbol and we want 4-level soft decision on each received bit, the code could be defined in terms of its transform over GF(64) in which $\beta = \alpha^9$ is an element of order 7. If the primitive polynomial $\alpha^6 + \alpha + 1 = 0$ is used to generate GF(64) then it is easy to verify that $\beta^3 + \beta^2 + 1 = 0$. As this is a primitive polynomial of order 3, the powers of β form the non-zero terms of a sub-field with 8 elements which remains closed under operations within GF(64). A code sequence can therefore be regarded either as a vector over GF(8) or as a vector over a sub-field of GF(64).

2.2 Calculation of Euclidean Distance

Having resolved the arithmetic problems, the decoding algorithm must find an error sequence which corresponds to the syndrome of the received sequence and results in an output sequence with q-ary values. For the decoding to be maximum likelihood, the output sequence must be such that the distance between the received sequence and the output sequence is minimized, taking into account soft decision information on the transmitted symbols. The distance metric to be used will therefore be a soft decision or Euclidean metric. For this to be possible algebraically, there must be a relationship between arithmetic operations and the Euclidean distance between symbols. The required relationship must depend on the algorithm to be used, but it is difficult to define an algorithm with no concept of how Euclidean distance can be measured. What has been done here is to propose a way of relating Euclidean distance on individual symbols to operations in the finite field, in the hope that this might provide a starting point for discovering an algorithm or heuristic that achieves either minimum distance decoding or some approximation to it. As yet no such method has been found.

It has been found that convenient mappings onto a binary channel do exist provided each bit is taken to represent a particular element of the sub-field used by the encoder. The mapping is based on the primitive polynomial used to generate the extension field $GF(q^m)$. The only condition that needs to be imposed is that this polynomial is of weight at least 2^m-1. The 2^m-1 non-zero received levels on a designated bit of each symbol are now set to represent terms, or linear combinations of terms, from the generator such that they sum to zero. For the other bits of the symbol, the received levels are mapped as before but successively multiplying by β.

For example, the three bits used to convey a symbol over GF(8) could be regarded as representing α^0, α^9 and α^{18} in GF(64) to provide 4-level soft decision on each bit. The field GF(64) is generated using $\alpha^6 + \alpha + 1 = 0$. Since this is of weight 3 we let the levels on the bit representing α^0 be designated 0, α, α^6 and α^0. For other bits we multiply by α^9 and α^{18}. The possible received values for the three transmitted bits would thus be as shown in table 1.

	bit 2	bit 1	bit 0
level 3	α^{18}	α^9	α^0
level 2	α^{24}	α^{15}	α^6
level 1	α^{19}	α^{10}	α^1
level 0	0	0	0

Table 1. Mapping of symbols from GF(64) onto three 4-ary symbols

All symbols of GF(64) can be created through linear combinations of the above. In every measurement of distance between two symbols, one of the symbols will take values from the smaller field. Because of the way the mapping was constructed, the sum of two symbol values yields a polynomial which can be used to represent the distance between the received and postulated values of each bit. Thus a received symbol α^{24}, 0, α^1 is at a distance α^{19}, α^9, α^1 from α^{18}, α^9, 0. If we compared α^{19}, α^9, α^6 with 0, 0, 1, exactly the same result would be obtained, as appropriate for a symmetric channel.

Note that the above example has provided 4-level (2-bit) quantization on each received bit. For the more usual 8-level (3-bit) quantization, the field for decoding would be GF(512) and a polynomial used to generate GF(512) would need to be of weight at least 7.

3 TRELLIS DECODING

3.1 Principles

A part of the objective of the above approaches has been to find algebraic techniques for decoding. This is still our ultimate objective but, for the purposes of checking the performance of soft decision decoding, a trellis decoding method has been implemented. Wolf [2] showed that the encoder for any block code could be represented by a trellis with at most q^{n-k} or q^k states. For a cyclic code, the trellis is regular except that the number of states grows over the first n-k symbols and reduces over the last n-k (assuming n-k < k). Even when the trellis is at less than maximum size, it is still a part of the full trellis, just as with a convolutional code the number of states that can be reached increases at the start of the sequence and reduces to 1 as the encoder is cleared at the end of the data.

The applicability of trellis decoding can best be seen if the encoder is modelled as a FIR filter whose impulse response is the generator sequence of the code and whose input is the data followed by n-k zero symbols. The code sequence is therefore a convolution of the data with the generator sequence. Decoding can therefore be done by maximum likelihood methods, i.e. computing the distance between the received sequence and all possible paths through the trellis. The number of paths to be considered can be limited by use of the Viterbi algorithm. During the last n-k stages when the input sequence is taken to be zero, the number of states reduces to one and the survivor sequence at that point determines the decoder output.

The number of paths to be computed over a single block can be calculated from the above considerations. Clearly the amount of computation required will vary through the block, but a realistic implementation might buffer each received block to be decoded whilst the next is being received. Thus the number of paths per received bit might be an appropriate measure of computation. The figures for a few simple codes are shown in table 2. If we consider that a figure of the order of 512 or 1024 would be considered to be the upper limit of feasibility for a convolutional code, we can see that full trellis decoding of the (15,11) code will not be feasible, but that the others may be possible.

q	n	k	total paths	paths/ bit
8	7	5	1680	80
8	7	3	5264	250.7
16	15	13	45600	760
16	15	11	7479840	124664

Table 2. Trellis decoding complexity for example RS codes

3.2 Theoretical Performance of Trellis Decoding

The expected performance of trellis decoding was evaluated by use of the usual formula based on the Union bound which upper bounds the bit error rate for convolutional codes; see, for example, Viterbi and Omura [3]. The weight distribution of the code is represented by a polynomial of the form

$$T(D,I) = \sum_{d,i} T_{d,n} D^d I^i$$

where d and n represent the input and output weights respectively of a code path and $T_{d,i}$ is the number of paths of that combination of input and output weights. The upper bound to the bit error rate is then

$$B.E.R. \leq \frac{1}{m}\sum_d A_d Z^d$$

where

$$A_d = \sum_i i T_{d,i}$$

m is the number of bits in each input symbol and Z is a value which depends on channel conditions. For a binary symmetric channel with bit error rate p, $Z = 2\sqrt{p(1-p)}$ and for unquantized soft decision decoding on an AWGN channel using binary PSK modulation, $Z = e^{RE_b/N_0}$ where R is the code rate and E_b/N_0 is the energy per information bit divided by the (single sided) noise power spectral density.

By enumerating all the codewords, the transfer function of the (7,3) RS code over GF(8) was found to be

$$\begin{aligned} T(D,N) = {} & D^6(3N + 6N^2 + N^3 + 6N^4 + 4N^5 + N^6) + \\ & D^8(3N + 19N^2 + 33N^3 + 25N^4 + 21N^5 + 12N^6 + 5N^7 + N^8) + \\ & D^{10}(3N + 3N^2 + 26N^3 + 48N^4 + 39N^5 + 23N^6 + 12N^7) + \\ & D^{12}(8N^2 + 17N^3 + 35N^4 + 48N^5 + 30N^6 + 11N^7 + 5N^8) + \\ & D^{14}(6N^3 + 11N^4 + 10N^5 + 11N^6 + 7N^7 + 3N^8 + N^9) + \\ & D^{16}(N^3 + N^4 + 4N^5 + 7N^6 + N^7). \end{aligned}$$

Hence the output bit error rate with maximum likelihood decoding is

$$BER \leq \frac{1}{3}(68Z^6 + 460Z^8 + 696Z^{10} + 744Z^{12} + 260Z^{14} + 76Z^{16})$$

For the (7,5) code, the paths are too numerous to give the full transfer function, but the formula for bit error rate is

$$BER \le \frac{1}{3}(1080Z^4 + 10292Z^6 + 44848Z^8 + 80988Z^{10} + 72952Z^{12} + 29548Z^{14} + 5664Z^{16} + 388Z^{18})$$

An alternative approach to estimating the hard decision performance is to assume that it should be similar to that of algebraic decoding. If it is assumed that all error patterns of weight greater than $t = (n-k)/2$ will result in an incorrect decoding and that, because of the non-systematic nature of the decoder, incorrectly decoded blocks have a bit error rate of 0.5, then an upper bound on the performance of algebraic decoding can easily be obtained.

The theoretical performance curves for the (7,3) and (7,5) codes are shown in figures 1 and 2. Both codes show an expected coding gain with soft decisions of around 2 dB at bit error rates in the region of 10^{-5} to 10^{-6}. It is seen that the predictions for hard decision trellis decoding are substantially inferior to those for algebraic decoding, particularly for the (7,5) code and at low values of E_b/N_0. It is assumed that the calculated upper bound for bit error rate of trellis decoding is not very tight in the circumstances of hard decisions, low distances and low E_b/N_0 and that therefore the algebraic performance is more likely to be representative. This latter surmise is supported by fact that the projected improvement from soft decision decoding is greater than would be expected on the basis of the asymptotic gains and by the simulation results discussed below.

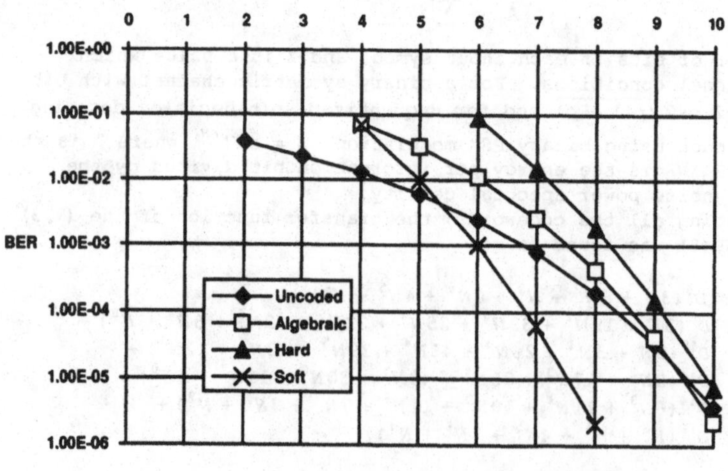

Figure 1. Performance of (7,3) RS code

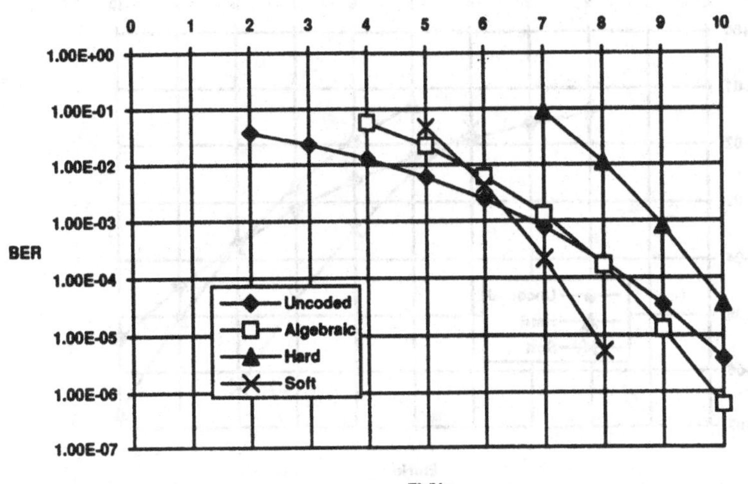

Figure 2. Performance of (7,5) RS code

For codes of length 15 the weights have not been enumerated, but hard performance can be estimated by calculating the performance of algebraic decoding. The difference between the asymptotic coding gains using hard and soft decisions can then be used to estimate soft decision performance. Soft decision improvements of 1.76 dB and 2.22 dB respectively have been assumed for the (15,13) and (15,11) codes. The results are shown in figures 3 and 4. The predicted soft decision coding gain for the (15,13) code is very similar to that for the codes of length 7 but the (15,11) code shows an expected soft decision gain of around 4 dB at 10^{-6} bit error rate. This code would therefore be of considerable interest for implementation if the complexity problems could be overcome.

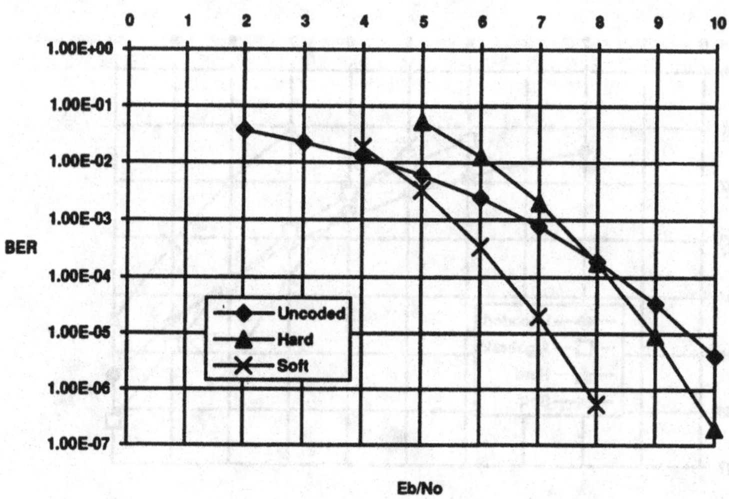

Figure 3. Performance of (15,13) RS code

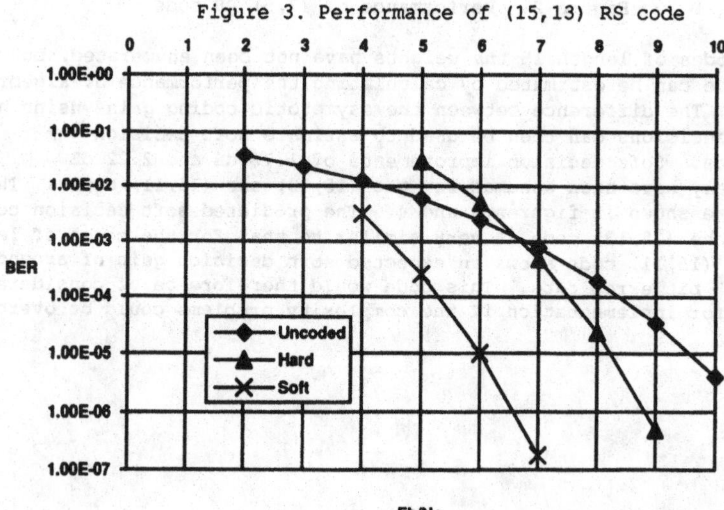

Figure 4. Performance of (15,11) RS code

4 REDUCED SEARCH METHODS

In order to extend the applicability of trellis decoding, reduced search methods may be considered. For convolutional codes, sequential decoding is well established, however the applicability of sequential decoding to trellis decoding of Reed Solomon codes is unclear. We have chosen to implement the SA(B) algorithm proposed by Aulin [4]. This was originally applied by the authors of that paper to Correlative Phase Modulation, where it worked well, and to Ungerboeck codes, where it did not. It was subsequently established [5] that the reasons for poor performance with Ungerboeck codes and other convolutional codes were related to the difficulties in resynchronizing with the correct code path after a decoding error. For our application, however, we do not have that problem since resynchronization will be achieved with the start of each new block.

The SA(B) algorithm can be described very simply. If in the trellis more than B paths are produced, only the best B paths are retained and extended at the next input symbol. If sufficient paths are kept, the ones which are lost or eliminated would be unlikely to figure in the final solution in full trellis decoding except in circumstances where even maximum likelihood decoding is unlikely to produce correct decoding. Determination of the appropriate value of B is done experimentally.

5 SUMMARY OF SIMULATION RESULTS

Some Monte-Carlo simulations have been undertaken to check the above predictions. This has been done by encoding randomly generated information sequences and simulating their transmission over a binary channel subjected to additive white Gaussian noise. The number of results obtained is so far small, particularly at lower bit error rates where long simulation runs are needed, and so only a few tentative conclusions are presented. For the length 7 codes, the algebraic decoding performance seems to represent a fairly tight over-bound on the bit error rate with hard decision decoding; even at bit error rates of around 10^{-2}, the simulated performance was within 0.5 dB of the algebraic decoding curve. At lower bit error rates the simulated hard decision curve appears to converge on the algebraic curve. The advantage of soft decision decoding appears to be maintained to bit error rates around 10^{-2} as well, 8-level soft decision decoding showing an improvement of approximately 1.5 dB and 2.0 dB for the (7,5) and (7,3) codes respectively which is close to the asymptotic values. This provides justification for the method used to estimate performance of the codes of length 15.

The effects of varying the value of B in the SA(B) algorithm have been studied on the (15,13) code, for which full Viterbi decoding would require B=256. It appears that as B is reduced the bit error rate shows negligible change until some threshold is reached, after which it degrades rapidly. Using soft decision decoding with an output bit error rate around 10^{-3}, the performance was maintained down to B=32, degraded slightly at B=16 and was significantly worse for lower values of B. Thus there appears to be a realistic chance of achieving good performance in more complex codes with manageable computation.

6 CONCLUSIONS

Some progress can be made in the problem of finding algebraic methods for soft decision decoding of Reed Solomon codes. A complete solution has not, however, been discovered.

The complexity of trellis decoding method limits the size of FIR encoder which can be modelled efficiently for full Viterbi decoding. Nevertheless the application of trellis decoding techniques is viable for some simple RS codes and can improve their performance significantly and therefore extend their applicability.

There appears to be a realistic chance that reduced search methods, such as the SA(B) algorithm could extend the viability of trellis decoding to more complex RS codes which offer improved coding gains.

REFERENCES

1. Blahut, R.: Theory and Practice of Error Control Codes, Addison Wesley, Reading, Massachusetts 1983.
2. Wolf J.K.: Efficient maximum likelihood decoding of linear block codes using a trellis", IEEE Transactions on Information Theory, IT-24/1 (1978), 76-80.
3. Viterbi A.J. and J.K. Omura: Principles of Digital Communication and Coding, McGraw-Hill, New York (1979).
4. Aulin, T.: Study of a new trellis decoding algorithm and its applications", ESTEC study report 6093/84/NL/DG (1985).
5. Aulin, T.: Application of the SA(B) detection algorithm to coded PSK modulation", ESTEC study report 7108/87/NL/PR (1991).

8

APPLICATIONS OF CODING

VLSI IMPLEMENTATION OF A FRACTAL IMAGE COMPRESSION ALGORITHM

R. Creutzburg, W. Geiselmann and F. Heyl
University of Karlsruhe, Karlsruhe, Germany

Abstract

In this paper we describe a VLSI implementation of a lossy image compression [1] algorithm given in [2]. The chip was designed using the full–custom VLSI CAD system ISIS and was performed in the hardware description language HDL of this system on transistor level [3, 4, 5]. The chip area is about $1\,\mathrm{mm}^2$ using a 2μ-CMOS technology.

Introduction

The efficient compression of image data is a very important problem in computer science and telecommunication. We can distinguish between the following two main strategies to compress image data, the lossless image data compression, and lossy image data compression. In most cases it is not necessary to store and reconstruct the full image information because the human visual system is often unable to recognize very small details.

A new and very interresting research field in image data compression are the fractal image data compression methods [6, 7, 2].

1 The fractal approach

Modern fractal geometry has been introduced by MANDELBROT in the late seventies [8]. Since that time it became a powerful tool in particle and signal analysis, statistics, physics, etc. In [6, 7, 2] simple fractal image compression algorithms have been introduced and in the following we will explain and extend some aspects of these algorithms.

In connection to the fractal geometric objects the fractal dimension of a set is often needed. Using a well-known method for the determination of the fractal dimension of a one-dimensional signal one can easily develop a method for data compression.

For the determination of the fractal dimension of a curve A a measure of variable length (often called a "yardstick") is used. The curve is approximated with this yardstick using a fixed length. After measuring the length of the curve using this fixed length another yardstick length is used to approximate the curve length. Roughly spoken, the limes "quotient of the length of the curve and the lenght of the yardstick" corresponds to the fractal dimension. More precise:

For each $\epsilon > 0$ (corresponding to half the length of the yardstick) let $N(A,\epsilon)$ denote the smallest number of closed balls of radius ϵ needed to cover A.

$$D := \lim_{\epsilon \to 0} \frac{\ln(N(A,\epsilon))}{\ln(\frac{1}{\epsilon})}$$

is called the <u>fractal dimension</u> of A if this limes exists.

1.1 Compression

For the compression of an image we start in the first image row. The grey values of the pixels can be viewed as a quasi-fractal one-dimensional signal to be determined. We put one end of our yardstick on the level defined by the grey level of the first pixel in the image row. Then we let the yardstick fall until its other end will be stopped by our imaginary edge line. Then we advance the yardstick to this new location and continue the process until the entire image will be covered.

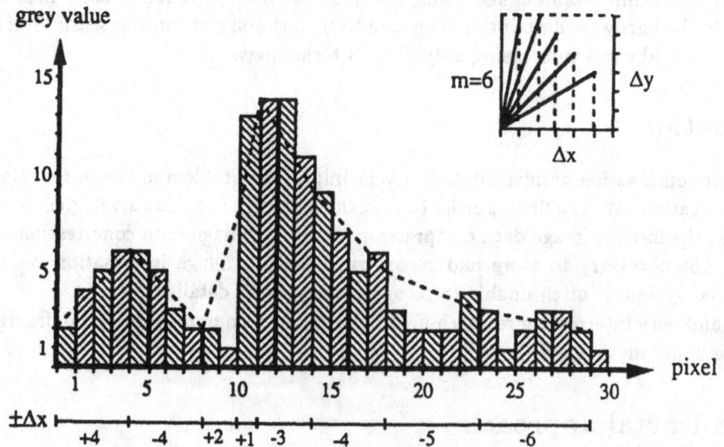

Figure 1: The scheme of the compression algorithm.

Consider an actual grey value c_i of a fixed point p_i that is considered. The corresponding horizontal distance Δx till the end point of the yardstick of length m is obtained using the following method (compare 2):

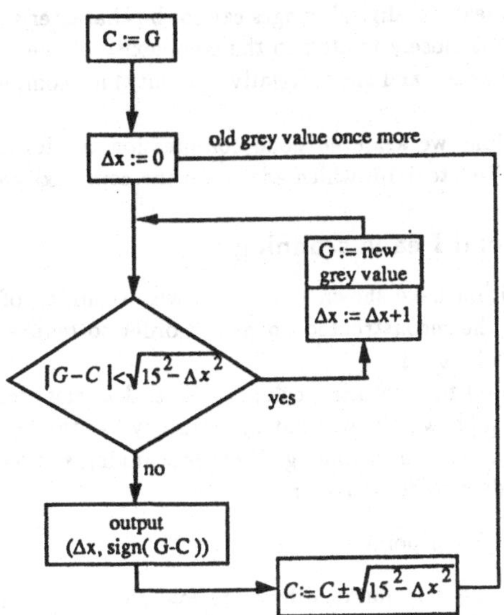

Figure 2: The structure of the algorithm.

1. $\Delta x := 0$,

2. if $|c_{i+\Delta x} - c_i| < \sqrt{(m^2 - \Delta x^2)}$ then set $\Delta x := \Delta x + 1$ and go to 2.

The sign is determined in the following way:

3. $s := sign(c_{i+\Delta x} - c_i)$.

From the fact that the transmission uses only integer values the problem of the accumulation of the rounding errors of the grey values arises.
Therefore, in each step a correction of the (next) grey value $c'_{i+\Delta x}$ has to be done:

4. $c'_{i+\Delta x} := c_i + s \cdot \sqrt{(m^2 - \Delta x^2)}$.

Here also the case is included where the grey value difference between two neighbor points is larger than the yardstick length. Here Δx equals 0 and the grey value is stepwise approximated until the necessary value is achieved.
In each step the values of Δx and s are transmitted. After that point number i is newly fixed:

5. $i := i + \Delta x$

and the same procedure is repeated until the end of an image row is achieved. Then follows the transmission of the initial grey value of the next image row, etc.

For the transmission of the compressed data it is recommended to transmit a number of bits according of the yardstick length for each horizontral distance. If the yardstick length is m then $\log_2[m + 1]$ bits plus 1 bit for the sign are necessary in every step. In experiments we used 15 bit for the yardstick length. So 5 bit per step are necessary for the transmission. The decompression is performed in analogy to the compression.

Note that, strictly speaking, digital images cannot be characterized as fractals. Moreover, the fractal dimension d is closely related to the complexity of the image. It would be close to unity for the smooth areas, and small (relative to unity) for complex, difficult to compress images.

In software simulations we achieved good compression results for satellite images with compression rates of 1:6 up to 1:16, which adapts to the complexity of the image.

1.2 Image arrays and Peano scanning

First tests of the algorithm have shown that a row-wise scanning of the image shows a lot of row-wise artifacts in the reconstructed image. In order to reduce these artifacts another scanning strategy has to be used.

In several tests we used the row-wise, column-wise and zigzag-scanning and the scanning along a Peano curve. In [9] it was shown that this strategy has the best correlation properties between neighbor point in the curve among all other scanning strategies.

The Peano curve is defined recursively:

- The curve of order 0 is a point.
- The curve of order $n+1$ is obtained by a fourfold copying and rotation of the curve of order n and a connection of corresponding parts of the curve.

The method is shown in 3 for the Peano curves of order 1, 2 and 5.

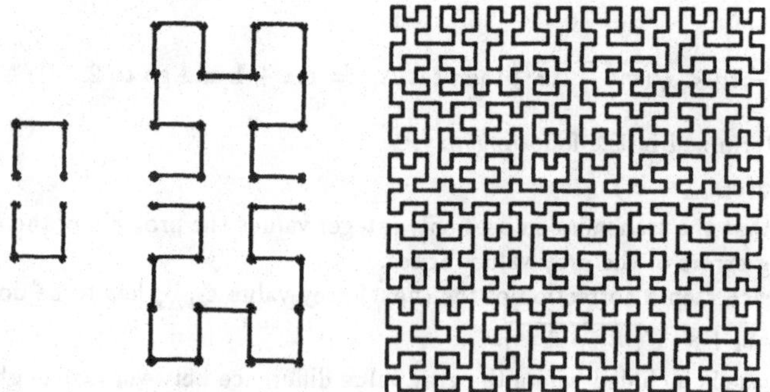

Figure 3: Peano scanning scheme.

2 Implementation

The compression algorithm of section 1.1 perfectly suits to a VLSI design. It has been implemented by us with the full custom layout system ISIS [3]. We have used well known standard modules as 4-bit manchester carry adders [5], we have modified and combined those elements for special problems as to add an 8-bit number to a 4-bit number and we have designed special modules where necessary.

2.1 Structure of the VLSI-design

The structure of the chip can be seen in Fig. 4. The algorithm in hardware works in the following way:

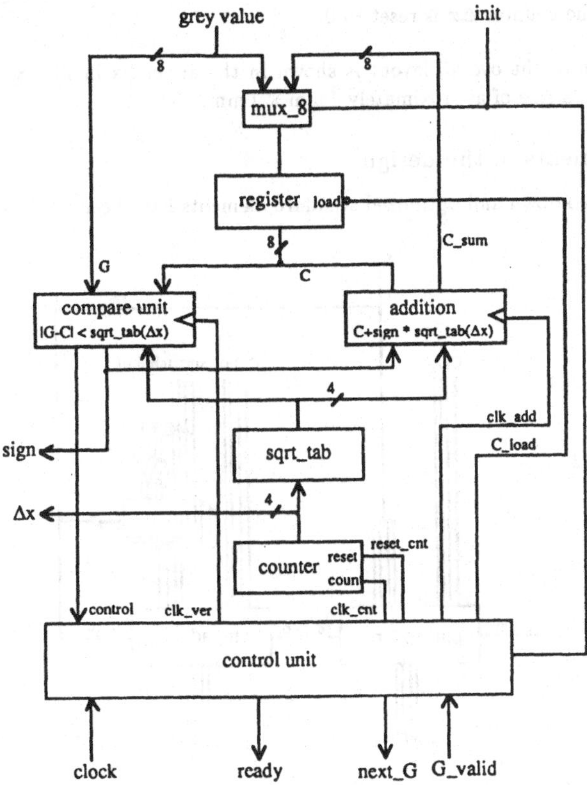

Figure 4: Structure of the VLSI design.

- The starting point C of the yardstick is stored in an 8-bit register.
- The value Δx is computed and stored by a 4-bit counter.
- The square root is "computed" by a small table.
- In each clock cycle the following calculation is done:
  The `compare unit` calculates `control` $= |G - C| < \sqrt{m^2 - \Delta x^2}$. According to this signal one of the two following actions is performed:

  "control = true": The actual grey value is close enough to C so that the yardstick does not intersect with this value. The counter is increased by one and the next grey value is loaded to the chip.

"control = false": The yardstick intersects with this grey value. The signal ready is set true by the control unit, the signals sign and Δx should be read as outputs of the chip.
The new starting point of the yardstick is calculated to $C + \texttt{sign} * \sqrt{m^2 - \Delta x^2}$ and the counter Δx is reset to 0.

The floorplan of the overall layout is shown in the appendix in Fig. 8. In a 2μ-CMOS technology it has a size of approximately 1 mm × 1 mm.

2.2 The elements of the design

As far as possible known and optimized standard elements have been used. The registers are

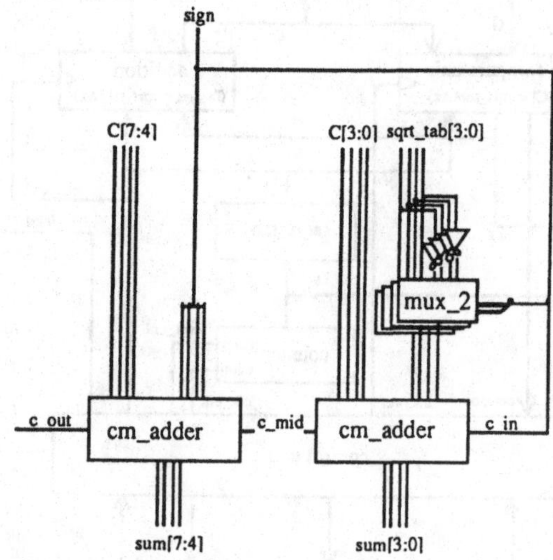

Figure 5: The structure of the 8 bit adder, consisting of two 4 bit adders.

build up by master-slave flip-flops (in [5] called "D flip-flops"). The 4-bit counter for Δx has been designed according to the synchronous counter of [5]. The addition (subtraction) of the 8-bit value C and the 4-bit value $\sqrt{m^2 - \Delta x^2}$ is realized by two manchester carry adders (see Fig. 5). According to the output signal sign of the compare unit the number $\sqrt{m^2 - \Delta x^2}$ is added or subtracted to C. Switching between addition and subtraction is realized by the 4 multiplexers in Fig. 5.

Calculating the square root $\sqrt{m^2 - \Delta x^2}$ for the 16 possible values of Δx can easily be realized by a table, see the left two columns of table 1. In hardware such a problem is usually solved by a ROM. For these special values the representation of the four output bits $y_0, \ldots, y_3$ in terms of Boolean functions of the input bits $x_0, \ldots, x_3$ (see table 1) gives the idea of a better implementation. Using this separation, this function can be realized in a much more efficient way by 2 ANDs, 1 OR, 1 XOR, 5 inverters and 10 multiplexers (see Fig. 6).

Δx	Δy	$x_3x_2x_1x_0$	$y_3y_2y_1y_0$	y_3	y_2	y_1	y_0
0	15	00 00	1111				
1	15	00 01	1111	1	1	1	1
2	15	00 10	1111				
3	15	00 11	1111				
4	14	01 00	1110				
5	14	01 01	1110	1	1	$\neg(x_1 \wedge x_0)$	$x_1 \wedge x_0$
6	14	01 10	1110				
7	13	01 11	1101				
8	13	10 00	1101				
9	12	10 01	1100	1	$\neg x_1$	x_1	$\neg x_0$
10	11	10 10	1011				
11	10	10 11	1010				
12	9	11 00	1001				
13	7	11 01	0111	$\neg(x_1 \vee x_0)$	$x_1 \oplus x_0$	$\neg x_1 \wedge x_0$	$\neg(x_1 \wedge x_0)$
14	5	11 10	0101				
15	0	11 11	0000				

Table 1: Function table for square root calculation, $\Delta y = \sqrt{m^2 - \Delta x^2}$.

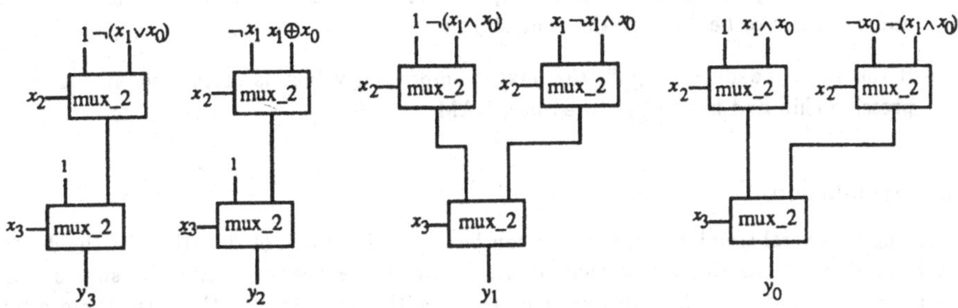

Figure 6: Calculation of the 4-bit number $\sqrt{m^2 - \Delta x^2}$ out of the 4 bits $x_0, \ldots, x_3$ using Boolean functions.

The `compare unit` evaluates the relation $|G - C| < \Delta y$ from the inputs G, C, Δy and has in addition the sign of $G - C$ as output. This module consists of 3 manchester carry adders, 1 multiplexer, AND and OR gates.

A method to integrate the chip described in the previous section into a system can be done in the following way (compare Fig. 7):

- The `control unit` sends the first grey value to the input of `compression`, by selecting the address of this value for the ROM. While the signal `init` is "high" this value is stored in the compression chip.

- After setting `init` to "low", the control unit has to wait until `next_G` demands for the next grey value.

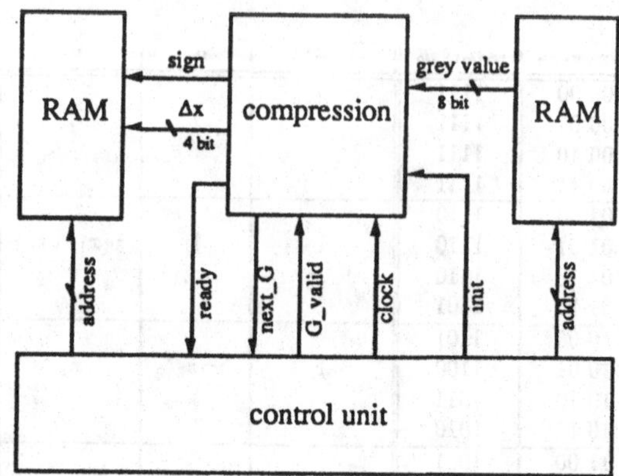

Figure 7: Integration of the compression chip into a system.

- Then **control** sends the address of the next grey value to the RAM and sets **G_valid** to "high" for one clock cycle (**next_G** will switch to "low"). If a (two dimensional) picture should be compressed according to the peano curve of section 1.2 this calculation of the addresses has to be done by the control unit.

- If the signal **ready** is "high", the next "compressed value" is calculated by the compression chip and has to be stored in a RAM

2.3 Simulation

According to several simulations, the chip can be run with a clock of 20 MHz. For the depth of the calculation, the worst case that occurs is within the **compare unit** the subtraction $G - C = 255 - 255$ and the following comparing with some Δy. In this case three 4-bit additions have to be calculated one after the other. This can be done within $40\,ns$ without problems. Using an asymmetric clock ($40\,ns$ "high" and $10\,ns$ "low") this results in a clock frequency of 20 MHz. The speed of the compression of a picture depends on the structure of the picture. The average over several examples is a rate of 0.8 bit per clock cycle. Thus the compression of a picture with 512×512 points needs $16\,ms$, therefore 60 of these pictures can be compressed within $1\,s$. Therefore, the method is well suited for real-time applications. Hence it is possible to integrate it in a single video camera.

Acknowledgements The authors thank *Armin Nückel* and *Hans Härtl* for their help in discussing the subject and during the VLSI design.

References

[1] Storer, J.A.: *Data Compression - Methods and Theory.* Computer Science Press: 1988.

[2] Walach, E.; Karnin, E.: *A fractal approach to image compression.* Proc. Int. Conference Acoust. Speech Signal Process. ICASSP'86, Tokyo 1986.

[3] ISIS Manuals; Racal-Redac Ltd, Tewkesbury, Gloucestershire, England; 1985.

[4] Mead, C.; Conway, L.: *Introduction to VLSI Systems.* Addison Wesley:1980.

[5] Weste, N.; Eshragian, K.: *CMOS VLSI Design.* Addison Wesley: Reading (MA) 1985.

[6] Creutzburg, R.; Ivanov, E.: *Fast algorithm for computing fractal dimensions of image segments.* in: Cantoni, V.; S. Levialdi; R. Creutzburg; G. Wolf (Eds.): *Recent Issues in Pattern Analysis and Recognition.* Lecture Notes in Computer Science **399**, Springer-Verlag: Berlin-Heidelberg 1989, pp. 42-51.

[7] Creutzburg, R.; Mathias, A.; Ivanov, E.: *Fast algorithm for computing the fractal dimension of binary images.* Physica A **185** (1992), pp. 56-60.

[8] Mandelbrot, B. B.: *The Fractal Geometry of Nature.* Freeman: New York 1988.

[9] Quinqueton, J.; Simon, J.C.: *On the use of a Peano scanning in image processing.* Issues in Digital Image Processing, Sijthoff & Noordhoff 1989, pp. 357-365.

A The floorplan of the chip

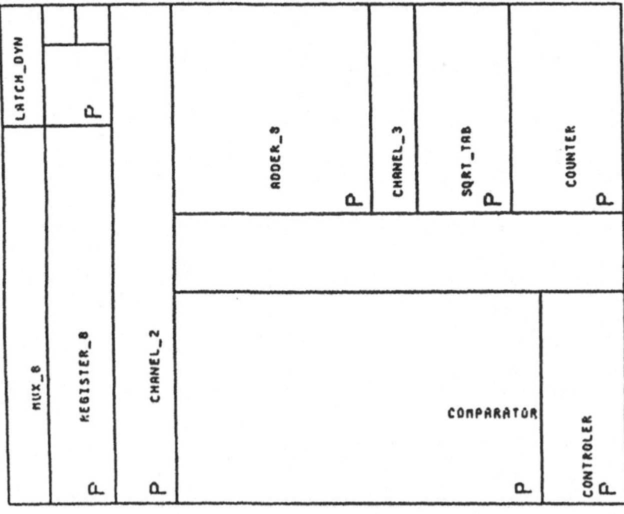

Figure 8: The floorplan of the overall layout. It has a size of approximately 1 mm × 1 mm.

Printed in the United States
By Bookmasters